CAMBRIDGE LIBRARY COLLECTION

Books of enduring scholarly value

Technology

The focus of this series is engineering, broadly construed. It covers technological innovation from a range of periods and cultures, but centres on the technological achievements of the industrial era in the West, particularly in the nineteenth century, as understood by their contemporaries. Infrastructure is one major focus, covering the building of railways and canals, bridges and tunnels, land drainage, the laying of submarine cables, and the construction of docks and lighthouses. Other key topics include developments in industrial and manufacturing fields such as mining technology, the production of iron and steel, the use of steam power, and chemical processes such as photography and textile dyes.

The Steam Engine

Thomas Tredgold (1788–1829) has been described as 'the most influential technical author of his generation and possibly of the nineteenth century'. His writings contributed greatly to the wider understanding of engineering, and it is his definition of civil engineering that the Institution of Civil Engineers wrote into their charter of 1828. Published in 1827, this work provides a historical survey and explanation of 'a masterpiece of human contrivance'. Tredgold breaks his subject down into ten sections, each covering areas such as the properties of steam, the differing means of harnessing its power, the history of the steam engine's invention and improvement, and the various applications of steam power. Containing many tables, formulae and line drawings, this thorough work complements Charles Frederick Partington's *Historical and Descriptive Account of the Steam Engine* (1822), which is also reissued in this series.

Cambridge University Press has long been a pioneer in the reissuing of out-of-print titles from its own backlist, producing digital reprints of books that are still sought after by scholars and students but could not be reprinted economically using traditional technology. The Cambridge Library Collection extends this activity to a wider range of books which are still of importance to researchers and professionals, either for the source material they contain, or as landmarks in the history of their academic discipline.

Drawing from the world-renowned collections in the Cambridge University Library and other partner libraries, and guided by the advice of experts in each subject area, Cambridge University Press is using state-of-the-art scanning machines in its own Printing House to capture the content of each book selected for inclusion. The files are processed to give a consistently clear, crisp image, and the books finished to the high quality standard for which the Press is recognised around the world. The latest print-on-demand technology ensures that the books will remain available indefinitely, and that orders for single or multiple copies can quickly be supplied.

The Cambridge Library Collection brings back to life books of enduring scholarly value (including out-of-copyright works originally issued by other publishers) across a wide range of disciplines in the humanities and social sciences and in science and technology.

The Steam Engine

*Comprising an Account of its Invention
and Progressive Improvement*

Thomas Tredgold

CAMBRIDGE
UNIVERSITY PRESS

CAMBRIDGE
UNIVERSITY PRESS

University Printing House, Cambridge, CB2 8BS, United Kingdom

Cambridge University Press is part of the University of Cambridge.
It furthers the University's mission by disseminating knowledge in the pursuit of
education, learning and research at the highest international levels of excellence.

www.cambridge.org
Information on this title: www.cambridge.org/9781108070287

© in this compilation Cambridge University Press 2014

This edition first published 1827
This digitally printed version 2014

ISBN 978-1-108-07028-7 Paperback

This book reproduces the text of the original edition. The content and language reflect
the beliefs, practices and terminology of their time, and have not been updated.

Cambridge University Press wishes to make clear that the book, unless originally published
by Cambridge, is not being republished by, in association or collaboration with,
or with the endorsement or approval of, the original publisher or its successors in title.

THE
STEAM ENGINE,

COMPRISING

AN ACCOUNT OF ITS INVENTION AND PROGRESSIVE IMPROVEMENT;

WITH AN

INVESTIGATION OF ITS PRINCIPLES,

AND THE

PROPORTIONS OF ITS PARTS FOR EFFICIENCY AND STRENGTH:

DETAILING ALSO ITS APPLICATION TO

NAVIGATION, MINING, IMPELLING MACHINES, &c.

AND THE RESULTS COLLECTED IN

NUMEROUS TABLES FOR PRACTICAL USE.

ILLUSTRATED BY TWENTY PLATES, AND NUMEROUS WOOD CUTS.

BY THOMAS TREDGOLD,
CIVIL ENGINEER;

MEMBER OF THE INSTITUTION OF CIVIL ENGINEERS; AUTHOR OF ELEMENTARY PRINCIPLES OF
CARPENTRY; A PRACTICAL TREATISE ON THE STRENGTH OF IRON, &c.

" It is certain, that of all powers in nature, heat is the chief."—BACON.
" The errors are not in the art, but in the artificers."—NEWTON.

LONDON:
PRINTED FOR J. TAYLOR,
AT THE ARCHITECTURAL LIBRARY, N°. 59, HIGH HOLBORN.
1827.

T. BARTLETT, PRINTER, OXFORD.

TO

THOMAS HOBLYN, ESQ.

FELLOW OF THE ROYAL SOCIETY,

VICE PRESIDENT OF THE SOCIETY

FOR

THE ENCOURAGEMENT OF ARTS, MANUFACTURES, AND COMMERCE, &c. &c. &c

THE AUTHOR

INSCRIBES

THIS WORK

DESCRIPTIVE OF

THE PRINCIPLES AND CONSTRUCTION OF THE

STEAM ENGINE;

WHICH, INVENTED AND PERFECTED BY THE ARTISTS OF BRITAIN, HAS RENDERED
HEAT AN INEXHAUSTIBLE SOURCE

OF

WEALTH AND PROSPERITY

TO THE

BRITISH EMPIRE.

PREFACE.

OF the various books published on that important and national subject the Steam Engine, there is not one in our own or any foreign language, which I consider as a fully satisfactory illustration of its principles; it is therefore only requisite for me to state this fact to render any apology unnecessary for the work I now offer to the notice of the Public. I have frequently and successfully claimed attention as an author; and in this case I hope to meet with equal success, and to shew by the labour and attention I have bestowed on this important subject, how highly I value the ostensible character I have acquired, and the extensive encouragement I have received.

It has been too common of late for mathematicians to complain of want of patronage, and to censure official authorities for not encouraging science, forgetting that research will always be estimated by its intermediate utility; and while they continue to confine their attention to abstract knowledge, while they do not devote a greater part of their time to its application to the wants and the welfare of society, they must be contented with a small share of those advantages which result from combining with practical skill, the power afforded by abstract reasoning. They should recollect that a Watt could have earned no fame, in an age nor in a country where the value of mecha-

b

nical power was unknown. In following the application of science to art, I have not, however, I hope been unsuccessful in adding also to the stores of pure science ; and, so far from being insensible to the value of abstract research, I wish it to be pursued with redoubled vigour by those who have spirit to break through the prejudices of existing systems, and study from nature : but it should be cultivated with a desire to promote the great end of human research, that is, the improvement of the condition of man ; otherwise the fantasies of the Greek philosophers might with equal force claim the student's regard.

I hope these remarks will tend to encourage those who pursue knowledge, whether with the energy of youth or the more steady enthusiasm of riper years ; and as all nature, so all art, must ever be the result of those immutable proportions and laws of action which it has pleased our Creator to impress on matter, its objects are truly boundless. Our imperfection consists generally in not being able to foresee all the circumstances which have an influence on the effects of causes ; but in proportion as we proceed in knowledge, we also acquire greater powers of perception : that which was at first difficult becomes easy, and the mind is often roused by the bright gleam of truth, breaking as it were accidentally upon a mass of obscure ideas, and rendering the true solution of the difficulty at once obvious ; and as my gifted countryman Emerson has remarked, " the labour and fatigue of seeking after it instantly vanishes."

I proceed now to give some idea of this work. It appears to be large for its object ; but, though confined to a single source of power, that power is gigantic, and involves so many new and important doctrines in mechanical science and practice, that it was impossible in justice to comprise it in less space. The work is in Ten Sections.

In the *First*, the history of the progressive improvement of the steam engine is traced, from the period of its first suggestion by the Marquis of Worcester, to its present state of high perfection.

The *Second Section* presents an analysis of the nature of steam and of other species of vapour; the laws of their combination with heat, and of their elastic force, density, and comparative power; with the principles of calculating their velocity when in motion, loss of force by cooling, &c. In this section it is shewn that water is of all other known fluids that best adapted for producing steam.

The *Third Section* treats of the laws of combustion, and of the effect of different species of fuel in producing steam; the proportions of fire places and chimneys of boilers, and the precautions necessary for their security and effect: the nature and application of safety apparatus is fully discussed. The section closes with a developement of the principles of condensing steam.

In the *Fourth Section*, the power afforded by a given quantity of steam, and all the methods of developing it, are illustrated both in a popular and scientific manner; and the theoretical defect of the rotary action of steam is investigated. The various modes of applying the power of steam are shewn, with a classification of engines; and the velocity and proportions which give a maximum of effect in engines, as well as the nature and office, and the power lost in working the air pumps of engines, are investigated.

The *Fifth Section* treats of the construction of the essentially different varieties of noncondensing steam engines; these engines are all of the high pressure kind, and the causes of loss of power, and means of employing steam to the best advantage, and the mode of calculating the power and proportion of the parts, are given in detail for each species.

The *Sixth Section* treats, in like manner, of the construction, proportions, power, and economy of condensing engines: in these sections, for the first time, those minute causes which affect the action of steam are not only stated, but are reduced to measure; and I trust in such a manner as to be most useful, both to those who wish to apply, and to those who wish to improve, the steam engine.

In the *Seventh Section*, the proportions and construction of the parts of steam engines are considered, as of cocks, valves, slides, pistons, stuffing boxes, &c.; also the modes of opening and closing valves, and the like, followed by a description of the different kinds of piston-rod guides, and an investigation of crank motions, and of the combinations for producing parallel motion. Also practical rules for the strength of the various parts of steam engines are added, and especially for boilers of different kinds.

The *Eighth Section* treats, First, of the modes of equalizing the action of the steam engine, as by fly wheels or counter weights. Secondly, of regulating the power of engines, as by valves, governors, regulators, &c. Thirdly, the method of ascertaining the state and intensity of the forces in engines, and the means of measuring their effective power. And, Fourthly, of the mode of working a steam engine.

The *Ninth Section* illustrates the application of steam power, to raising water, to the drainage and business of mining, to impelling machinery for manufacturing and for agricultural purposes, and its application to land carriage by means of railways.

The *Tenth Section* is on steam navigation; and the stability of vessels, their resistance to motion in fluids, the means of propelling them, and the modes of proportioning the power to the effect, are investigations altogether new; and of necessity so, for the theory of the resistance of fluids hitherto taught in schools, is erroneous and cannot be applied. I have therefore endeavoured to explain the methods of my own researches in popular rather than strictly scientific discussions, reserving for a separate work the full developement of my views on this important branch of science.

The tables will be useful in practice, and the plates are accompanied by descriptions, so as to render them of easy reference, and also to enable me to refer to the parts of the work which they tend to illustrate.

I am indebted to the friendly assistance of some of my professional brethren for access to information, which otherwise I could not have obtained: in a few instances, their favours arrived too late, except for my own satisfaction in finding that they conformed to the principles laid down in this treatise; of Mr. Bevan's interesting experiments on the resistance of boats I have given only part, because the others were evidently affected by the limited section of the canal. One of the plates (XIII.) was furnished by Mr. White, Engineer, and a few of the others are selected from the very accurate plates drawn by Clement, and published in Partington's History of the Steam Engine ; the rest are engraved from my own drawings, and are aided by a great number of wood engravings on the pages.

My great object has been to lead the reader to study the principles of the steam engine, and to furnish him not only with materials for study, but also with methods of reasoning, and in sufficient variety to enable him to examine any new case likely to occur ; and in proportion to the care and pains he bestows on the inquiry, he will feel the advantage of the few steps I have taken in this interesting and important subject.

I shall conclude in the language of Sir Isaac Newton, on a greater occasion, "I heartily beg that what I have here done may be read with candour, and that the defects I have been guilty of upon this difficult subject may not be so much reprehended as kindly supplied, and investigated by new endeavours of my readers."

THOMAS TREDGOLD.

16 Grove Place,
    Lisson Grove, London.
        August 13, 1827

CONTENTS.

———

xii   CONTENTS.

SECT. II.—OF THE NATURE AND PROPERTIES OF STEAM, ITS ELASTIC FORCE, EXPANSIVE FORCE, AND POWER OF MOTION.

CONTENTS. xiii

Sect. III.—Of the Generation and Condensation of Steam and the Apparatus for those Purposes.

c

SECT. IV.—OF THE MECHANICAL POWER OF STEAM, AND THE NATURE, GENERAL PROPORTIONS AND CLASSIFICATION OF STEAM ENGINES.

SECT. V.—OF THE CONSTRUCTION OF NONCONDENSING ENGINES.

CONTENTS. XV

Sect. VI.—Of the Construction of Condensing Engines.

Sect. VII.—Of the Proportions, and the Construction of the Parts of Steam Engines.

CONTENTS.

SECT. VIII.—OF EQUALIZING THE ACTION, REGULATING THE POWER, MEASURING THE USEFUL EFFECT, AND MANAGING THE STEAM ENGINE.

CONTENTS.

TABLES.

MEASURES, WEIGHTS, &c. USED IN THIS WORK.

Temperature is measured by degrees of Fahrenheit's scale, of which the freezing point is 32°, and the boiling point 212°.

Heat is measured by the degrees the same quantity of heat would increase the temperature of a given quantity of water at 60°, with the barometer at 30 inches.

Mechanical power is measured by the elementary horse power, as settled by Mr. Watt. A horse power is = 33,000 lbs. raised one foot high per minute, or = 550 lbs. raised one foot high per second; and a day's work of a horse is this power acting eight hours.

This horse power is, in French measures, 4661 kilogrammes raised one metre high per minute.

The pound is the avoirdupois pound, = 7000 troy grains, = ·4535 French kilogrammes.

The foot is = ·3048 French metres.

An atmosphere is 30 inches of mercury = ·762 French metres.

THE STEAM ENGINE.

SECTION I.

AN ACCOUNT OF THE INVENTION AND PROGRESSIVE IMPROVEMENT OF THE STEAM ENGINE.

ART. 1.—WHEN an efficient mechanical power is produced by the generation, or generation and condensation, of the steam or vapour of any liquid, the combination of vessels and machinery for that purpose, is called a Steam Engine. This engine was for a considerable time after its invention called a Fire Engine, and not improperly, for the active agent is heat or fire. The liquid almost universally employed for obtaining steam is water, but it may be obtained from alcohol, ether, and other fluids; fortunately, however, water, the most easily procured, is equal if not superior to any other.

2.—That the application of heat would generate steam from water, and that the steam so generated would isssue with much force from a small aperture in the vessel employed to generate it in, must have been known at a very early period. The eolipile, and some other similar instruments for illustrating natural phenomena, were well known among the Egyptians, Greeks, and Romans. Vitruvius, who wrote during the reign of Augustus Cæsar, refers to the eolipile as an illustration of the effect of heat in producing winds;* but he clearly had no idea of steam being rendered useful as a mechanical power. Philibert de l'Orme proposed placing an eolipile over a fire as a means of impelling smoke up a chimney,† and several applications of this instrument are described in the works of Solomon de Caus, Brancas, Van Drebbel, and various other writers, the greater part of whom are mentioned by M. Montgéry,‡ an author, who has been at considerable trouble to shew that the invention of the steam engine is not of English origin.

* Vitruvius, lib. 1. cap. vi.   † Traité d'Architecture, folio, Paris, 1567.
‡ Notice historique sur l'invention des machines à vapeur.

3.—But unless it be shewn that an engine had been actually invented, and was un-doubtedly applicable to some of the purposes for which the steam engine is now employed, and for which alone it has become valuable, it appears to be mere trifling to search for authorities, and absolutely unworthy of occupying the time or attention of a man of real science. The blast of an eolipile is certainly not a mode of employing steam capable of producing the species of useful effect which is obtained by a steam engine, and, as a proof of its inefficiency, the same principle of action (that is by impulse) has never been rendered applicable to produce mechanical power for useful purposes in a steam engine.

It is not my object, therefore, to inquire when it was first ascertained that steam has force; but, to endeavour to trace the history of its suggestion in a practical form, and of its application in the arts and manufactures; to develope the various changes and improve-ments the steam engine has received; and to shew, among the host of projectors, those who have really advanced our knowledge, either regarding the principles, the construction, or the arrangement of this powerful prime mover.

It is easy to perceive that I have assigned myself a difficult task, but it is equally evident that if it be accomplished, in a judicious and candid manner, it will form a valuable addi-tion to an interesting branch of mechanical science; hence, I am encouraged to proceed, and trust to leave my reader with an impression, that I have been just in deciding between the claimants of the invention of each of the parts of the steam engine.

1663. *Marquis of Worcester, died* 1667.

4.—The idea of employing the impulsive force of the eolipile, seems to be the only one which had been formed for using steam as a source of motion before the time of the Marquis of Worcester; and he, in a little work entitled " A Century of the Names and Scantlings of Inventions," undoubtedly describes a method of employing the pressure of steam for raising water to great heights.* His work was first published in 1663, and under the sixty-eighth invention we have the following name and scantling :—

" LXVIII. *A Fire Water Work.*—An admirable and most forcible way to drive up water by fire, not by drawing or sucking it upwards; for that must be, as the philosopher calleth it, *infra sphæram activitatis,* which is but at such a distance. But this way hath no bounder if the vessels be strong enough; for I have taken a piece of a whole cannon, whereof the end was burst, and filled it three-quarters full of water, stopping and screwing up the broken end, as also the touchhole, and, making a constant fire under it; within twenty-four hours it burst and made a great crack; so that having a way to make my vessels so that they are strengthened by the force within them, and the one to fill after the

---

* Another engine, which the marquis terms a " Water-commanding Engine," seems to have been the one for which he obtained an act of parliament, allowing him the monopoly of the profits arising from its use.

other, I have seen the water run like a constant fountain stream forty feet high. One vessel
of water rarefied by fire driveth up forty of cold water.   And a man that tends the work is
but to turn two cocks, that, one vessel of water being consumed, another begins to force
and refill with cold water, and so successively, the fire being tended and kept constant;
which the selfsame person may likewisely abundantly perform in the interim between the
necessity of turning the said cocks."

    This description puts it beyond a doubt that the Marquis of Worcester knew that steam,
heated in a close vessel, acquires an immense degree of force, and that this force could be
effectively applied to raise water.   The effect of condensation he does not appear to have
been at all acquainted with, and therefore, his mode of operation must have been exceedingly simple, and probably, of the nature exhibited in the annexed figure:—Where B is
the boiler; C, one of the vessels with a pipe to deliver the water to an elevated cistern D.

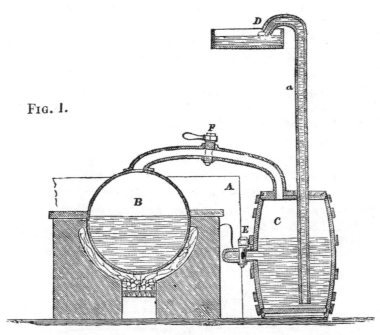

Fig. 1.

Now suppose the vessel C to be supplied from a cistern of cold water A by a pipe, so that
it would be filled on opening the cock E, and this cock being closed; if, when the steam
in the boiler is of sufficient strength, the cock F be opened, the pressure of the steam on
the water in C would cause it to ascend from C, through the pipe a into the cistern D.
The vessel C being emptied and the cock F being shut, it would refill with water on opening the cock E.   Another vessel C and its cocks and pipes are necessary to complete the

species of water engine indicated by the description, and these may be on the other side of the boiler.

Such a mode of raising water would be most expensive from the quantity of condensation when the steam came in contact with cold water, but it was fully capable of producing the quantity of effect mentioned, for it is only equivalent to raising twenty cubic feet of water or 1250 lbs. one foot high by one pound of coal, or about the 200th part of the effect of a good steam engine. Hence, it appears, that to the Marquis of Worcester must be ascribed the first invention and trial of a practical mode of applying steam as a prime mover, and of applying it to one of those great purposes for which it has been so useful to society.

1683. *Sir Samuel Morland, died* 1695.

5.—From a part of a manuscript in the Harleian collection in the British Museum, it appears that a mode of raising water by steam, similar to that of the Marquis of Worcester's, was proposed, among other methods, to Louis XIV, of France, by Sir Samuel Morland. It contains no description of the method he intended to employ, but there is sufficient to indicate that its author was not without knowledge of his subject.

The title of the part which treats of the power of steam is, " The Principles of the New Force of Fire, invented by Chev. Morland in 1682, and presented to his most Christian Majesty, 1683;" and these principles are explained as follows:—" Water being converted into vapour by the force of fire, these vapours shortly require a greater space (about 2000 times) than the water before occupied, and sooner than be constantly confined, would split a piece of cannon. But being duly regulated according to the rules of statics, and by science reduced to measure, weight, and balance, then they bear their load peaceably (like good horses) and thus become of great use to mankind, particularly for raising water, according to the following table, which shews the number of pounds that may be raised 1800 times per hour, to a height of six inches, by cylinders half filled with water, as well as the different diameters, and depths of the said cylinders."

Cylinders.		Weight of the Load to be raised.
Diam. in feet.	Depth in feet.	
1	2	15 lbs.
2	4	120 —
3	6	405 —
4	8	960 —
5	10	1875 —
6	12	3240 —

These numbers are obviously proportional to the capacity of the cylinders.

The table is continued in the original to shew the effect of a number of cylinders of the largest of the above sizes, each one being capable of raising 3240 lbs.

Morland has given the increase of volume, which water occupies in the state of vapour at common pressures, so nearly, that we may suppose it to be the result of experiment, while his allusion to the force of steam being sufficient to burst a cannon, and his proposal of the method to a foreign prince, render it probable that he was not a stranger to the volume the Marquis of Worcester had published twenty years before.

Morland's researches seem to have had little influence on the progress of the practical application of steam.

6.—In 1695, Dr. Papin suggested the idea of employing the expansion and contraction of steam to form a partial vacuum under a piston for raising water, and making the pressure of the atmosphere on the upper side of the piston the moving power.* The real authors of the atmospheric engine were very likely indebted to this suggestion, but neither Papin himself, nor his rival, Savery, discovered how to turn this suggestion to advantage. Indeed, it was proposed in a form which was not practicable, the fire was to be alternately applied to, and removed from, the cylinder, and the expansion of the water in it, by heat, was to raise the piston, and its contraction, by cooling, when the fire was removed, was to cause a partial vacuum, and, consequently, the descent of the piston was to be produced by the pressure of the atmosphere. If such a scheme was ever tried, the result must have been sufficiently discouraging for Papin to abandon it and adopt a new one, which it will be found he actually did, after seeing an engraving of Savery's engine.

1698. *Thomas Savery.*

7.—These projects were speedily followed by a direct practical application of the steam engine to raising water, for which " letters patent," were granted to Captain Thomas Savery, in July 1698, (these being the first on record granted for a steam engine,) and, Dr. Robison says, it was " after having actually erected several machines," of which, Savery gave a description in a pamphlet he published in 1699,† called " The Miner's Friend," which was republished with additions in 1702.

In June 1699, Captain Savery exhibited a model of his engine before the Royal Society; and the experiments he made with it succeeded to their satisfaction.‡ It consisted of a

* Phil. Trans. Abridg. IV. 155, 1697.

† Robison makes it 1696, but this does not appear to be correct. Switzen's date 1699, is taken as likely to be the right one, from his System of Hydrostatics II. 326.

‡ Phil. Trans. Abridg. IV. 198, 1699.

furnace and boiler B; from the latter two pipes, provided with cocks C, proceeded to two steam vessels S, which had branch pipes from a descending main D, and also to a rising

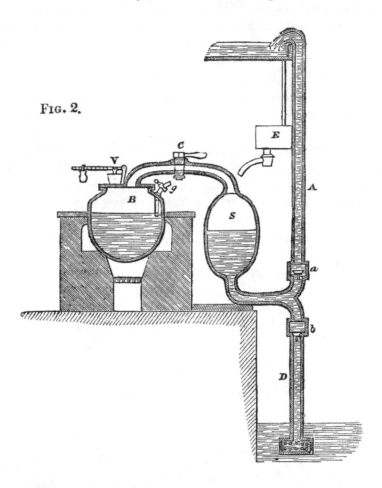

FIG. 2.

main pipe A; each pair of branch pipes had valves a, b, to prevent the descent of the water raised by the condensation or by the force of steam.   Only one vessel S is shewn, the other being immediately behind it.   One of the steam vessels being filled with steam, condensation was produced by projecting cold water, from a small cistern E, against the vessel; and into the partial vacuum made by that means, the water was forced up the ascending main D, by the pressure of the atmosphere from a depth of about twenty feet; and, on the steam being let into the vessels again, the valve b closed and prevented the descent of the water, while the steam having acquired force in the boiler, its pressure caused the water to raise the valve a, and ascend to a height proportional to the excess of the elastic force of the steam above the pressure of the air.

Captain Savery afterwards simplified this engine considerably by using only one steam vessel. To prevent the risk of bursting the boiler, he applied the steelyard safety valve V; invented by Papin for his digester. The cocks were managed by hand; and, to supply the boiler with water, he had a small boiler adjoining to heat water for the use of the large one, and thus prevent the loss of time, which must have occurred on refilling it with cold water.

Several engines for raising water appear to have been erected according to Savery's plan, and to have succeeded tolerably where the water had not to be raised more than forty feet, but this was not sufficient for mines where a new and powerful machine was most wanted.

The new principles, introduced into the steam engine by Savery, consist of the use of condensation in the steam vessel by cold applied externally. He also used a method of supplying the boiler with hot water, contrived a mode of ascertaining the quantity of water in the boiler, by inserting the cock g, called a gauge cock, and applied the safety valve of Papin's digester as a means of preventing accidents.

The defects of his engine are obvious. A cold vessel and cold fluid must at each operation condense, and, therefore, waste a great quantity of steam; and the height to which water could be raised, unless by the use of such powerful steam as to render it dangerous, was too limited to be applicable to mining purposes. Its effect would, however, be vastly superior to that of the Marquis of Worcester's. Whether Captain Savery did or did not know of the previous schemes, his claims to original invention are certainly considerable, and to his enthusiasm and talent, we undoubtedly owe the first effective steam engine.

1698. *Dr. Dennis Papin.*

8.—Dr. Papin, professor of mathematics at Marbourg, whose former project I have noticed, (art. 6,) is said to have made many experiments on raising water by the force of fire in 1698, by the order of Charles, Landgrave of Hesse; and in 1707, he published a small treatise on the subject in which he ascribes to the landgrave, the whole merit of the first idea of a steam engine. Papin's trials in 1698, whatever they were, did not end in producing any thing in an useful shape; and, while he candidly acknowledges that Savery's scheme was not borrowed from any thing done in Germany, it appears that he did not follow up his experiments, till after he had seen an engraving of Savery's engine, in June 1705; a pretty conclusive argument, that no satisfactory results had been arrived at in these experiments, and there is a wide distinction between unsuccessful experiments and invention.

To do justice to the claims of Papin, it will be sufficient to describe his engine in its most improved state, and as he gives it after knowing what Savery had effected. It con-

sisted of a boiler B, provided with a safety valve *v;* and a cylinder G H, connected to the
boiler by a steam pipe S.   The cylinder was closed at the top, and contained a floating

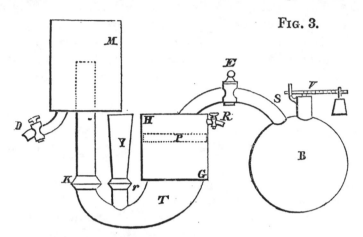

Fig. 3.

piston P; and the base of the cylinder terminated in a curved tube T, which ascended
into a cylinder M, the bent tube had a pipe, Y, from a reservoir of water communicating
with it, and it was provided with a valve at *r.* Now suppose the cylinder G H, to be filled
with cold water by the pipe Y from the reservoir, and the boiler to contain strong steam;
by opening the cock E, the steam would be admitted, and, pressing on the floating piston
P, cause the water to ascend into the cylinder M; its return is prevented by the valve K,
and the steam cock E being shut and the cock R opened, to let the condensed steam escape
at the pipe R, the water from the reservoir refills the steam cylinder through the pipe Y,
and it is ready for repeating the operation.   The water raised to be directed to any useful
object by the pipe D.[*]

A reference to the Marquis of Worcester's plan will shew that Papin did no more than
repeat his experiments.   The scheme of adding to the effect by the introduction of red hot
irons into the cylinder G H, is too absurd to insert; but it is in some measure redeemed
by the suggestion that the water raised by the engine might be applied to drive a water
wheel; thus giving the idea of a steam engine being applicable to impel machinery.

9.—In 1699, Mr. Amontons published a description of a machine, designed to be
moved by the spring of air when expanded by heat, and afterwards condensed by contact
with cold water.[+] The continual access of heated air to water would ultimately render the
air saturated with vapour, but even then it would not be more than an air engine, and a
very indifferent one, being exceedingly complex.

[*] Belidor's Archi. Hydrau. II. p. 328.
[+] Prony's Nouvelle Archi. Hydrau. II. p. 89, (note,) where it is described.

1705. *Thomas Newcomen.*

10.—The trials of Savery's engines made known their defects, yet evidently strengthened the idea that steam could be effectively applied to raise water; and the immense expense of raising water from deep mines, so embarrassed their proprietors, that there were most powerful incentives at that period to engage in further researches on the subject. To this stimulus we are indebted for another construction of the steam engine by Thomas Newcomen, a smith, of Dartmouth, who, in conjunction with John Cawley, a plumber, of the same place, and Captain Savery, obtained letters patent for the invention in 1705.* The novelty of this construction consists entirely in condensing the steam below an air-tight piston, in a cylindrical vessel having an open top; and the idea was very probably taken from the project of Papin in 1695, (see art. 6.:) for it appears that Newcomen was in correspondence with Dr. Hook on the subject, to whom the speculations of Papin were well known; but the mode of effecting the object was entirely different from Papin's. It consists in admitting steam below a piston; and, at first, the steam was condensed by applying cold water to the outside of the cylinder; but injection of cold water by a jet into the interior was soon found to be a more effective method, and is said to have been disco-vered by accident.† The following is a description of the engine, as far as it was improved by Newcomen. B represents the boiler with its furnace for producing steam, and at a small height above the boiler is a steam cylinder, C, of metal, bored to a regular diameter, and closed at the bottom; the top remaining open. A communication is formed, between the boiler and the bottom of the cylinder, by means of a short steam pipe S. The lower aperture of this pipe is shut by the plate *p*, which is ground flat, so as to apply very accu-rately to the whole circumference of the orifice. This plate is called the regulator, or steam cock, and it turns horizontally on an axis *a*, which passes through the top of the boiler, and is fitted steam-tight; and has a handle *b* to open and shut it.

* Switzer says, on report, that Newcomen was as early in his invention as Savery. Sys. of Hydros. ii. 342.

† Desaguliers' Experimental Philosophy, ii. p. 533. The piston was kept tight by a quantity of water on the top of it; and as they were working by condensing from the outside, they were surprised to see the engine make several strokes very quickly, and found that it was owing to a hole in the piston letting down water to condense the steam. This suggested the idea of injection.

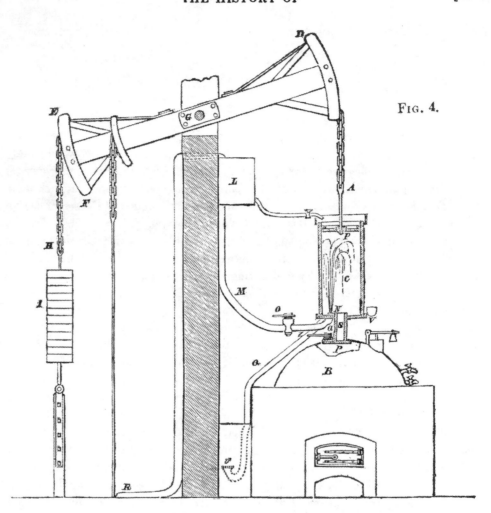

Fig. 4.

A piston P is fitted to the cylinder, and rendered air-tight by a packing round its edge of soft rope, well filled with tallow, to reduce the friction, and its upper surface is kept covered with water to render it steam-tight. The piston is connected to a rod P A, which is suspended by a chain from the upper extremity D of the arched head of the lever, or working beam, which turns on the gudgeon G. This beam has a similar arched head E F, at its other end, for the pump rod H, which receives the water from the mine. The end of the beam to which the pump rod is attached, is made to exceed the weight and friction of the piston in the steam cylinder; and when the water is drawn from such a depth, that the pump piston is too heavy for this purpose, counterpoise weights must be added at I, till the piston will rise in the steam cylinder at the proper speed. At some height above the top of the cylinder is a cistern L, called the injection cistern, supplied with water from the forcing pump R. From this descends the injection pipe M, which enters the cylinders through its bottom, and terminates in one or more small holes at N. This pipe has at O a cock, called the injection cock, fitted with a handle. At the opposite side of the cylinder, a little above its bottom, there is a lateral pipe, turning upwards at the extremity, and provided with a valve at V, called the snifting valve, which has a little dish round it to hold water for keeping it air-tight.

There proceeds also from the bottom of the cylinder a pipe Q, of which the lower end is turned upwards, and is covered with a valve v; this part is immersed in a cistern of water called the hot well, and the pipe itself is called the eduction pipe. To regulate the strength of the steam in the boiler, it is furnished with a safety valve, constructed and used in the same manner as that of Savery's engine, but not loaded with more than one or two pounds on the square inch.

The mode of operation remains to be described. Let the piston be pulled down to the bottom of the steam cylinder, and shut the regulator or steam valve p. Then the piston will be kept at the bottom by the pressure of the atmosphere. Apply the fire to the boiler till the steam escapes from the safety valve, and then on opening the steam regulator, the piston will rise by the joint effect of the strength of the steam, and action of the excess of weight on the other end of the beam. When it arrives at the top of the cylinder, close the regulator p, and by turning the injection cock O, admit a jet of cold water, which condenses the steam in the cylinder, forming a partial vacuum, and the piston descends by the pressure of the atmosphere, raising water by the pump rod H from the mine. The air which the steam and the injection water contain, is impelled out of the snifting valve V, by the force of descent, and the injection water flows out at the eduction pipe Q; and by repetition of the operations, of alternately admitting steam and injecting water, the work of raising water is effected.

These operations were done by hand, till a boy, named Humphrey Potter, contrived to attach strings and catches to the working beam, for opening and shutting them while he was at play;* after which, more permanent appendages were added to answer the purpose, and the engine became a step nearer to a self regulated machine.

The engine in this simple and efficient state was termed the Atmospheric Engine. It was brought to this degree of perfection about 1712; and such engines were erected in various places.

The novelty of this engine is chiefly in its mechanism; but as this mechanism produces

* Desaguliers' Experimental Philosophy. ii. 533.

all the difference between an efficient and an inefficient engine, I am inclined to set a higher value upon it than on the fortuitous discovery of a new principle. To point out what is actually due to Newcomen would be difficult, and for want of evidence we must be content with examining the state of the engine. The admission of steam below an air-tight piston, attached to the impelled point of a lever properly counterpoised; its rapid condensation by injection of water, which is essential to gain effect; and the mode of clearing the cylinder of air and water after the stroke; are all in addition to the principles and mechanism before in use: and these are wholly due to Newcomen or those connected with him.

1718.   *Henry Beighton, F.R.S.* died 1743.

11.—The arrangement of the parts of the atmospheric engine, the mode of fixing, and the mechanism for opening and shutting the valves, were greatly improved by Mr. Henry Beighton, an Engineer, of Newcastle-upon-Tyne. He also seems to have been the first to reduce the calculation of the powers of engines to a regular system, and published a "Table of the Dimensions and Power of the Steam Engine" in 1717, which has been found to accord with practice;[†] and he directed the construction of several large engines. He also remarked the fact of steam heating a very large proportion of water in condensing, and communicated to Dr. Desaguliers some experiments on the bulk of steam formed by a given quantity of water, the result of which was erroneously stated in consequence of a singular mistake in the calculation; and it is also obvious, that the mere quantity of water, and bulk of the cylinder, could not possibly give the result he expected, even on the supposition that the cylinder was maintained at 212° during the experiment.[‡] I cannot leave

---

* Dr. Hutton remarks, it is probable that Mr. Beighton died in 1743 or 1744, as it appears that he conducted the Ladies' Diary for the Stationers' Company, from 1714 to 1744 inclusively; discharging that trust with such satisfaction to the company, that they permitted his widow to enjoy it for many years afterward, by employing a deputy to compile that very useful annual little book. In that almanack, for the year 1721, Mr. Beighton inserted a curious table of calculations on the steam engine. Phil. Trans. Abrid. vii. p. 442.

† Desaguliers' Course of Experimental Philosophy. ii. 534.

‡ In Mr. Beighton's experiment (Desaguliers' Ex. Phil. ii. 533,) made on the steam engine, to know what quantity of steam a given quantity of water produces, he found by several trials with a divided steelyard on the safety valve on the top of the boilers at Griff and Wasington, that when the elasticity of the steam was just one pound on a square inch, it was sufficient to work the engine; and that about five pints in a minute would feed the boiler, as fast as it was consumed in producing steam for the cylinder at sixteen strokes per minute. Griff's cylinder held 113 gallons of steam every stroke, hence 113 × 16 = 1808 gallons = 14464 pints, therefore five pints of water produced 14464 pints of steam; consequently, 1 pint would produce 2893 pints of steam of that density and temperature it had in the cylinder at the termination of the stroke; but this temperature and density not being ascertained, the experiment does not shew the bulk corresponding to the atmospheric pressure; for the elastic force in the boiler differs considerably from that in the cylinder.

the memory of Beighton without the remark, that though he was not distinguished by the novelty of his views, yet, the sound knowledge he had of science, seems to have been of more real advantage to those who sought benefit from the steam engine, than the undirected efforts of his predecessors.

1720. *Leupold.*

12.—About this period various writers gave notices of the different engines that had appeared, but those who added nothing, either in theory, experiment, or construction, it would be as tedious as useless to notice. But in this class, Leupold, the industrious German, collector of mechanical inventions, ought not to be placed, he having given the first sketch for a high-pressure engine with a piston; it is further remarkable as having a four-passage cock for the admission and emission of steam.

The scheme of Leupold is simple; over a boiler B, he placed two cylinders C C, fitted with steam-tight pistons, *p p*. A four-way steam cock S, is placed between the boiler and

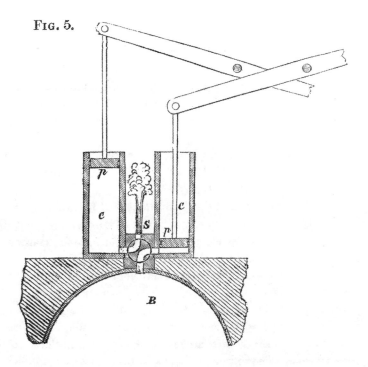

FIG. 5.

cylinders, so as to alternately admit steam into one cylinder and let it out from the other. The piston, by the admission of strong steam from the boiler below it, is raised, and depresses the other end of a lever connected to the rod of a plunger of a pump, which causes the water to rise through the pipe, and by the alternate action of the steam in the two cylinders, a continual stream of water is raised. Thus the first rude notice of the principle of employing high-pressure steam under a piston was given.

13.—It does not appear that any thing was added to the existing knowledge of the steam engine by Dr. Desaguliers, though, from his fondness for experimental philosophy, we are led to expect he would have taken an important place in reducing to fixed principles the phenomena he was daily called upon to witness, and make known to the world. This was not the case; and, for historical information, he was evidently too prejudiced in favour of particular individuals to allow him to detail facts with candour and fairness; therefore, the matter he collected, in his Experimental Philosophy, is no further valuable, than by making the state of the engine at that period, and part of the researches of Beighton, known.

1736. *Jonathan Hulls.*

14.—The atmospherical steam engine, as improved by Beighton, began to be very generally adopted in the coal works and copper mines; and it does not seem to have required any great stretch of invention to direct such an efficient power to other purposes, besides that of raising water.

The first attempt, however, on record, was one to apply steam to navigation, and was made by Jonathan Hulls, who, on the 21st of December, 1736, obtained a patent, for what may strictly be considered a steam boat.

The letters patent, and a description of this boat, illustrated by a plate, was published in a tract, by Hulls, in 1737, under the following title:—" A Description and Draught of a new invented Machine for carrying Vessels or Ships out of or into any Harbour, Port, or River, against Wind or Tide, or in a Calm." As the origin of the invention of steam boats has been strongly contested, this pamphlet, which it is now very difficult to obtain on account of its rarity, has been brought forward to prove that Jonathan Hulls was the first person who suggested the power of steam as a means of propelling paddle wheels. His mode of converting the reciprocating motion of the engine into a rotatory one, is less simple than the crank, but it appears to have been the first attempt, and was done in the following manner: Let a, b, c, be three wheels on one axis, and d, e, two wheels loose on another axis A, with ratchets. so as to move the axis only when they move forwards; f, g, h, are three ropes, and P is the piston of the engine. When the piston descends, the wheels a, b, c, move for-

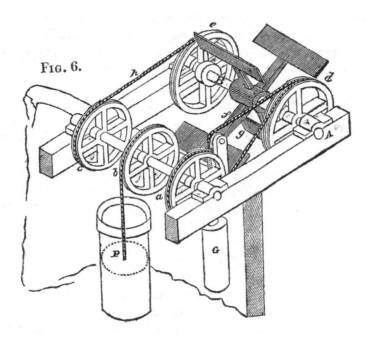

FIG. 6.

ward, and the ropes *g*, *h*, cause the wheels *e*, *d*, to move; the wheel *e* forward, and the wheel *d* backward, and the latter raises the weight G, which moves the wheel *d* forward during the ascent of the piston; consequently, the axis A B, with the paddle wheels, would be constantly moved round in the same direction, and by an equable force. This is certainly a beautiful contrivance for rendering so irregular a first mover equable, and considering the object it was intended for, it is not a complex arrangement; for besides equalizing the power, it gives a means of increasing or diminishing the velocity in the ratios of the diameters of the wheels. The pamphlet of Hulls bears evidence of being the work of an ingenious and well informed mind; and we must regret the causes which prevented his views meeting the encouragement they merited.*

1739. *Bernard Belidor, born* 1698, *died* 1761.

15.—Belidor, so eminently distinguished as a writer on the theory and practice both of civil and military engineering, treated of the steam engine in 1739, and undoubtedly presented the most accurate information then existing in France on the subject.† He gives a

---

* Hull's pamphlet may be seen at the British Museum, or the London Institution ; and many civil engineers have succeeded in adding it to their own collections.

† Architecture Hydraulique, tom. ii. p. 300—331.

slight sketch of its history; and infers, from his inquiries, that the three nations of Europe most advanced in the pursuit of knowledge, each furnished a man of science to participate in the glory of the important discovery: that Papin in Germany, Savery in England, and Amontons in France, were each occupied in studying the means of making use of the action of *fire* for moving machines; but the first suggestion of the idea, in an intelligible form, he acknowledges to be due to the Marquis of Worcester. Belidor closes his historical notice by remarking, that all the fire engines that had been constructed out of Great Britain, had been executed by English workmen; and then proceeds to describe the atmospheric engine at Fresnes near Condé, in that minute, clear, and practical manner, which renders his writings so valuable. To the theory of the action of steam Belidor added nothing, and the formulæ he has given for calculating the load to be lifted by an engine are neither very simple nor accurate; like those of Beighton they apply only to the statical equilibrium of the machine.

1741. *John Payne.*

16.—The first direct experiment to determine the density of steam was made by John Payne.* His process was well devised, but wanted the addition of a thermometer. He took a copper globe twelve inches in diameter, having two cocks fitted to it, and a small valve. The vessel thus prepared was hung over a large vessel, in which water was rarefied into steam, and by a pipe the steam was admitted at one of the cocks into the globe, and the other being also open, the steam being allowed to blow through, forced out the air that was in the globe, and supplied its place; when both cocks were suddenly shut, and the globe taken down and hung over a vessel of cold water with the lower cock immersed in water. The cock was opened under water, on which the water rushed violently into the globe till it had supplied the vacuum, when the cock was again shut, and the globe, with the water, was put in the scales, and found to weigh 713 oz. which taken from 727 oz. the whole weight before, there remains only fourteen ounces the difference, from which he inferred that all the air was nearly excluded out of the globe by the steam. He again excluded the air out of the globe with steam as before, and both the cocks being closed, with the globe full of steam, he put the globe in the scales, and it weighed 202·5 oz. He then opened one of the cocks and let in the air, and by adding weight in the other scale it was found to weigh 203 oz, which shewed that the weight of the air the globe contained, was ·5 oz. or 218·75 grains. The globe being filled with steam as before, and condensed with cold water on the outside of the globe, and the metal again made very dry, and the air let into the globe, the water from the condensed steam was found to weigh ninety-six grains. It is worthy of remark here, that this gives the density

* Phil. Trans. vol. xli. p. 821. or Ab. vol. viii. p. 518.

of steam at 212° to that of air at 60°, as 96 : 218·75, or as 0·44 : 1.   The true density of steam at 212°, is nearly as 0·48 : 1.

The globe filled with steam, as before, only now, he, not knowing the effect of temperature, continued the globe longer with the steam passing through it, by which it acquired a greater degree of heat, for he had found by these experiments, that the least degree of cold less than the steam, would condense a part of it again into water, and hence, the quantity could not be ascertained which would exclude the air out of a given space, which was the chief end of the experiment.   In this experiment he succeeded in excluding the air with less steam; for, on weighing the globe, when the steam was condensed, the air let in, and all cold, it was found that the weight of the water condensed from the steam was only about forty-eight grains, which filled, when converted into steam, 925 cubic inches of space, so as to exclude nearly all the air.   From which he concluded that one cubic inch of water will form 4000 inches of steam.   To admit of comparison the temperature should have been observed, as I have little doubt that the steam was so rarefied by heat as to cause this result.

17.—Mr. Payne also attempted to introduce a new mode of generating steam; his apparatus consisted of a cast iron vessel of the figure of a frustrum of a cone, its diameter at the bottom being four feet, with a semi-globular end of copper of about five feet and a half diameter.   In the inside a small vessel was inserted, which Payne calls a *disperser*, which vessel had pipes round the sides fixed to it; the bottom rested on a central pin, on which it revolved, so as to spread the water it received from above, through an iron pipe.   The end of this pipe passed up through the head, and was enclosed very tightly, but so as to be easily moved with a circular motion, so that the water might be dispersed or showered round on the sides of the red hot cone, or ignited vessel, in a very exact manner.   From experiment he states that a pot or vessel, of the size and shape here mentioned, will, being kept to a dark red heat and the water regularly dispersed, expand 6·5 cubic feet of water into steam in an hour.   And that, by experiments made at Wednesbury and Newcastle-on-Tyne, 112 lbs. of pit coal will by this method expand twelve cubic feet of water into steam. This is near the truth of what may be done; but the method has no advantages, and the apparatus soon fails.   It is a duty, however, to an ingenious man to record his attempts to establish useful truths even when he fails in his object.   It shews the state of knowledge at the time on these subjects; and it saves others from repeating useless experiments. The mode of generating steam we have just described has been more than once revived lately.

18.—The engine of Savery had hitherto required the attendance of a person to open and shut the cocks.   This defect appears to have been first removed by Gensanne, a

Frenchman, who contrived a self-acting apparatus for the purpose, in 1744; and afterwards De Moura, a Portuguese, sent a model of another method to the Royal Society, which is ably described by Smeaton, in the Transactions for 1751.* His general description is sufficient for the purpose of shewing how the action is obtained. The engine consists of a receiver with a steam and an injection cock. It has a suction and a forcing pipe, each furnished with a valve, and a boiler, which may be of the then common globular shape, and, having nothing particular in its construction, a description of it will not be necessary; also the rest of the parts already mentioned being essential to every machine of this kind, a further account of them may be dispensed with. What is peculiar to this engine is a float within the receiver, composed of a light ball of copper, which is not loose in it, but fastened to the end of an arm made to rise and fall by the float, while the other end of the arm is fastened to an axis: and, consequently, as the float moves up and down, the axis is turned round one way or the other. The axis is made conical, and passes through a conical socket, which last is fixed to the side of the receiver. On one of the ends of the axis, which projects beyond the socket, is fitted a second arm, which is also moved backwards and forwards by the axis as the float rises or falls. By these means, the rising or falling of the surface of the water within the receiver communicates a corresponding motion to the outside, in order to give the proper motions to the rest of the apparatus which regulates the opening and shutting of the steam and injection cocks, and serves the same purpose as the plug frames, &c. in Newcomen's engine.

1751. *Francis Blake, F. R. S.*

19.—A paper on the best proportions for steam engine cylinders, by Mr. Francis Blake, was published in 1751;† which merits attention both as one of the first steps in theoretical inquiry respecting the proportions of engines, and on account of the result he obtained. It is evident, he remarks, from the principles of mechanics, that the contents of the cylinder remaining the same, the quantity of water discharged at each lift will in all cases be equal; and this equality is obtained by only adjusting the distance of the centre of the piston from the fulcrum of the beam. It will be granted also, that the excess of the column of atmosphere above that of water, is equivalent to a weight on the piston, driving it to a depth of about five feet within the cylinder; by the present construction acceleratedly, till friction, and resistance from the uncondensed steam which remains in the cylinder even after the injection, and is increased in elasticity while its bounds are diminished, shall equal the accelerative force: and that then again the piston may be retarded the rest of the way. But, independent of friction, we can, notwithstanding this diminution of force by the

---

* Phil. Trans. Vol. XLVII. p. 436, or Abridg. Vol. X. p. 252.
+ Phil. Trans. Vol. XLVII. p. 197, or Abridg. Vol. X. p. 187.

remainder of steam within the cavity of the cylinder, demonstrate the ratio of the velocities, and the times of descent of the pistons, in cylinders of unequal altitudes, to be exactly
the same as if the resistance were nothing, whence we shall without difficulty arrive at some
conclusion in this matter.   Let M N be the working part of a steam engine cylinder of the
usual height, equal in diameter to a shorter one $m\,n$ and the rarefaction in both of them
being supposed the same, A Q$=a\,q$, R Q$=r\,q$, and A R$=a\,r$, may represent the excess of
the atmosphere's weight above the column of water, the resistances to the pistons from the
remainder of steam and the effective force respectively.   Make $a\,k$: A K $::a\,n$: A N. and

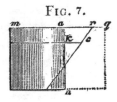
Fig. 7.
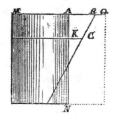

at all similar positions, the resistance $b\,c$ of $m\,n$, and force $k\,c$ on its piston, will be equal to
the resistance B C of M N, and force K C on its piston; and (by Newton's Princip. prop.
39.) in the descent of bodies, we have $\sqrt{a\,k\,c\,r}$: $\sqrt{A\,K\,C\,R}$ :: celerity in k : celerity in
K.   But these areas being evidently as the corresponding parallelograms $k\,q$ and K Q,
and these again as their heights, the celerities generated are in the subduplicate ratio of
$a\,k$ to AK, as if the resistance had been invariable.

To apply this to steam engines, if T W be a cylinder of equal content with the cylinder
M N, the quantity of water delivered by both will, as observed above, be the same at each
lift; but the cylinder T W is no higher than $m\,n$, and their rarefactions are supposed equal,
therefore, by what has been proved with regard to the times, the time of the piston's descent
in T W, will be to that of the piston's descent in $m\,n$ :: $\sqrt{E\,W}$: $\sqrt{A\,N}$; whence, in any
given time, the short cylinder T W will perform more than the longer one M N of equal
content, and that in the ratio of their diameters; for as $\overline{T\,E^{2}}\times\overline{E\,W}=\overline{M\,A^{2}}\times\overline{A\,N}$; and
$\overline{E\,W}$: A N :: $\overline{M\,A^{2}}$: $\overline{E\,T^{2}}$, therefore, $\sqrt{E\,W}$: $\sqrt{A\,N}$ :: M A : T E.   And he further
remarks, the friction is diminished with the slowness of the motion, because the periphery of
the piston increases in a less ratio than its area.

The result of his whole reasoning is in favour of a short cylinder, and it must be allowed
to be ingenious; but the proper question is, What form of cylinder will enable us to do the
most work with the least steam? and not the most work in the least time with a cylinder
of a given capacity.   (Sect. IV.)

Mr. Blake also investigated the relation between the power and resistance which gives a maximum effect in a given time when the motion accelerates from rest, both when the force is uniform, and when variable, increasing as the distance.* (See Sect. IV.)

1757. *Keane Fitzgerald, F. R. S.*

20.—It was natural to expect that the atmospheric engine being now in considerable use, the means of saving fuel would be considered in places where it was expensive. Mr. Keane Fitzgerald, in 1757,† proposed, with this object, to agitate the water in the boiler by a stream of air, on Dr. Hales' plan for evaporating; not perceiving the difference between forming steam and accelerating evaporation. But in consequence Dr. Hales applied to him respecting working ventilators for mines by steam engines, and a rotary motion being necessary to that end, Fitzgerald contrived one to render the steam engine applicable to the purpose. The method he adopted nearly resembled, in principle, that before contrived by Hulls for his steam boat,(art. 14.;) but instead of regulating it by a weight, Fitzgerald proposed to use a *fly wheel;* and remarks, that the steam engine by such means may be applied to corn mills, raising coals, &c. Fitzgerald also shewed the impropriety of the then usual mode of constructing the working beam with its axis below its centre of gravity, and altered the place of the axis of the engine beam of the York water-works engine, with much advantage to its effect.

1758. *William Emerson, born 1701, died 1782.*

21.—A brief but clear description of the atmospheric engine was published by Emerson in his Mechanics, with the mode of computing its power; as far as statical equilibrium between the power and resistance is concerned. He also in his Miscellanies gives a solution of a problem, which has for its object to determine the relation between the power and resistance when the effect is greatest. It may be stated as follows :—In a steam engine there is given the effective pressure of the atmosphere upon the piston, and the length of the stroke, to find the water to be drawn at a stroke, so that the greatest quantity shall be drawn in a given time, supposing the force uniform, and the arms of the beam of equal length. Emerson's solution differs from Blake's (art. 19,) in taking the whole time of the ascending and descending strokes into the account; and in not considering the moving power as a gravitating mass of matter. It is, therefore, more strictly applicable to the question, though still not perfectly so, as the space, not the time, should be given. (See Sect. IV.)

22.—The celebrated practical engineer, James Brindley, attempted to improve the

---

* Phil. Trans. Vol. LI. p. i. or Abridg. Vol XI. p. 317
† Phil. Trans. Vol. L. p. 53, and 157.

construction of the steam engine boiler by forming it of wood and stone, and inserting a fire place and chimney of cast iron in the internal part of the boiler, so as to surround both as far as possible, on all sides, by the water of the boiler. This plan he expected would render more of the heat of the fuel effective; and, therefore, obtained a patent, in 1759, for the arrangement. That it was founded on mistaken views of the nature of combustion, and of the quantity of the loss of heat, would not be difficult to prove, (see art. 190.) and accordingly never was adopted in general practice.

1762.   *Dr. Joseph Black, born* 1723, *died* 1799.

23.—The relation between the quantity of fuel and the effect of steam in an engine became now an important subject; but the different quantities of heat combined with the same body according as it was in a solid, a liquid, or gaseous state, or with different bodies at the same temperatures, had not yet been determined, or rather the fact was not distinctly known, and, therefore, crude opinion must have directed the wisest, as it now directs the ignorant man in his attempts to improve the steam engine. To Dr. Black we owe the first investigation of the combination of heat with bodies in the solid, liquid, and gaseous state, which he began to teach publicly in 1762, the heat so combining with them, he shewed was insensible to the thermometer, and hence he called it *latent heat*.

The quantity of heat required to convert boiling hot water into steam, he found exceeded five times the quantity which made water boil. Dr. Black also shewed that different bodies required different quantities of heat to produce the same change of temperature, and denoted the property by the phrase, capacity for heat; the term now usually employed is specific heat. (See Sect. II. art. 70.)

The principles of managing confined fires, and the nature and effect of fuel, were also taught by Dr. Black.

In the inquiries respecting heat he was followed by Dr. Irvine and Dr. Crawford, who made experiments to determine the specific heat and latent heat of various substances.

1765.   *John Smeaton, F. R. S. born* 1724, *died* 1792.

24.—Smeaton was not of a cast of mind likely to seize the views of Dr. Black, and turn them to account in managing the action of steam; his talent was chiefly confined to improving the construction and proportions of existing machines, by selecting the best methods known, and making experiments; accordingly we find he designed a portable atmospheric engine to make trials upon in 1765; and these experiments he was preparing for in 1769.*   Smeaton afterwards directed the erection of several large atmospheric engines,

---

* Reports, Vol I. p. 223. and II. 338.

and brought them to a degree of perfection which has not been exceeded in later times. I propose briefly to follow through the most interesting of his inquiries; commencing from his portable engine. This seems to have been the first attempt to make an engine capable of being removed from place to place. The fire place was formed entirely within the boiler; and, in the place of an ordinary beam, a wheel 6·2 feet in diameter, with a chain, communicated the motion from the piston to the pump rod.

The diameter of the cylinder was eighteen inches, its area, in circular inches, 324 inches; and allowing seven pounds to the inch, which such a cylinder, he remarks, would very well carry, we have 2268 lbs. The number of strokes per minute is stated to be ten of six feet each; hence, the effect is $2268 \times 10 \times 6 = 136080$ lbs. raised one foot, or four horses' power; he reckoned it equivalent to six horses; and, therefore, his value of the horse power is 22680 lbs. raised one foot high per minute, instead of the usual standard 33000 lbs.

Respecting fuel, he remarks, it has been found by experience that a two feet cylinder requires 174 lbs. of Newcastle coal per hour; which, reduced in the ratio of the capacity, gives ninety-seven pounds and a half per hour for the eighteen inch cylinder, or a four horse engine, according to the common application of fire; but he had reason to think an engine constructed like his would not require above sixty-five pounds per hour for a four horse engine.

The fire place was of a spherical figure, of cast iron, and entirely within the boiler; the coals were introduced by a large pipe from the outside of the boiler to the fire place, and the smoke passed off by a curved pipe with an iron funnel to promote a sufficient draught. The ashes fell through a pipe covered by a grate eighteen inches diameter, the whole being joined to the boiler by proper flanches, and always covered with water. In so short a flue the force of the fire cannot be wholly exhausted within the compass of the boiler, therefore the curved pipe was surrounded by a copper vessel adapted to its shape, into which was brought the feeding water, that it might be raised to a greater degree of heat than if brought immediately from the hot well into the boiler. It is also obvious that by this arrangement, the coolest part of the water comes in contact with the flue, to take the heat from the smoke before it ascends the chimney. The bars of the grate were cast into a loose ring capable of being taken out, and replaced when occasion required. On the large scale also Smeaton's boilers were admirably adapted for generating steam, little inferior to any that have been since contrived.

In a report to the London bridge water works, in 1771, Smeaton proposed to regulate the power of the engine by the injection, whereby the engine-keeper would be enabled, while the engine was working, to vary the quantity in proportion to the column to be lifted, and avoid the ill effects arising from a variation of the column, and save fuel.

Smeaton's first effective introduction of the improvements resulting from his experiments on the engine, seems to have been in the early part of 1774;* and by these improvements he appears to have reduced the expenditure of fuel about one-third. In 1775, he designed the Chase Water Engine, the cylinder of which was seventy-two inches in diameter, and the stroke nine feet. Its power was equivalent to the exertion of 108 horses, and its consumption of fuel was estimated at 1136 lbs. of Newcastle coal per hour. At its full power it was proposed to make nine strokes per minute, but to be regulated by the cataract to four strokes and a half per minute. The construction of the beam, and other parts of the engine are sufficiently curious to entitle it to the strict attention of the student of the subject.†

There seemed to be few practical circumstances that escaped Smeaton's inquiry respecting the atmospheric engine; and he drew up for his own use a table of the proportions of the parts for different sized engines, which still exists in the collection of his papers, which was purchased by Sir Joseph Banks. But the most important of his researches relate to the load upon the piston, on which he remarks he had found engines calculated to carry a load, varying from under five pounds to upwards of ten pounds to the square inch, those lightly loaded being expected to go with the greatest velocity, so that an engine carrying five pounds to the inch must go with double the velocity of one loaded to ten pounds, the cylinders being of equal area, in order that the effects of the power might be equal. He further adds, in engines, however, as in other machines, there is a maximum which, without new principles of power, cannot be exceeded : bad proportions of the parts and bad workmanship, may make an engine fall short, in any degree, of what it should do, but its maximum cannot be exceeded by the most accomplished artists.

Experience had, however, in some degree, directed, to a mean burden. The original patentees (Newcomen, & Co.) from some of their first performances, laid it down as a rule to load the piston, so as but little to exceed eight pounds to the inch; but, on more experience, they diminished that load, and amongst the best engines previous to Smeaton's own time the load was made about seven pounds to the square inch. And he further states that any proportion will do if the parts be properly proportioned, but, from a long course of very laborious experiments, he had fixed his scale near upon, but somewhat under eight pounds to the inch, including raising the injection water.

The labours of Smeaton shew the imperfect state of mechanical science as applied to practice in a remarkable degree. He actually designed an engine to be erected at Long Benton to raise water for turning a water wheel to draw coals from a pit;‡ and in 1781 proposed one of Boulton & Watt's engines to be erected for raising water for driving a corn mill ;§ using such arguments as these in support of his opinion.   " It is to be apprehended

* Reports, Vol. II. p. 337.   † Smeaton's Reports, Vol. II. p. 350.
‡ Smeaton's Reports, Vol. II. p. 435.   § Reports, Vol. II. p. 378.

that no motion communicated from the reciprocating beam of an engine, can ever act with perfect equality and steadiness in producing a circular motion like the regular efflux of water in turning a water wheel; and much of the good effect of a water mill is well known to depend upon the motion communicated to the millstones being perfectly equable and smooth; the least tremor or agitation takes off from the complete performance. Secondly, all the engines he had seen were liable to stoppages, and so suddenly, that in making a single stroke the machine is capable of passing from nearly its full power and motion to rest; for whenever the steam gets lowered in its heat below a certain degree, for want of renewing of the fire in due time or otherwise, the engine is then incapable of performing its functions. In the raising of water, (a business for which the fire engine seems peculiarly adapted,) the stoppage of the engine is of no other ill consequence than the loss of so much time, but in the motion of millstones grinding corn, such stoppages would have a particularly bad effect."

It was certainly not a gratifying circumstance to Smeaton to find that his tedious inquiries had been rendered nearly useless by a new mode of operation. To find that his cautious system of analysis was not in all cases the best mode of rendering the powers of nature useful to man. Yet if it were his labours on the steam engine alone on which his fame rested, there would be sufficient to command our esteem and respect. Its further improvements, its close cylinder, its double action, undoubtedly owed much of their perfection to the use of those modes of construction applied by Smeaton to the air pump.

1766. *John Blakey.*

25.—Though there are so many circumstances in the mode of action which reduce the effect of an engine of Savery's construction, these defects seem only to hold out an inducement to speculative men to attempt to remove them, and among these Blakey was one of the most sanguine. He obtained, in 1766, a patent for a new mode of constructing Savery's engine by using two receivers, one placed over the other, with a pipe of communication between them. The contact of the steam and water was to be prevented by a stratum of oil, forming a species of fluid floating piston. He further proposed to admit air to occupy the place between the steam and water so as to prevent condensation during the process of forcing: both methods inferior to the floating piston of Papin. Blakey had, however, sufficient art to persuade the public that he had made a valuable discovery, and to get Ferguson, the Lecturer, to shew off its advantages by a steam fountain.* In practice his method was found to be worthless.

In generating steam Blakey seems to have first proposed cylindrical tubes for boilers,

---

* Ferguson's Lectures, Vol.I. p. 312.

the description of which he published in 1774.　He was also the author of a pamphlet entitled, " A short Historical Account of the Invention, Theory, and Practice of Fire Machinery," printed in London in 1793, which was chiefly filled with short notices of his own labours on the subject, now of no interest.

1769.　*James Watt, L. L. D. F. R. S. born* 1736, *died* 1819.

26.—The commencement of the researches of Mr. Watt appears to have been in 1764, two years after Dr. Black began to teach his doctrines regarding heat.　Mr. Watt began by making experiments on the elastic force and bulk of steam; and gradually developed those principles which form the basis of his valuable improvements on the steam engine, but he did not so far mature his plans as to apply for a patent till 1768, which was enrolled in 1769.　The specification is brief and not illustrated by figures, hence I will give it entire; and then distinguish the principles and methods of construction which had not been anticipated.

Mr. Watt's patent of 1769, was for his " Methods of Lessening the Consumption of Steam, and consequently of fuel in Fire Engines," and his specification is as follows:— First, " That vessel in which the powers of steam are to be employed to work the engine, which is called the cylinder in common fire engines, and which I call the steam vessel, must, during the whole time the engine is at work, be kept as hot as the steam that enters it; first, by enclosing it in a case of wood, or any other materials that transmit heat slowly; secondly, by surrounding it with steam or other heated bodies; and, thirdly, by suffering neither water, nor any other substance colder than the steam, to enter or touch it during that time.—Secondly, In engines that are to be worked wholly or partially by condensation of steam, the steam is to be condensed in vessels distinct from the steam vessels or cylinders, although occasionally communicating with them; these vessels I call condensers; and, whilst the engines are working, these condensers ought at least to be kept as cold as the air in the neighbourhood of the engines, by application of water or other cold bodies.—Thirdly, Whatever air or other elastic vapour is not condensed by the cold of the condenser, and may impede the working of the engine, is to be drawn out of the steam vessels or condensers by means of pumps wrought by the engines themselves, or otherwise.—Fourthly, I intend in many cases to employ the expansive force (pressure) of steam to press on the pistons, or whatever may be used instead of them, in the same manner as the pressure of the atmosphere is now employed in common fire engines.　In cases where cold water cannot be had in plenty, the engines may be wrought by this force of steam only, by discharging the steam into the open air after it has done its office.—Fifthly, Where motions round an axis are required, I make the steam vessels in form of hollow rings or circular channels, with proper inlets and outlets for the steam, mounted on horizontal axles like the wheels of a water mill; within them are placed a

E

number of valves that suffer any body to go round the channel in one direction only : in these steam vessels are placed weights so fitted to them, as entirely to fill up a part or por_ tion of their channels, yet rendered capable of moving freely in them by the means herein- after mentioned or specified. When the steam is admitted in these engines between these weights and the valves, it acts equally on both, so as to raise the weight to one side of the wheel, and by the re-action on the valves successively, to give a circular motion to the wheel, the valves opening in the direction in which the weights are pressed, but not in the contrary; as the steam vessel moves round it is supplied with steam from the boiler; and that which has performed its office may either be discharged by means of condensers, or into the open air.—Sixthly, I intend in some cases to apply a degree of cold not capable of reducing the steam to water, but of contracting it considerably, so that the engines shall be worked by the alternate expansion and contraction of the steam.—Lastly, Instead of using water to render the piston or other parts of the engines air and steam-tight, I employ oils, wax, resinous bodies, fat of animals, quicksilver, and other metals, in their fluid state.

" Be it remembered, that the said James Watt doth not intend that any thing in the fourth article shall be understood to extend to any engine where the water to be raised enters the steam vessel itself, or any vessel having an open communication with it."*

The great and valuable improvement described in this specification is that of condensing in a separate vessel; and it necessarily involved a method of clearing the condenser of air and water. The application of the principle could be rendered perfect only by keeping the cylinder as hot as the steam, and the condenser as cold as it could be done with eco- nomy; and the methods proposed by Mr. Watt for accomplishing these objects are at once novel and efficient.

The idea of using steam pressure was not new, not even when applied to a piston, (see art. 12,) but the application of it in a close cylinder by means of a stuffing box, such as Smeaton had applied to the air pump, was a new mode of construction, which, it may be inferred, was intended in the engine specified though it is not described. The scheme of a rotary steam vessel, or steam wheel, was also first made public in this specification, though in a very imperfect manner.

27.—Mr. Watt's steam wheel not answering on trial, his next object seems to have been to convert the reciprocating motion of the piston rod into a rotary one. Methods for this purpose had been contrived by Hulls, and Fitzgerald; and patents had been ob- tained for similar ones by Stewart in 1769, and by Washborough in 1778, also by Steed in 1781, for the simple crank motion.

Notwithstanding the existence of these methods, Mr. Watt obtained a patent for five

* Robison's Mech. Phil. Vol. II. p. 119. Repertory of Arts I. 217. 1794.

others in 1781, one of which was the sun and planet wheel motion, which he used for some time on account of the crank being Steed's patent.

In 1782 Mr. Watt obtained another patent, embracing various methods of applying steam. First, for an expansive steam engine, with six different contrivances for equalizing the power; secondly, the double power steam engine, in which the steam is alternately applied to press on each side of the piston, while a vacuum is formed on the other; thirdly, a new compound engine, or method of connecting together the cylinders and condensers of two or more distinct engines, so as to make the steam which has been employed to press on the piston of the first, act expansively upon the piston of the second, &c.; and thus derive an additional power to act either alternately or conjointly with that of the first cylinder; fourthly, the application of toothed racks and sectors to the end of the piston or pump rods, and to the arches of the working beams, instead of chains; fifthly, a new reciprocating semi-rotative engine, and a new rotative engine or steam wheel.

By the double engine the same cylinder was rendered capable of doing double the quantity of work in the same time, the steam pressure acting, and condensation taking place, both during the ascent and descent of the piston. Simple as this change appears, after being made, it is attended with many striking advantages; it renders the power nearly uniform, diminishes the proportion of cooling surface, a less boiler is necessary, and it reduces the bulk and weight of the engine.

Of the modes of regulating the power of steam engines, the most effective were, first, by limiting the opening of the regulating valves which admit the steam to act on the piston, and letting it continue so open during the whole length of the stroke; secondly, by letting them open fully at first, and shutting them completely when the piston has proceeded only part of its stroke; or, thirdly, by the use of a throttle valve placed in the steam pipe, which, acting in the same manner as the floodgate of a mill, admits no more steam than gives the desired power.

The second of these methods of regulating the power of the engine is the best, and forms the basis of what is called Watt's Expansive Engine; which renders available more of the power of the steam than when the piston is acted upon by the whole force of the steam through the whole length of the cylinder. This principle is said to have been adopted in an engine at Soho manufactory, and some others, about 1776, and in 1778 at Shadwell water works; but it was not made public till the above patent was obtained in 1782. Whereas the same principle had been made known the preceding year by Hornblower, though applied in a different manner; and of two discoverers of the same principle, with different modes of application, that one should have the merit who first makes it known to the public,—for inventions are rarely withheld except through interested motives; and when a secret is kept for an individual's benefit all claim to priority of invention ceases.

28.—There yet remained another step to complete the mechanism of the double engine, viz. a guide for the piston rod, and this appears to have been first accomplished, in 1784, by the invention of the parallel motion. This is an ingenious combination of levers, one point of which describes a line nearly straight, and to this point the piston rod is connected, so that its rectilineal movement causes the beam to vibrate. This Mr. Watt secured by patent in 1784, together with a new rotative engine, in which the steam vessel was to turn upon a pivot, and be placed in a dense fluid, the resistance of which to the action of the steam was to cause the rotative motion; an improved method of applying the steam engine to work pumps or other alternating machinery, by making the rods balance each other; a new method of applying the power of steam engines to move mills which have many wheels required to move round in concert; a simplified method of applying the power of steam engines to the working of heavy hammers or stampers; a new construction and mode of opening the valves, with an improved working gear; and a portable steam engine and machinery for moving wheel carriages.

Mr. Watt obtained a patent, in 1785, for a method of constructing furnaces, in which the best principles the philosophy of the period could furnish, are applied to elicit the heat and consume the smoke of fuel. He also applied to the steam engine the conical pendulum as a governor, the steam gauge, condenser gauge, and a useful little instrument for ascertaining the state of the steam in the cylinder, called an indicator.

29.—The only part of the theory of the action of steam which Mr. Watt attempted to investigate from first principles is the power it affords by expansion, and this is done imperfectly. The proportions and mode of construction he adopted, seem to have been obtained wholly from trial, and the natural consequence was a long delay before he brought his engines into use; for though his facility of invention must have been considerable, his means of judging of the value of his combinations were very limited: indeed, he appears to have possessed no other certain test than the expensive and slow mode of proving each by making models and trial machines. On the ground of not having been remunerated for these expensive trials he got the term of his patent extended to 1800; and in consequence, with his partners, made a rapid fortune. He devoted a considerable portion of the latter part of his life to chemical philosophy, and particularly its application to the arts. As an author he has contributed on the steam engine only some historical notices of his own inventions, a few corrections of Dr. Robison's article in his Mechanical Philosophy, and notes to it containing his experiments on the latent heat and elastic force of steam, not given in detail to the world till they were rendered unnecessary by the inquiries of others.

30.—The share which Boulton had in the improvement and introduction of the steam engine must not be forgotten, for, as it has been remarked by Baron Dupin, " Watt's engine was, when invented by him, but an ingenious speculation, when Boulton, with as much courage as foresight, dedicated his whole fortune to its success." He did not hesitate even when Smeaton declared his conviction that it could never be generally applied as a useful agent.   Besides, Boulton rendered no small service to Watt and to Great Britain when, by his extraordinary talents in manufactures and commerce, he exempted his partner from all the cares of life, from all commercial speculations, and from all those difficulties which are the inevitable consequences of great enterprise in trade.   Boulton did still more; he triumphed over all those interests and prejudices which necessarily arose in the beginning to retard the success of the new steam engines and their application. " Men," continues Dupin, " who devote themselves entirely to the improvement of industry, will feel in all their force the services that Boulton has rendered to the arts and mechanical sciences, by freeing the genius of Watt from a crowd of extraneous difficulties which would have consumed those days that were far better dedicated to the improvement of the useful arts."

31.—A curious apparatus for trying the elastic force of different vapours was invented by T. H. Zeigler, which he describes in a memoir published at Basle in 1769, with tables of the results of his experiments, but it appears he had not taken care to free his apparatus from air before the trials; they are therefore useless.

1781. *Jonathan Hornblower.*

32.—In 1781, a patent was obtained by Mr. Hornblower, for a mode of applying the expanding power of steam.   For when steam is confined on one side of a piston, and a partial vacuum is formed on the other, the steam will move the piston till its force be in equilibrium with the friction and uncondensed steam; and as much power as is communicated during this motion is in addition to the ordinary effect of steam pressure. To gain power in this manner, Mr. Hornblower used two vessels in which the steam was to act, and which, in other steam engines, are called cylinders, employing the steam after it had acted in the first vessel to operate a second time in the other, by permitting it to expand itself, which he did by connecting the vessels together, and forming proper passages and apertures, whereby the steam might at proper times go into and out of the said vessels.*

The effect would be nearly the same as that derived from cutting of the steam before the piston arrives at the end of its stroke, as was afterwards done by Mr. Watt, (art. 28.) but has the decided advantage of being a more equable method of employing steam power ; and in large engines the construction of Hornblower's is also superior, because strong steam can be used in a small cylinder with less risk : he, however, does not appear to have intended to use powerful steam; and he could not use his invention because the improved mode of condensation remained the right of Boulton and Watt.

* Repertory of Arts, Vol. IV. 861. 1796.

Like Watt, many other engineers imagined that a rotary motion might be communicated with advantage by the direct action of steam; and two combinations for that purpose were made by Hornblower. His first was an ingenious but complicated machine for which he had a patent in 1798.* The second was more simple; it was secured by a patent in 1805. It consists of four vanes revolving in a cylinder round its axis. The vanes are like those of a smoke jack, but of thickness sufficient to form a groove in their edges, to hold stuffing for the purpose of making them steam-tight in their action. They are mounted on an arbor which has a hollow nave in the middle. Into this nave the tails of the vanes are inserted, and each opposite vane affected alike by having a firm connection with one another; so that if the angle of one of the vanes with the arbor be altered, the opposite one will be altered also, and the opposite ones are set at right angles to each other; so that when a vane is flatly opposed to the steam, the opposite vane will present its edge to it, and thus they are constantly doing in their rotation on their common arbor; so that the steam acts against the vane on its face for about a quarter of a circle, or ninety degrees, in the cylinder where it is destined to act; and as soon as it has gone through the quarter of the circle, it instantly turns its edge to the steam, while at the same instant another vane has entered the working part of the revolution, and the rotation proceeds without interruption. This engine was to be furnished with the condenser and discharging pump of Watt, but Hornblower added what he considered an improved method of discharging the air from the condenser.

It is easy to shew that the friction, and other sources of loss of power, are much greater in the rotary than in the rectilineal action of steam, while the loss by rendering a reciprocating motion rotary is very small; (Sect. IV and VII.) but I notice this as one of the most simple combinations proposed for a rotary engine.

33.—A series of experiments on the elastic force of steam from 32° to 212° was published by Mr. Achard in 1782. He also examined the elastic force of the vapour of alcohol; and observed, that when steam and alcohol vapour were of equal elastic force, the temperature of the latter was about thirty-five degrees lower, but that the difference of temperature was not constant; it seemed to be greater or less as the elastic force was greater or less.

1782. *Marquis De Jouffroy,*

34.—The idea of employing the steam engine to propel vessels, which had been suggested by Hulls, (art. 14,) was first tried in practice by the Marquis De Jouffroy, who, in 1782, constructed a steam boat to ply on the Saône, at Lyons; it was 140 feet long

---

* Repertory of Arts, Vol. IX. 289. Old Series.

and fifteen feet wide, and drew 3·2 feet of water. He made several experiments with it, and it was in use fifteen months on the Sáône.*

35.—In 1785 M. Perronet gave a very full description, in the French Encyclopædia, of an atmospheric engine erected near Saint Guilain, in Hainault; this description is remarkable for its clearness, and practical details, and not less so from its being introduced by stating Papin to be the inventor of the steam engine in a most unqualified manner; though admitting the first to have been constructed in England.

1788. *Patrick Miller.*

36.—About this time various competitors for the application of steam navigation appeared, (1785—88;) in America, two rivals; James Rumsey of Virginia, and John Fitch of Philadelphia. In Italy the application of steam power to vessels was proposed by D. S. Serratti, and in Scotland by Mr. Miller of Dalswinton, who afterwards on the sight of a model of a steam carriage invented by Mr. William Symington of Falkirk, was so much pleased with the model that he desired Mr. Symington to make him a small steam engine, to work a twin or double boat on Dalswinton Loch. The engine having been accordingly executed, and put on board the boat, the experiment was made at Dalswinton in the autumn of 1788, and it succeeded so well, that Mr. Miller commissioned Mr. Symington to purchase a gabert, or large boat, at Carron, and to fit up a steam engine on board of it, to make a trial on a larger scale. Every thing being completed, the trial was made on the Forth and Clyde canal, in the summer of 1789, Messrs. Miller, Stainton, Taylor, &c. being on board, and the result answered their most sanguine expectations; but most unaccountably, after having thus established, at a considerable expense, the practicability of employing the power of steam in navigation, Mr. Miller seems to have neglected it entirely.†

37.—The theory of the steam engine still made small progres, though it excited some degree of attention.

Bossut had described an atmospheric engine in the first edition of his Hydrodynamique, in 1771, with some formula for its statical equilibrium; in the edition of 1786, he investigated the proportion of counterweight, but for a particular case, and not including the actual circumstances of the moving forces.

38.—A rotary engine was proposed in 1789 by Cooke,‡ and a patent was obtained for one in 1790, by Bramah and Dickinson,‖ and for another in 1791 by Sadler.§ The

* Dictionaire de Physique, art. Chaloupe à Vapeur.
† Short Narrative of Facts relative to Steam Navigation, Edin. Phil. Journal.
‡ Repertory of Arts, Vol. III. p. 401. 1795.   ‖ Idem Vol. II. p. 73.
§ Idem Vol. VII. p. 170.

peculiar construction of all these engines I need not describe, as the principle of a rotary engine will be shewn to be attended by a loss of effect, which mechanical combinations cannot remove. (See Sect. IV.)

39.—Bramah and Dickinson's patent included three varieties; the most simple of which is designed with pistons sliding in an eccentric wheel, the steam to enter at S; and the opening to the condenser being at C, the pressure causes the smaller wheel to revolve, and the pistons to slide in it. All the varieties are specimens of that beautiful

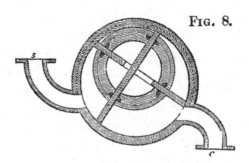

FIG. 8.

style of executing machinery which Bramah contributed so much to introduce in this country, and which has been carried to such high perfection by his pupil, the celebrated Maudslay.

1790. *Bettancourt.*

40.—Chev. Bettancourt, who was employed by the Spanish government to collect models of hydraulic machines, made a series of experiments on the force of vapour of water and of alcohol, at different temperatures. They are more accurate than those which were at that time before the public; but still had not that precision which is necessary to develope the laws of the force of vapour. He made a model of the double acting engine, with a new mode of forming the valves; and, Prony says, from merely seeing the exterior of a double acting engine when at work.*

1790. *R. Prony.*

41.—M. Prony is the author of one of the most extensive of the French works on the steam engine; it forms a part of his " Architecture Hydraulique," which commences in the first volume, and occupies nearly the whole of the second.

* Archi. Hydraulique, Vol. I. p. 574.

M. Prony begins with the properties of caloric, and the tables of Bettancourt on the force of vapour; and from the latter constructs empirical formula for calculating the force of vapour at different temperatures. These are not a little complex considering their want of conformity with experiment. He then proceeds to the description of engines as then constructed, and their parts; which are illustrated by plates having figures on a large scale. When he arrives at the parallel motion, the nature of the curve described by the extremity of the piston rod is very fully investigated, with tables to shew its variation from a straight line for a given range in the curve. It is followed by the proposal of a method for determining the diameter of the steam cylinder, which is little better than telling the artist to guess at it, and correct his guess by an intricate formula. The part on the steam engine terminates with a calculation of the effect produced by a given quantity of fuel, where the time of combustion is certainly erroneously introduced.

The rest of the volume is occupied by an analytical investigation of empirical formula for the expansive forces of elastic fluids and vapours at different temperatures, which has been rendered wholly useless by later researches having shewn the experiments to be inaccurate

It is remarkable that Prony had not acquired a knowledge of the advantage of steam acting expansively; though when his second volume appeared, it had been fifteen years a contested discovery in England. Of his labours it may be said, that they afford the strongest evidence that mere mathematical talent is not sufficient for the promotion of mechanical science, otherwise the principles of the steam engine would not have remained to be investigated.

1795. *John Banks.*

42.—Mr. Banks, in a work on mills, published in 1795, has treated of the maximum of useful effect in atmospheric steam engines. He considers the space, or length of the stroke, the given quantity, in which his investigation differs from those of Blake and Emerson. He has, however, by considering the atmospheric pressure as a gravitating weight, failed in giving a correct solution.

One of his problems includes the weight of the moving parts of the engine; and he adds some useful practical formulæ for the statical equilibrium of engines for raising water, with examples.

In 1803, Mr. Banks gave some rules for the strength of engine beams, both for wood and cast iron; and also a description of a gauge for determining the state of rarefaction in the cylinders and condensers of steam engines, in principle the same as the common barometer; and differing from the ordinary condenser gauge by having a cistern instead of a syphon for the mercury. His rules for the strength of beams are to find the relation between the

F

pressure and breaking weight, and to let the breaking weight exceed the pressure by six, eight, or ten times.*

1797. *Edmund Cartwright.*

43.—The simple and neat combination of Cartwright next claims attention, and on more grounds than one. He attempted to condense the steam by means of cold applied externally to the condenser; it consisting of two metal cylinders lying one within the other, and having cold water flowing through the inner one and enclosing the outer one. By this construction a very thin body of steam is exposed to a very great quantity of cooling surface. And, by placing the valve to change the steam in the piston, a constant communication is at all times open between the condenser and the cylinder, so that whether the piston ascends or descends, the condensation is always taking place.

One of the chief objects of this arrangement was the opportunity it afforded of substituting ardent spirit or alcohol, either wholly or in part, in the place of water, for working the engine. For as the fluid with which it is worked is intended to circulate through the engine without mixture and very little loss, the using alcohol, after the first supply, it was expected, would be attended with little or no expense. The power obtained from alcohol, it was then imagined, would require only half the fuel which was necessary to obtain the same power from water; (see Sect. IV.) and Cartwright proposed, in some cases, to apply this engine to a still, to obtain mechanical power by the distillation of ardent spirit, so as to save the whole of the fuel.† How he was to keep the engine in a workable state, and yet obtain a pure spirit, neither he nor his friends seem to have considered.

In order to reduce the friction of the piston, which, when fresh packed in the common way, lays a very heavy load upon the engine, Cartwright made his solely of metal, and expansive; by this method he further expected some advantage from saving of time and expense in the packing, and from the piston fitting more accurately, if possible, the more it was worked. (See Sect. VII.) Cartwright was very desirous of simplifying all the other parts of his engine, having only two valves, and those are as nearly self-acting as may be. Cartwright's engine is represented in the annexed figure. It is a single acting engine, and A is the cylinder; B, the piston; I, the pipe which conducts the steam to C, the condenser, which is a double cylinder, the steam passes between the inner and outer one into the pump D, which returns the condensed fluid back into the boiler, through E, the air box, with *e* its valve.

As the pipe from the pump, through which the condensed fluid is returned into the

---

* Power of Machines, p. 103.   † Phil. Mag. Vot. I. p. 3.

boiler, passes through the air box, what air or elastic vapour may be mixed with the fluid rises in the box, till the ball which keeps the valve *e* shut, falls, and suffers it to escape.

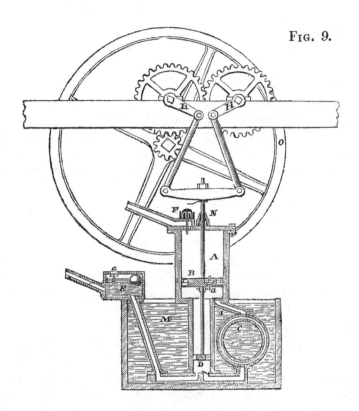

FIG. 9.

F is the steam valve; *a* the piston valve; H, H, two cranks, upon whose axles are two equal wheels working in each other, for the purpose of giving a rectilinear direction to the piston rod ; and M is the cistern that contains the condensing water.   The metallic pistons he formed of metal rings as shewn by the section of the piston, which by springs are forced outwardly against the surface of the cylinder, so that the piston may adapt itself to any inequality in its form.   The piston rod is also made steam-tight by a metallic box, constructed in the same manner as shewn at N.   O is the fly wheel for regulating the motion of the engine.

The metallic piston is the only part of the engine which was really new in principle, and for its invention we are undoubtedly indebted to Cartwright; and though we cannot say any other part is new except in arrangement, we admire the appearance of simplicity and originality which distinguish his design, even knowing that both theory and practice forbid us attempting to use the methods he proposed.

Cartwright included in his patent, a rotary engine, which is simple in appearance, but in reality involves many difficulties in construction, besides the loss of effect which must necessarily follow from steam acting on a rotary piston.

1797. *John Curr.*

44.—A work containing the proportions of the parts of atmospheric engines as they were executed in 1797, with brief technical directions for constructing them, illustrated by plates shewing the parts on a large scale, was published at Sheffield, by John Curr.* It contains no general description of the engine, and he assigns no reasons for any of the proportions he has given, except when speaking of the pressure on the piston, he says, that when the pressure was increased from seven to eight pounds and a half per square inch the engine did less, and also when reduced to 6·1 lbs. it did somewhat less; and he does not recommend a greater load than six and a quarter or six and a half pounds.† The engine had a sixty-one inch cylinder, and made twelve strokes, of eight feet and a half each, per minute. The consumption of coals was ten hundred weight of small coal, or sleck, per hour. The power of the engine would be nearly equal to fifty-four horses' power, and as the ratio of coal to sleck is about as three is to four, it is equivalent to about 840 lbs. of coal per hour. And at this ratio, one pound of sleck raises 97,600 lbs. of water one foot high, and one pound of coals 130,000 lbs. one foot high.

45.—In 1797, an engine on Savery's principle was described by William Nicholson in his Philosophical Journal, which Mr. Kier had erected in 1793. It acted wholly by condensation; the steam vessel being raised somewhat above the height to which the water was to be raised. It had a provision for letting in a small portion of air between the steam and the water, and the construction was extremely simple and judicious. The boiler was seven feet long, five feet deep, and five feet wide, and it consumed six bushels (522 lbs.) of good coal in twelve hours in its best state, and seven in its worst state. Under these circumstances it made ten strokes per minute, and raised seventy cubic feet of water twenty feet high in a minute.

According to this statement, in the best state of the engine, eighty-seven lbs. of coal were consumed in two hours, or 120 minutes, and 1400 cubic feet of water raised one foot high per minute; or $1400 \times 120 = 168,000$ cubic feet by eighty-seven pounds of coal,

---

* The Coal Viewer and Engine Builders' Practical Companion.
† Idem. p. .40

which multiplied by sixty-two pouuds and a half, the weight of a cubic foot of water, and divided by eighty-seven, gives 120,000 lbs. for the load raised one foot by one pound of coals, which is about one half the effect produced by an engine with a piston and Watt's condenser, and less than the effect of the common atmospheric engine as used for the coal mines.

An attempt was also made to improve Savery's engine, by condensing in a separate vessel, by John Nancarrow; but the nature of the engine does not permit of this being applied with much effect.

1799. *Matthew Murray, died* 1826.

46.—In the construction and improvement of some of the parts of engines, much was done by Mr. Murray, of the firm of Fenton, Murray, and Wood, of Leeds. These improvements were made the subjects of patents, and though it appeared that some of them had been before used by Boulton and Watt, they did not become publicly known till Mr. Murray obtained patents for them.

In his patent of 1799, in order to save fuel, Murray proposed to place a small cylinder with a piston on the top of the boiler, connected to a rack, by means of which the force of the steam within the boiler opens or closes the damper fixed on an axis in the chimney, thus increasing or decreasing the draught of the fires, so as to keep up a regular degree of elastic force in the steam. Mr. Murray also thought some advantage would be gained by placing the steam cylinder in an horizontal, instead of a vertical position, with a view of rendering the engine more compact than the usual construction; he also adopted a new method of converting the reciprocating motion of the piston rod to a rotary one of equal power, by means of a property of the rolling circle, and shewed how to fix the wheels for producing motion alternately in perpendicular and horizontal directions.*

47.—Mr. Murray's patent of 1801, was for six different objects:—First, for a method of constructing the air pump.—Second, for a method of packing stuffing boxes, &c. by bringing the moveable parts of each in immediate contact, which prevents the piston rod receiving any oblique pressure, by the lid being screwed down more upon one side than the other.—The third and fourth methods relate to the construction and motion of the valves.—The fifth was a method of connecting the piston rod to the parallel motion.—And the last for the construction of fire places, by which the smoke arising from the fire was to be consumed, and in which he had been anticipated.

48.—Another patent was obtained by Mr. Murray, in 1802, for a portable engine, but

* Repertory of Arts, Vol. XI. p. 311. Old Series.

as it included some of the methods for which Messrs. Boulton and Watt had patents, it was at their instance repealed in the following year.

1799. *William Murdoch.*

49.—Mr. Murdoch, a partner in the firm of Boulton Watt and Co, obtained in 1799 a patent for some new methods of construction, which consist of a mode of boring the metallic cylinders and pumps more equably by means of an endless screw, worked by a toothed wheel; and a method of simplifying the construction of the steam vessel and steam case, in engines formed on Mr. Watt's plan, by casting the steam case of one entire piece, to which the cover and bottom of the working cylinder are to be attached. He also proposed to cast the cylinder and steam case in one piece of considerable thickness, and bore a cylindric interstice between the steam case and steam vessel, leaving the two cylinders attached at one end, and to close the other by a ring of metal. Another improvement included in the patent was, a plan for simplifying the construction of the steam valves or regulators of the double engine, by connecting together the upper and lower valves, so as to work with one rod or spindle. The tube which connects them being hollow serves as an eduction pipe to the upper end of the cylinder, and a saving of two valves is effected; and, lastly, he adds a scheme for a rotary engine consisting of two toothed wheels working in an airtight vessel, which he imagined would work with considerable power. Mr. Murdoch's modes of moving the valves have added much to the simplicity and neatness of the double engine, and to his skilful superintendance the steam engine owes many of its perfections; and its success in Cornwall was greatly aided by his activity, integrity, and resources for overcoming the difficulties which the drainage of the mines presented.

1801. *Dr. John Robison, born 1739, died 1805.*

50.—Dr. Robison, to whom the mixed mechanical sciences are so much indebted for a more judicious combination of theory with practice than is to be found in any preceding author, and for treating them in a more popular style, seems to have bestowed much attention on the principles and construction of the steam engine. His analytical knowledge was ample for the purpose, and the access of the friend of Watt to practical data must have been easy in proportion as Watt was the liberal friend of science; therefore, much is expected when we take up the volume which contains the articles of Dr. Robison on the steam engine.

The first article contains a rather diffuse statement of the physical properties of steam. The phenomena of boiling, and the effect of pressure in altering the temperature necessary for ebullition, and the popular doctrines of latent heat, are fully stated. It also contains a series of experiments on the elastic force of the steam of water, and of alcohol, (see art. 95&104;) and we have only here to remark on them, that they were not made with sufficient accuracy even to establish the justness of some of his own views on the subject; also, the rule for the elastic force of steam derived from these experiments, and stated to be "sufficiently exact for practical purposes" is very far from being so, and has had a little effect in misleading some of the engineers who have ventured to speculate on the improvement of steam engines. But, on the whole, Dr. Robison's is the best article on steam I have seen.

The article on the steam engine consists of the history, mixed with detailed descriptions, of the engines of Savery, Newcomen, Watt. &c.; and such theoretical discussion as he has given is also blended in the same mass. In the historical portion the memory of Papin is not quite so respectfully treated as we could have wished, and the circumstance of Watt being the private friend and countryman of the author has not been without its effect on the historian. In other respects Dr. Robison has been impartial. In description there is a want of system, but he is full and particular; and he has been of unknown value in giving information to the competitors of Boulton and Watt, and in furnishing matter for minor writers. In theory he has reprinted the speculations of Bossut respecting the best velocity for atmospheric engines, with some additions, and Watt's mode of computing the pressure on the piston of the expansive engine; but neither of these inquiries are conducted in such a manner as to be of use to engine makers.

The reputation of Dr. Robison has given much additional value to his articles on the steam engine; hence their effect has been unparalleled, and if we find little of novelty in his labours, it was no small favour to have the scattered knowledge on the subject collected with so much skill, and treated with so much clearness and good taste.

51.—A modification of Watt's manner of constructing boiler fire places was contrived in 1800, by Messrs. Roberton, of Glasgow, which is more convenient in practice though the same in principle. (See Sect. III.) They also attempted to make the steam which escapes by the sides of pistons useful in adding to the effective power of engines. But the complexity and expense of apparatus to obtain so small an increase of power, renders this and some other expedients of that time of little if any value.

1801. *Joseph Bramah, born* 1749, *died* 1814.

52.—The rotary engine, the joint product of Messrs. Bramah and Dickinson, has already been noticed, (art. 39.) In 1801 Mr. Bramah obtained a patent for a new mode of applying the four-passaged cock to steam engines, with some other variations in their construction.

The four-passaged cock he made to turn continually in the same direction, and yet produce the same effect as by turning it backwards and forwards, but by turning constantly the same way the wear is rendered more equable, and consequently the combination is more durable.

He also adjusted the movements so as to give, at the proper time, as instantaneous and free a passage to the cylinder and condenser as possible; and formed the apertures so that the cone might be pressed equally into its seat by the force of the steam.

The mining and manufacturing interests of the country felt most severely the injurious effect of the exclusive privileges that had been granted to Boulton and Watt, without those restrictions which ought to guard public rights when a power of monopoly is renewed; and their claims to those privileges were strongly contended against, and the imperfections of their specification exposed with enthusiastic warmth, by Mr. Bramah, in a pamphlet published in 1797. Indeed the fortunate idea of condensing in a separate vessel, which in Watt's single engine is the only essential part in saving of fuel beyond what Smeaton had accomplished, would undoubtedly in a short time have occurred to some other person, and mines must have been drained at a more economical rate, long before that monopoly ceased. The progress of the public good should never be retarded for individual interest, and therefore monopoly should never be renewed except so that any other person may, at a fair and at a fixed rate of licence, join in it.

53.—A series of tables for the proportions of the cylinders of atmospheric engines, to produce a given effect, were published, in 1801, by Mr. Thomas Fenwick, whose employment in the management of coal works near Newcastle gave him a good opportunity of knowing what would answer best in practice.

He infers from some experiments that the whole friction of the atmospheric engine is about four pounds per square inch, on the area of the piston, and on account of the frequent bad effects attending designing an engine with too small an allowance for excess above its ordinary work, he makes his computations at five pounds and a half effective power for each square inch of piston.

In a later edition of his work he gives tables for an improved atmospheric engine with a separate condenser in which the ratio of the effect is as 17 : 10, when the same sized cylinder is used. The saving of fuel he does not mention, as at coal works it is not considered of much importance; for if the first expense of an engine be small, and its operation simple and efficient, it is of more value to a coal owner than a finer piece of machinery.

1801. *John Dalton.*

54.—At this period a knowledge of the nature and properties of vapours began to become important in chemical science, in meteorology, and in other branches of natural philosophy, and therefore a wholly different class of writers engaged in the investigation, which had made so little progress in the hands of mechanical people. The first chemist who distinguished himself by attempting a full investigation of the theory of vapour was Mr. John Dalton. He made an accurate series of experiments on the expansive force of steam at temperatures lower than 212º,—made experiments and ascertained various phenomena relative to the expansion of gases, the mixture of air and vapour, the nature of evaporation and of combustion. And though he failed in his attempt to reduce any of these to general laws, yet he gave such an impulse to the inquiry as rendered it one of universal research among chemical philosophers. The importance of Dalton's inquiries, and even their connection with the theory of the steam engine, did not appear, at first, to be much noticed. The idea that Watt had done every thing possible to be done respecting the power of steam had stopped inquiry among men of science, and left the manufacturers and capitalists of the country, who were wishful to encourage improvement, to be guided by vain and ignorant projectors, or ruined by pretending knavery.

1802. *William Symington.*

55.—In 1801, Mr. Symington was encouraged to proceed with a steam boat, by Thomas, Lord Dundas, of Kerse, who wished that one might be applied to drag vessels on the Forth and Clyde Canal in place of horses, and accordingly a series of experiments on a large scale, which cost nearly £3000; were set on foot in the year 1801, and ending in 1802. The boat Mr. Symington made was for towing, and it had a steam cylinder twenty-two inches in diameter, and four feet stroke. A complete model of it, with a set of ice-breakers attached, may be seen at the rooms of the Royal Institution in London. This tow-boat proved to be very much adapted for the intended purpose, but no direct practical application of steam power to this object resulted from it.

1802. *Trevithick and Vivian.*

56.—The idea of a high pressure engine had occurred to Leupold, (art. 12.) and to Watt, (art. 26.) but neither of them had reduced their notions to practice, and it was not till 1802 that this simple mode of applying steah was brought into use by Messrs. Trevithick and Vivian.* Their object seems to have been to form a simple and portable engine for cases where water was scarce, or where gaining the whole effect of the fuel was of less consequence than moving a cumbrous load of matter.

* Repertory of Arts. Vol. IV. p. 241. New series

G

Indeed their high pressure engines were intended chiefly for propelling of carriages upon rail-roads, and when used for this purpose the boiler was composed of cast iron, of a cylindrical form; mounted horizontally upon a frame with four wheels, the cylinder of the engine being placed vertically within the boiler near to one end. The piston rod moved a cross head, between two guides, and by a connecting rod descending from each end of the cross head to two cranks the motion was communicated to the wheels of a carriage; a fly wheel in this case is not required because the momentum of the carriage supplies its place.

The first trial of this species of moving power for carriages took place on a railway at Merthyr Tidvil in 1805. Its use was not at that period followed up, but it is now with some slight modifications extensively employed on rail-roads.

Several projects for trifling variations in the construction of engines, and for methods of applying fuel, appeared about this time, but none of either sufficient novelty or importance to claim particular attention.

The nature and application of heat had been so well illustrated by Rumford, and many of its more recondite properties so ably developed by Leslie, that there seemed to be little reason to expect any material improvement beyond the best mode then in practice. The cylindrical boilers which Blakey projected, and Rumford had tried, were again remodelled by Woolf; but in his practice we find he has reverted to methods nearly like those of Rumford, instead of continuing to follow his own. The steam engine itself had also apparently obtained its most simple and efficient form, except in the eyes of those who expected to use its direct rotary action. The fact however was otherwise, for by a most simple change of a previous combination it had to be materially improved.

1804. *Authur Woolf.*

57.—The mode of condensation invented by Watt being now public property, and the term of Hornblower's patent having expired, Mr. Woolf adopted the arrangement of the latter, with the alteration of using high pressure steam in the small cylinder, and employing the condensing apparatus of Watt. But a change of the working force of the steam would have been too slight a ground to have claimed a patent upon, and therefore he commences his specification with a claim of the discovery of a new law of the expansibility of steam. This law of expansibility he stated, with much confidence, as the result of experiments; but no doubt he had deceived himself. His assumed law of expansion, is, that steam generated at any number of pounds above the pressure of the atmosphere will expand to an equal number of times its volume, and still be equal in elastic force to the pressure of the atmosphere, the temperature being unaltered. Hence steam generated at forty pounds on the square inch was expected to expand to forty times its bulk, and yet be equal to the elastic force of the atmosphere. But it is a well known law of the expansion of fluids that the temperature being constant, the bulk is inversely as the pressure; and calling the pressure of the atmosphere fourteen pounds, we have 14 : 14+40 : : 1 : 4, nearly. Therefore steam generated at fifty-four pounds on the square inch, or forty pounds above the pressure of the atmosphere, would expand only to four instead of forty times its volume. (See art. 120.) And though Woolf's assertions were so directly opposed to the laws of the constitution of elastic fluids, they have found their way as undoubted experimental truths into works which ought to have high claims to respectability; and it should be a lesson of care to authors, unless they have no higher wish than to reprint advertisements.

The employment of high pressure steam to act expansively by means of a double cylinder, gives the utmost degree of power in the most equable manner, and with the most safety. Hence either for machinery engines, or mine engines, it seems the most economical mode of obtaining power. I object to strong steam on account of its danger, but my readers may not have like apprehensions. Woolf's other patents are for projects of little if any value.

58.—It would be an omission to pass without notice the exertions made by Oliver Evans about this period to get into use the high pressure steam engine. His scheme for employing it had not at first many supporters, and he had some rivals. His engine differs little from that of Trevithick and Vivian in construction, but from a work called "The Abortion of the Young Steam Engineer's Guide" it appears, that the expansive force of the steam was to be employed. The "Abortion" is a curious work, it betrays that strange mixture of absurd speculation and indistinct perception of truth, which distinguishes the generality of enthusiastic projectors, and is valuable only to those who can select by means of previous knowledge or experience. A volcanic steam engine, and the idea of employing the force of solar heat by means of a burning glass to work an engine, are among his projects.

59.—The claims of our American brethren to improvement, and to judicious construction and application, are however much stronger than those of our continental neighbours; and of American claims we have reason to speak with pride rather than with other feeling. British genius and industry have not been extinguished by transplanting to another climate. It is true that many of the projects they have yet formed are rather extravagant than novel, being seldom founded on the sober reasoning of science. Time will, however, check this evil, and we may expect them to hold that rank in the new world which Britain has held with such honour for some centuries in the older portion. The chief object of their engineers has been to render steam useful in navigation; and considering the importance to America of navigating her immense rivers, it is not surprising that the application of the power of steam to propelling vessels should by persevering efforts have been first carried into successful practice in that continent. This was achieved through the activity and zeal of Mr. Fulton, who appears evidently, however, to have derived much of his knowledge of the subject from what was done in Scotland. The first American steam boat that completely succeeded was launched at New York, on the 3rd of October, 1807, fitted with a steam engine made by Boulton and Watt for Mr. Fulton in 1804,* and soon afterwards this vessel plied between that city and Albany, a distance of 160 miles.

60.—The successful introduction of steam navigation in Britain we owe to Mr. Henry Bell, who, in 1811, built a steam vessel according to his own plans, with a forty feet keel, and ten feet six inch beam, fitted it up with an engine and paddles, and called it the Comet, because it was built and finished the same year that a large comet appeared.

Since that time the progress of steam navigation has been exceedingly rapid, and has had a most beneficial influence on the trade of the country.

61.—An almost innumerable quantity of schemes for improvements on the steam engine have been crowded on the public eye within the last ten years, but except a few for improvements in construction, of small importance, there has been nothing done that is worthy of detaining the reader to notice, towards either the improvement of the engine, or of the mode of generating steam, so as to increase the power of a given quantity of fuel in the steam engine.

62.—Some valuable experiments on the elastic force, bulk, and latent heat, of steam, made by Mr. John Southern in 1803, were published by Mr. Watt, and the experiments of Dr. Ure and Mr. P. Taylor on the elastic force of steam, have led to a considerable advance in theoretical investigation. The improvements in the manufacture of steam engines have also been important; but we have no reason to expect any material increase of its power; it seems to have reached its limit, and we might equally hope to add strength to a man or a horse. New modes of applying the power of steam may be devised, and new objects may be found to which it may be applied with advantage, and its theoretical principles will become more generally and more perfectly known.

It may also be found that the vapour of some other substance may be used with advantage, in certain cases, instead of that of water; of this, however, there is not much hope; and my reasons for this opinion will be shewn in treating of the properties of vapour, (art. 115.) Probably some other source of power will be discovered which will divert the at-

* Fifth Report on Holyhead Roads' Steam Boats. Mr. Watt's letter p. 210.

tention of projectors, and the only one in nature which appears unemployed by man seems to be that of the electric fluid; how far it may be rendered useful is a matter of curious inquiry, and dangerous in proportion to its power and our ignorance of its nature.

63.—Some idea of the rapid progress of the application of steam power may be formed from the circumstance, that, in the year 1789, the first steam engine was erected in the town of Manchester: before that time the manufactories were dispersed throughout the remotest districts, as they depended chiefly upon falls of water for power, the more expensive one of animal force being the only remaining expedient. The engines of Watt produced the most complete revolution in this respect. The factories were transported from the most wild and inaccessible places to towns and cities; and furnished with the means of uniting, under the same roof, the various branches of the manufacture, so that the raw material is now, with astonishing rapidity, converted into the most perfect cloth.

In Glasgow, the first steam engine, erected for spinning cotton, was put up in January, 1792, at Scott and Co's. cotton mill, near Springfield. This was seven years after Boulton and Watt put up their first steam engine for spinning cotton in Messrs. Robinsons' mills, at Papplewick, in Nottinghamshire.

The number of steam engines in Glasgow and its neighbourhood in 1825, as collected by Mr. Cleland, is

	Number of engines.	Horse power.
In manufactures . .	176 . .	2970
In collieries . . .	58 . .	1411
In stone quarries . .	7 . .	39
In steam boats . . .	68 . .	1926
In Clyde iron works . .	1 . .	60
Total	310	6406

The average horses' power of the engines is 20$\frac{7}{}$.

The steam engines in Great Britain and Ireland, employed in the year 1817, in the manufacture of cotton yarn, amounted to more than 20,000 horses' power, and such has been the advantages resulting from the application of machinery, that one person can produce more yarn in a given time, than 200 could have produced about sixty years ago.

In the iron, woollen, and flax manufactures, the beneficial effects from employing the steam engine have been equally important.

The total extent to which steam power is applied in Great Britain, is estimated, by Baron Dupin, to be equivalent to the power of 320,000 horses in constant action. To this immense command of power our country owes much of its commercial prosperity, besides a vast addition to the comforts and conveniences of life.

The increased employment of steam has, however, in no instance been so great as in its application to navigation in Britain.   A solitary steam boat navigated the Clyde in 1811; in 1825, fifty-one steam boats plyed on that river; and from the first successful trial in 1811 up to 1822, the number of steam vessels in Britain increased to about 140, with a power equivalent to the exertion of 4700 horses, and a tonnage of 16,000 tons.

64.—In concluding this historical sketch it is of some importance to remark, that the whole tends to prove that the steam engine, in the highest state of perfection it has yet attained, is entirely of British origin.   The remark extends to the discovery of physical principles, as well as of mechanical combinations.   No new principle, nor no new combination of principles, has yet been derived from a foreign source; the most perfect of foreign steam engines being professedly copied from British ones, and not unfrequently manufactured by British workmen.

SECTION II.

OF THE NATURE AND PROPERTIES OF STEAM, ITS ELASTIC FORCE, EXPANSIVE FORCE, AND POWER OF MOTION.

ART. 65.—NATURAL bodies exist in three states, the solid, the liquid, and the gaseous. The state of many of them may be changed; thus, water may be in the solid state, as ice, in the liquid, as water, in the gaseous, as steam; and these changes take place only under particular degrees of heat and pressure: but there are some gaseous bodies which cannot be reduced to the liquid form by the means we at this time are acquainted with; though there has been so much accomplished as to render it tolerably certain, that all the gases known would be reduced to liquids, were they exposed to sufficient pressure and reduction of temperature.

66.—Those gases which are not changed into liquids, by the ordinary changes of temperature and pressure, are called *permanent gases*.

The gases which condense into liquids, by the common changes of temperature and pressure, are called *vapours*, or *steams*; I intend to use the term steam in preference to vapour.

67.—Heat is diffused through all the bodies in nature, whether they be in the state of solids, liquids, or gases, and it constantly tends to an equilibrium; so that, when by any means it is accumulated in particular substances, a portion is quickly given off to the surrounding bodies, to bring the whole to one common temperature. On the other hand, where bodies have been deprived of a portion of it, heat is given off to them by, or heat passes to them from, the surrounding bodies, to restore the equilibrium.

68.—When there is an equilibrium of heat, or the adjoining bodies are of the same temperature, if it be destroyed by the introduction of a fresh quantity of heat, different bodies will be found to absorb different quantities of the new portion of heat in restoring the equilibrium. The peculiar quantity which each body absorbs, under the same circumstances, is denominated the *specific heat* of that body. In comparing the specific heats of bodies, that of water at 60° is considered to be unity, and therefore becomes a measure of all the rest.

69.—The property of bodies to hold different quantities of heat at the same temperature, is sometimes called *capacity for heat;* but this term should be applied only to the whole quantity of heat in a body, otherwise it becomes the same as specific heat. Hence, when I speak of the capacity of a body for heat, it must be understood as applied to the whole quantity of heat the body contains.

70.—The dimensions of bodies are enlarged when heat is poured into them, and they contract when it is taken from them. And the natural consequence of bodies absorbing different quantities of heat to cause an equal change of temperature, is, that they do not all expand nor contract alike by the change.

The incontrovertible fact, that different substances have different capacities for heat being established, another necessarily presents itself, which, though it could not possibly escape observation, has seldom been properly applied. It is, that in every chemical change we effect, we are altering the capacities of bodies for heat, and, consequently, deranging the equilibrium of heat; for the products differ in their capacity from the ingredients.

71.—By the mere addition of heat many solids assume the form of liquids, and liquids the gaseous state. On the other hand, gases, by an abstraction of heat, become liquids, and liquids solids. But even this change of state is accompanied by a change of capacity. The capacity of steam for heat is greater than that of water; for steam requires an additional quantity of heat, and that heat which is required to expand the particles of a liquid to the distance they are apart in the state of steam, does not affect the thermometer. That is, when a given quantity of water, heated to 212°, is converted into steam of the same temperature, the heat necessary to produce the change, from water to steam, would raise the temperature of about six times as much water from the mean temperature to 212°.

72.—The heat absorbed by steam or vapour, during its formation, is called *latent heat;* it is, however, a term which conveys a false notion of the state of heat in bodies, for the heat is not latent, it is simply a difference of quantity, and not of quality; and some term that would convey a more accurate idea of the phenomena would be better.

73.—The heat cmbined or disengaged by a change of the state of a body, called latent heat, is measured in the same manner as specific heat; that is, by the quantity of heat necessary to raise the temperature of water one degree at 60°.

It was the eminent Dr. Black who first discovered, (in 1762,) that a change of state in natural bodies requires a certain addition or diminution of heat; and that the quantity is different for different bodies, and also different according to the nature of the change.

The importance of this discovery to general science is great, and its finest practical application is to the principles of the steam engine.

74.—The additional heat in the vapour or steam of any liquid, is not very easily determined, but since the discovery of Dr. Black, experiments have been made by several philosophers, distinguished for their accuracy and skill in such delicate researches. The method adopted by Dr. Black is simple and easily tried, but not accurate. When a vessel containing water is placed on a fire, the water gradually becomes hotter till its temperature reaches 212°, but after that its temperature does not increase. The water is flying off in steam, and the heat not raising the temperature higher, as we know it would do if the vessel were closed, we must conclude, that the heat which would be communicated to the water in a close vessel combines with the steam in an open one, and yet does not increase the temperature of that steam to more than that of boiling water. To ascertain the quantity of heat which is combined with steam, Dr. Black put some water in a tin plate vessel upon a red hot iron. The water was of the temperature 50°; in four minutes it began to boil, and in twenty minutes it was all boiled off. During the first four minutes it had received 162° or 40½° per minute. If we suppose that it received as much per minute during the whole process of boiling, the heat which entered into the water, and converted it into steam would amount to 40½° × 20 = 810°. This 810 degrees of heat is not indicated by the thermometer, for the temperature of steam is only 212°; therefore Dr. Black called it latent heat.* But the result is obviously inaccurate, because steam is formed during the heating of the water to the boiling point, and the vessel is losing heat from its surfaces in unequal quantities, and the effect of the fire is also unequal, being less as the heat of the water increases.

75.—The heat required to form steam may be more accurately determined by condensing the steam by a cold fluid, and the heat communicated to the fluid by a given weight of steam gives the additional quantity of heat contained in the steam. Mr. Watt made various trials in this manner in 1781, and those on which he placed the most reliance gave 950° for the additional heat in the steam of water.†   Count Rumford, Mr. Southern. and Dr. Ure made experiments on this principle.‡

---

* Dr. Thomson's System of Chemistry, Vol. I. p. 101.

† Watt's Notes on Robison's Mech. Phil. Vol. II. p. 7.

‡ This mode of conducting the experiment has sometimes led to erroneous results through a want of attention to the mode of calculation. The quantity of heat being measured by the specific heat of water, let W be the weight of water used to condense the steam, and t its temperature after the steam has been condensed in it ; t' being the quantity its temperature is raised. Also let w be the weight of steam, and s its specific heat when condensed ; and x the whole heat required for its formation into steam.

Then the heat communicated to the water by the steam, is as its weight multiplied by the rise of temperature, or $W . t$

Also water may be heated in Papin's digester to 400° without boiling ; because the steam is forcibly compressed, and prevented from making its escape. If when heated to 400° the mouth of the vessel be suddenly opened, part of the water rushes out in the form of steam, but the greater part still remains in the form of water, and its temperature instantly sinks to 212°; consequently, 188° of heat have suddenly disappeared. This heat must have been carried off by the steam. Now as about one-fifth of the water is converted into steam, that steam must contain not only its own 188°, but also the 188° lost by each of the other four parts; that is to say, it must contain 188° × 5, or about 940° of heat.

76.—The experiments of Dr. Black are not greatly different from the result obtained by Schmidt, for the latter found the heat of steam to be 5·33 times the heat which is required to boil water of the temperature 32°, the barometer being at 29·84 inches.* This is the best mode of expressing the heat, for there is reason to believe that the specific heat of water is not the same for every rise of temperature. But to reduce it to the usual measure in degrees; there are 180° between the boiling and freezing point, hence $180 \times 5{\cdot}33 = 959{\cdot}4°$ for the additional heat of steam.

77.—Mr. Southern, and Mr. W. Creighton, in 1803, made some experiments by condensing steam with a considerable degree of care ; the steam being generated at different temperatures and pressures. The pressure, temperature, heat of formation, and bulk of the steam, from a cubic inch of water, are shewn in the following table:

The condensed steam has the temperature t after the operation, and as its whole heat was $x.w$ before condensation, it must be $t.s.w$ after. Therefore

$$W\, t' = x\, w - t\, s\, w, \text{ or } \frac{W\, t'}{w} + t\, s = x =$$

the measure of the heat that will form the weight of steam w.

If T be the temperature of the steam before condensation, and s' its specific heat ; then

$$\frac{W\, t'}{w} + t\, s - T\, s' =$$

the heat of conversion into steam ; and it appears from experiment to be nearly a constant quantity for the same liquid.

But it is usual to suppose the specific heat of equal weights of the steam and the liquid which forms it to be the same, and then

$$\frac{W\, t'}{w} - s\, (T - t) =$$

the heat of conversion.   And for water where $s = 1$, it becomes

$$\frac{W\, t'}{w} + t - T =$$

heat of conversion.

* Nicholson's Philosophical Journal, Vol. V. p. 208, octavo series.

Pressure in inches of mercury.	Temperature.	Heat required to form the steam.	Bulk of steam from one cubic inch of water at 60°.	Bulk calculated from the first experiment.
40	229°	1157°	208	1208
80	270°	1244°	588	635
120	295°	1256°	404	427

If from the whole heat we deduct the difference of temperature, we have 1157°, 1203', and 1190°; whence it appears that the heat to form steam is nearly a constant quantity when the temperature is the same, being independent of the density.

Therefore the most convenient mode of expressing the quantity of heat is that adopted by Mr. Southern, which consists in ascertaining the constant quantity of heat required to be added to the actual temperature of the steam to give the whole heat necessary to form it. This quantity is

$1157 - 229 = 928°$; $1244 - 270 = 974°$; and $1256 - 295 = 961°$; and the mean 954°.

In another set of experiments, made under the same pressures and temperatures, the quantities of heat required in addition to the temperature were 942°, 942° and 950°,[*] the mean being nearly 945°, and the mean of both sets 949°. In this set of experiments an allowance was made for the heat communicated to the vessel, in the former set none was made.

These experiments are valuable because they afford a proof that the additional heat of steam is either accurately or nearly a *constant quantity*.

78.—And they also shew that the *bulk* or volume of steam is *inversely as the pressure*, when the temperature is not altered. For as 80 : 40 : : 1208 : 604, which added to the expansion would be 635, nearly; and 120 : 40 : : 1208 : 402, and adding the expansion it is 427, nearly; and conversely the *density* is *directly as the pressure*: the experiments being quite as near as could be expected in so extremely delicate an operation.

79.—Count Rumford obtained a higher result: and, from his known skill in such inquiries, much confidence may be placed in his experiments. The heat was measured by means of the temperature communicated to a copper vessel filled with water, which he called his calorimeter. Within this calorimeter a thin serpentine pipe, of copper, con-

* Robison's Mechan. Phil. Vol. II. p. 160—166.

tained the steam to be condensed; hence the fluids did not mix together, and loss by the escape of vapour was prevented.

The water which the calorimeter contained was of a lower temperature than that of the room by 5° or 6°; and when the thermometer of the calorimeter announced an augmentation of temperature of 10° or 12°, an end was put to the experiment.

The water produced by the condensation of the vapour in the serpentine, was carefully weighed, and from its quantity, as well as from the heat communicated to the calorimeter, the heat developed by the vapour in its condensation was determined.

As a small part of the heat communicated to the calorimeter, was produced from the cooling of the water, condensed in the serpentine pipe after the vapour had been changed into water, an account was kept of this heat. It was supposed that the water at the moment of condensation was at the temperature of 212°, being that of boiling water; and it was determined by calculation, what part of the heat communicated to the calorimeter must have been owing to the boiling water.

In making this calculation, Count Rumford remarks, no " account was taken of the difference in the capacity of water for heat, which depends on its temperature: this is but imperfectly known; and besides, the correction which would have been the result, could not but have been very small."

The following are the details and results of two experiments, made on the 21st of January, 1812. The duration of each of the two experiments was from ten to eleven minutes. The water had been boiled for some time to drive out the air which it contained before the steam was directed into the serpentine pipe of the calorimeter.

Temprrature of the room.	State of the calorimeter, equal in specific heat to 42909 grains of water.			Quantity of vapour condensed into water.	Heat of conversion of the water into vapour in degrees.
	Temperature at the beginning.	Temperature at the end.	Elevation of its temperature.		
				Grains.	
61°	55½°	67½	12¼	457	1029·3°
62¼°	57°	67¼	10¼	377	1052·3°
				Mean.	1040.8°*

* Philosophical Mag. Vol. XLIII. p. 65.

The result of the second experiment being compared by our formula, (note to art. 75,) we have

$$\frac{42909 \times 10\frac{1}{4}}{377} + 67\frac{1}{4} = 1262\cdot5,$$

from whence, deducting 212° on the supposition that the specific heat of steam is equal to that of water, we have 1050·5 for the constant quantity of heat for conversion into steam; the very small difference between this and Count Rumford's result, arises from the fractions neglected in reducing the French to English weights.

80.—Count Rumford also made experiments on the quantity of heat developed in the condensation of the vapour of alcohol; the results of these experiments were less regular than those of the experiments made with water, as might have been expected, but they were nevertheless sufficiently uniform to give the quantity of heat with considerable certainty.

The vapour which is extricated from spirit of wine when boiled, varies a little with the intensity of the fire used in boiling it; he took care therefore to note the time which was taken in every experiment, in order to be able to judge, by comparing the quantity of vapour condensed, with the time employed to form it, of the intensity of the heat employed to boil the liquid. In the following table will be found the details and results of five experiments made on the same day, (January 21, 1812,) with alcohol of different degrees of strength. The specific heat of the calorimeter and the water it contained, was always equal to that of 42909 grains of water, and the thermometer employed was that of Fahrenheit.

Specific gravity of the alcohol employed.	Time employed in the experiment.	Temperature of the apartment.	State of the calorimeter.			Quantity of alcohol condensed in the calorimeter.	Heat of conversion of the liquid into vapour.
			Temperature at the beginning.	Temperature at the end.	Elevation of its temp.		
						Grains.	
81763	4¼ min.	61°	56°	66½°	10½°	875	479·92°
84714	8 —	60½°	55½°	65½°	10°	755	500·03°
85342	7 —	61°	54¼°	68¼°	14¼°	1079	499·54°
85342	5 —	61°	56°	66½°	10½°	805	476·83°
85342	6½ —	64°	57°	71½°	14½°	1102	499·65°
						Mean.	491.13

On determining, by calculation, the quantity of water which may be heated *one degree*, by the heat developed in the condensation of the vapour, he took care to keep an account of the difference of the capacity of water for heat from that of alcohol.[*]

The result of Count Rumford's calculation is nearly the same as by the formula, (art. 75, note,) when we assume the specific heat of the alcohol vapour and liquid to be the same, and equal to 58. Thus from the second experiment

$$\frac{42909 \times 10}{755} + (\cdot 58 \times 65 \cdot 5) = 606 \cdot 3,$$

from whence, deducting $173 \times \cdot 58$ for the heat due to the temperature of the vapour, we have 506°, nearly, for the heat of conversion from liquid to vapour. The Count's number is 500·03.

Count Rumford also ascertained that the vapour of sulphuric ether afforded only about half the heat in condensation that alcohol afforded; or one-fourth of the heat furnished by condensing the steam of water.

81.—Important as a knowledge of the heat of conversion into vapour is, it was not further investigated till 1817, when Dr. Ure made a few experiments on different bodies.[†] His mode of experiment was exceedingly simple. The apparatus consisted of a glass retort of very small dimensions, with a short neck, inserted into a globular receiver, of very thin glass, about three inches in diameter. The glass was fixed steadily in the centre of 32340 grains of water, at a known temperature, contained in a glass basin. Of the liquid, whose vapour was to be examined, 200 grains were introduced into the retort, and rapidly distilled into the globe by the heat of an Argand lamp. The temperature of the air was 45°, that of the water in the basin from 42° to 43°, and the rise of temperature occasioned by the condensation of the vapour, never exceeded that of the air by four degrees. As the communication of heat is very slow between bodies which differ little in temperature, the air could exercise no perceptible influence on the water in the basin during the experiment, which was always completed in five or six minutes. A thermometer of great delicacy was continually moved through the water, and its indications were read off, by the aid of a lens, to small fractions of a degree.

The distillation was rapidly performed, and, we are assured by Dr. Ure, that in the numerous repetitions of the same experiment the accordances were excellent. The following table gives the mean result, the last column being calculated by the formula, of (note to art. 75.)

[*] Philosophical Mag. Vol. XLIiI. p. 67.
[†] Philosophical Transactions for 1818.

Liquid.	Specific gravity.	Temperature of the water in the basin.			Boiling point.	Heat of conversion into vapour.
		At the beginning.	At the end.	Difference.		
Water	1·000	42·5°	49·°	6·5°	212°	942·°
Alcohol	0·825	42·	45·	3·	175°	425·5°
Sulphuric ether	0·7	42·	44·	2·	112°	302·6°
Oil of turpentine	0·888	42·	43·5	1·5	316°	146·0°
Petroleum	0·75	42·5	44·	1·5	306°	150·0°
Nitric acid	1·494	42·	45·5	3·5	165°	517·0°
Ammonia	0·978	42·	47·5	5·5	140°	840·0°
Vinegar	1·007	42·5	48·5	6·		870·0°

The quantity of water of which the specific heat would be equivalent to the heat absorbed by the vessels, Dr. Ure has not given, but in assuming it to be about 1660 grains we shall be not far distant from the truth. Hence we have 32340 + 1660 = 34000 for the water equivalent to the specific heat of the cooling apparatus; and by the formula, (art. 75, note,)

$$\frac{34000 \times 6\cdot5}{200} + 49 - 212 = 942° \text{ for water.}$$

$$\text{And} \quad \frac{34000 \times 3}{200} + \cdot65 \, (175 - 45) = 425\cdot5 \text{ for alcohol.}$$

The others being calculated in the same manner afford the results in the last column, taking the specific heat from the usual tables. Through an oversight in calculation, Dr. Ure's numbers in the Phil. Trans. are erroneous.

82.—A further correction might be applied for the quantity of steam remaining in the retort, and the loss of heat in the operation. Dr. Ure, has, in a recent correction for loss of heat, made the heat of conversion of water into steam 1000; and under the impression that Count Rumford's are the most accurate experiments on the subject, I am inclined to think this number about right. If for these sources of loss we make a further allowance in Dr. Ure's experiments of the specific heat of 2000 grains of water, we shall have

$$\frac{36000 \times 6\cdot5}{200} + 49 - 212 = 1007;$$

and correcting the rest of the numbers in this manner the following are obtained.*

* Ure's Dictionary of Chemistry, art. Caloric.

	Equal weights.	Equal volumes.		Equal weights.	Equal volumes.
Water into steam	1007·0	1007o	Petroleum into vapour	165o	124o
Alcohol into vapour	455·5o	375o	Nitric acid into vapour	552o	830o
Sulphuric ether into vapour	322·6o	227o	Ammonia into vapour	895o	875o
Oil of turpentine into vapour	161·o	143o	Vinegar into vapour	930o	936o

Having followed through the best information hitherto laid before the public, on the heat required to produce steam, our next object must be to convert it into a form more directly useful for our purpose. For the quantity of heat which converts a liquid into vapour, requires the additional facts of the volume of the vapour, and its elastic force to render it valuable.

Of the Elastic Force of Steam.

83.—To obtain a rule for determining the force of steam at any temperature, or the temperature corresponding to any given force, we must have recourse to a rule found by trial from the best experiments; it is not a satisfactory method, but we have no other means of arriving at a rule in a case where the real causes of variation are not understood. We still however may gain some assistance, from previous reasoning, in forming our conclusions. In the first place, the index of the power representing the law of variation, must be of such a simple kind as to render it probable that it is the true one. Hence the index 5·13 employed by Mr. Southern,[*] is not likely to represent the law of nature; Mr. Creighton's index 6;[†] or Dr. Young's which is 7,[‡] are, either of them, more likely to be accurate. The true equation may be very complex, but this is not probable, and while we are ignorant of its nature, and can represent the results sufficiently near for practical use, by one index, it is best to adopt the simplest form, and particularly when it is equally as likely to be the true one as one of a more complex kind. In any attempt to find the index by the usual method of differences, the errors of experiment will have too great an influence.

84.—Secondly. It appears probable, that there is a degree of cold at which steam

* Robison's Mechanical Phil. Vol. II. p. 172.   † Phil. Mag. Vol. LIII. p. 266,
‡ Natural Phil. Vol. II. p. 400.

cannot exist,* and this must be the case when it is condensed by cold, till the cohesive attraction of the particles exceeds the repellant force of the caloric interposed between them, and the change from an elastic fluid to a solid may then take place without the intermediate stage of liquidity. This physical circumstance enables us to fix another element in the calculation; for there must be a temperature when the force is nothing.

85.—Thirdly. The greatest possible force of steam must next be considered, for we are certain that onr formula must be in error if it exceeds that limit. Suppose a given quantity of water, a cubic inch for example, to be confined in a close vessel which it exactly fills; and that in this state it is exposed to a high temperature. Then, as the bulk when expanded is to the quantity the bulk is increased by expansion, without change of state, so is the modulus of elasticity of water of that temperature to the force of steam of the same density as water. If our rule therefore gives steam a greater force than this at the same density and temperature, it must be erroneous. With these limitations we must in a considerable degree be guarded against error, and the method followed is next to be explained.

86.—Let f be the elastic force of steam, in inches of mercury, and t the corresponding temperature, and let a be the temperature below which the elastic force is 0. Consider f the abscissa, and $t + a$ the ordinate of a curve, of which the equation is $Af = (t + a)^n$, whence the coefficient

$$A = \frac{(t + a)^n}{f}.$$

Let the abscissa increase to f', and the ordinate to $t' + a$; then

$$\frac{(t + a)^n}{f} = \frac{(t' + a)^n}{f}; \text{ or } \frac{\log. f' - \log. f}{\log. (t' + a) - \log. (t + a)} = n.$$

Now if these points be near one extremity of the range of experiment, and two other points be taken near the other extremity, then

$$\frac{\log. f''' - \log. f''}{\log. (t''' + a) - \log. (t'' + a.)} = n, \text{ and consequently}$$

$$\frac{\log. f''' - \log. f''}{\log. f' - \log. f} = \frac{\log. (t''' + a) - \log. (t'' + a)}{\log. (t' + a) - \log. (t + a)}.$$

From four results of Mr. Southern's experiments, on steam from water, we find that $a =$

---

* An interesting paper on this subject by Mr. Faraday renders it equally so that the limit is different for different vapours; my formula had led me to the same conclusion, hence, it has another property justified by experience. See Phil. Mag. Vol. LXVIII .p. 344.

I

100 very nearly satisfies the conditions; and this value of a being inserted, we find $n = 6$ and $A = 177$, or its logarithm $= 2\cdot247968$.

Therefore, for water,

$$f = \left(\frac{t + 100}{177}\right)^6; \text{ or } t = 177 f^{\frac{1}{6}} - 100.$$

In logarithms,

$$\log. f = 6\left(\log.(t + 100) - 2\cdot247968\right).$$

87.—If the expansion of confined water, when its temperature is raised to 1150 degrees of heat, be $0\cdot9693$ of its bulk, the force necessary to confine it to its bulk at 60°, when exposed to a heat of 1150°, the modulus of water being 22,100 atmospheres at 60°, would be about 6925 atmospheres.* Our rule gives for the force of steam at that temperature and density

---

* The expanding power of heat, and the decrease of the modulus of elasticity, must be in the same ratio ; and most probably both vary as the square of the central distances of the atoms, and consequently as the $\frac{2}{3}$ power of the volume. Hence, if e be the expansion, the original bulk being unity, and m the modulus, it must be

$$\frac{m}{(1 + e)^{\frac{2}{3}}} =$$

the modulus at any expansion e ; and consequently (by art. 85.)

$$1 + e : e :: \frac{m}{(1 + e)^{\frac{2}{3}}} : \frac{m\,e}{(1 + e)^{\frac{5}{3}}} =$$

the force of compression capable of retaining the fluid in its original state of density.

The expansion varies as the expanding power of heat, and as the temperature, hence, it will be as the $\frac{5}{3}$ power of the temperature ; and it must be 0 at 40° ; consequently, $A(t - 40)^{\frac{5}{3}} = e$, and as from 40° to 212°, it is found to be 04333, we have $\frac{5}{3} \log.(t - 40) - 5\cdot089091 = e$. The agreement of this formula with experiment is shewn in the following table.

Temperature.	Expansion by formula.	Expansion by experiment.	Temperature.	Expansion by formula.	Expansion by experiment.
40°	0·00	0·00	400°	0·1484	
64°	0·00159	0·00133	800°	0·5155	
102°	0·00791	0·0076	1150°	0·9693	
212°	0·0433	0·04333	1171°	1·0000	

In the equation for the force at 1150 degrees of temperature, we have

$$\frac{m\,e}{(1 + e)^{\frac{5}{3}}} = \frac{22100 \times \cdot9693}{(1\cdot9693)^{\frac{5}{3}}} = 6925$$

atmospheres.

4137 atmospheres; and in the uncertainty both as to what the actual expansion of water would be in such high temperatures, and the decrease of its modulus, it is more prudent to be within that beyond the limit. But at, or near the temperature 1150° the rule will cease to be of any use because then it is simply the expansive power of compressed water; and it varies as the quantity water expands by a given change of temperature.

Having thus far explained the methods by which the rules have been obtained, it only remains to give them the most simple form for use, with illustrative examples.

88.—RULE I. To find the force of steam from water in inches of mercury the temperature being given.

Add 100 to the temperature, and divide the sum by 177; the sixth power of the quotient is the force in inches required.

Example. To find the force of steam for the temperature 312.°

$$\frac{312 + 100}{177} = 2\cdot3277.$$

Raise this to the sixth power and it gives 159 inches for the force of the steam in inches of mercury.

Or by logarithms. Add 100 to the temperature, and from the logarithm of this sum, subtract 2·247968; and six times the difference is the logarithm of the force in inches of mercury.

Example. To find the force of steam for the temperature 250°.

$$
\begin{array}{rr}
\text{Log. } 250 + 100 = 350 \text{ is} & 2\cdot544068 \\
\text{Subtract constant log.} & 2\cdot247968 \\
\hline
& 0\cdot296100 \\
& 6 \\
\hline
\text{Log. of force in inches of mercury} = \text{log. } 59\cdot79 = & 1\cdot776600 \\
\hline
\end{array}
$$

89.—RULE II. The force of the steam of water being given to determine its temperature.

Multiply the sixth root of the force in inches by 177, and subtract 100 from the product, which gives the temperature required.

Example. Let the force of steam be eight atmospheres, equal 240 inches of mercury, to find its temperature.

The sixth root of 240 may be easily found by a table of squares and cubes, by first finding its square root, and then the cube root of the square root. Thus the square root of 240 is 15·492, and the cube root of 15·492 is 2·493; hence, (2·493 × 177) − 100 = 341·20. Mr. Southern's experiment gives 343·6.

Or by logarithms. Add one sixth of the logarithm of the force in inches to 2· 247968, the sum is the logarithm of 100 added to the temperature.

Example. Let the force of steam be equal to sixty inches of mercury, which is nearly fifteen pounds on the square inch above the pressure of the atmosphere, to find its temperature.

$$\begin{array}{rl}
\text{Log. 60 is} & - - - \quad 1\cdot778151 \\
\hline
\text{and one sixth is} & - \quad \cdot296358 \\
\text{constant log.} & - - \quad 2\cdot247968 \\
\hline
\text{Log. } 350\cdot2 & - - \quad 2\cdot544326 \\
\hline
\end{array}$$

from which subtract 100, and it gives 250·2° for the temperature. Mr. Southern's experiment gives 250·3°.

90.—When sea water is employed, as it boils at a different temperature, the force of the steam is different. The correction in the rules is easily made by finding the constant number which corresponds to a force of thirty inches of mercury, at the boiling point, with different degrees of saturation with salt. Many of the people employed about boat engines, are not yet aware that there is a difference between the temperature of steam from common water, and that from salt water, when the force is the same. I will shew in another place (Sect. IV.) the effect this has on the power of the steam engine, but at present our object is to determine the force of the steam. Mr. James Watt was the only person who had made experiments on the steam of salt water; they were made in 1774.* He does not give them as being very accurate ones, but they are sufficient to establish the fact that there is a difference; and Mr. Faraday has lately had occasion to satisfy himself, on the same point, by various experiments.†

91.—The following table gives the boiling points of solutions of different salts in water.

---

* Robison's Mechanical Phil. Vol. II. p. 34.
† Quarterly Journal of Science. Vol. XIV. p. 440.

Name of salt.	Dry salt in 100 parts by weight of the solution.	Boiling point.	Authority.
Acetate of soda	60	256°	Griffiths.*
Nitrate of soda	60	246	————
Common salt	37	226	My trials.
Muriate of soda	30	224	Griffiths.
Ditto		222·35	Achard.†
Sulphate of magnesia	57·5	222	Griffiths.
Sulphate of lime	45	220	————
Alum	52	220	————
Sulphate of iron	64	216	————
Sulphate of soda	31·5	213	————
Ditto		217·6	Achard.

92.—According to the analysis of Dr. John Murray 10,000 parts sea water of the specific gravity 1·029‡ contain

$$
\begin{aligned}
\text{Muriate of Soda} \quad & 220\text{·}01 = \tfrac{1}{46}\text{·}\\
\text{Sulphate of Soda} \quad & 33\text{·}16 = \tfrac{1}{301}\text{·}\\
\text{Muriate of Magnesia} \quad & 42\text{·}08 = \tfrac{1}{238}\text{·}\\
\text{Muriate of Lime} \quad & 7\text{·}84 = \tfrac{1}{1275}\text{·}\\
\hline
& 303\text{·}09 = \tfrac{1}{33}\text{·}
\end{aligned}
$$

Or 1 part of sea water contains ·030309 parts of salts = $\tfrac{1}{33}$ of its weight.

93.—Now as the salts do not rise with the steam, the water in a boiler supplied with sea water becomes gradually more saturated, and after a certain time begins to deposit salt, if the means that have been invented for that purpose be not employed to prevent it. (See Sect. III.) And even then a certain degree of a saturation must be allowed to take place. The following table, with the constant numbers for different degrees of saturation will serve to illustrate this matter. The boiling point of water appears to be increased one degree by each addition of 2·6 parts to the proportion of common salt in 100 parts of water; at least so nearly that this regular law does not materially differ from the mean results of my experiments, which were made with a considerable degree of care. But it is difficult to make them on account of the degree of saturation constantly varying during the experiment.

* Quarterly Journal of Science, Vol. XVIII. p. 90.
† Thomson's Chemistry, Vol. II. p. 14.   ‡ Phil. Mag.

Proportion of salt in 100 parts by weight.			Boiling point.	Constant number.	Constant log.
Saturated solution } 36·37	=	$\frac{12}{33}$	226·	185·	2·267031
33·34	=	$\frac{11}{33}$	224·9	184·3	2·265563
30·3	=	$\frac{10}{33}$	223·7	183·6	2·263956
27·28	=	$\frac{9}{33}$	222·5	183·	2·262343
24·25	=	$\frac{8}{33}$	221·4	182·3	2·260859
21·22	=	$\frac{7}{33}$	220·2	181·6	2·259234
18·18	=	$\frac{6}{33}$	219·	181·	2·257604
15·15	=	$\frac{5}{33}$	217·9	180·4	2·256104
12·12	=	$\frac{4}{33}$	216·7	179·7	2·254461
9·09	=	$\frac{3}{33}$	215·5	179·	2·252812
6·06	=	$\frac{2}{33}$	214·4	178·3	2·251296
Sea water 3·03	=	$\frac{1}{33}$	213·2	177·6	2·249496
Common water		0	212	177·	2·247968

94.—The next point is to compare the formula with experiment, and we will commence with Mr. Watt's experiments on salt water. The water was nearly saturated with salt. It was more free from air than common water, but it parted with difficulty from that it contained. The results compared with the formula for saturated salt are shewn in the following table.

Watt's Experiments on the Steam from salt Water.

Temperature.	Forces in inches of mercury.		Temperature.	Forces in inches of mercury.	
	Watt's observations.	Formula for saturated solution.		Watt's observations.	Formula for saturated solution.
46°	0·01	0·24	195·5	15·34	16·64
85	0·58	1·00	201·5	17·16	18·77
113	1·72	2·33	207	19·34	20·92
139	3·54	4·66	210	21·8	22·18
160	6·27	7·72	212	22·74	23·05
169	8·12	9·47	216	24·6	24·87
180	10·85	12·04	218	25·52	25·84
187	12·67	13·01	220	26·5	26·84

In these as in all the early experiments on the force of steam, the force is less than it ought to be at low temperatures.

Mr. Watt's experiments on pure water afford a like discrepancy, as will be found by comparing the following table of results taken at random out of his series.*

Watt's Experiments on pure Water.

Temperature.	Force in inches of mercury.	
	Watt's observations.	Formula for pure water.
55°	0·15	0·45
118	2·68	3·59
180	14·73	15·67
225	37	38·32
240	49	50·24
261	68	72·0
272·5	82	86·89

The explanation offered by Mr. Watt himself is not sufficient to account for the difference except in the lower temperatures.

He supposes the stationary barometer must have had its scale placed ·2 of an inch too low; and if so, the same addition would be required to the forces in the preceding table on salt water. These tables, however, are not selected for minute accuracy, but to shew the important fact that the force of the steam of water depends on the temperature of the liquid which produces it, or which is in contact with it. For this they are sufficiently correct, and it is a circumstance which affects its elastic force both in the boiler and in the condenser; and is peculiarly interesting to those concerned in steam vessel engines. The temperatures not being the same the comparison is not so easy, but at 180° the force of salt water is 10·85; that of pure water 14·73 inches: at 212° salt water has a force of 22·74; pure water 29·56.

95.—The experiments made by Professor Robison were tried in a similar manner; and

* Robison's Mechan. Phil. Vol. II. p. 32—34.

as a method the same in effect was used by Bettancourt, whose results agree extremely well with Robison's, the description of it may be useful.

Professor Robison's apparatus for determining the force of steam.—This apparatus, in the first trials, consisted of a small digester of copper, A B C D, in the figure; the top of which had a thermometer inserted through the centre, and a loaded valve at V; with a third hole for inserting a barometer tube S G F, to ascertain the force at lower temperatures than 212°. The force at the higher temperatures was measured by the steelyard on the valve, and a plug was inserted in the place of the tube S G F, but the results with the valve were irregular and not satisfactory. Hence, the glass tube M N K, having a cistern L for mercury, was adapted to the hole in the digester, and instead of measuring the force by the valve, it was measured by the ascent of the mercury in the tube M N. The digester was heated by a lamp.

FIG. 10.

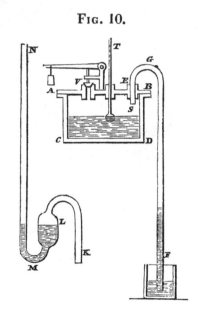

To determine the pressure at temperatures below 212°, the tube S G F was inserted as in the figure, and a basin of mercury provided at F. The lamp being applied, the water in the digester produced steam till it issued at both the valve and the pipe F, so as to expel the air; the lamp being removed, and both the valve and tube being closed, the latter by immersing it in the mercury, the mercury rose in the tube F G as the apparatus cooled, and the heights corresponding to different temperatures were noted; like observations were made as it reheated.

To determine the pressure at higher temperatures with the apparatus, the end K of the tube M N K was inserted at E, and as the temperature increased, the pressure of the steam in the cistern L caused the mercury to ascend, and consequently afforded a means of measuring the force of the steam.

The objection to this mode of trial is that the temperature of the mercury must be continually changing during the trial, and steam must either be condensing or generating on its surface during the time of observation. At each observation the temperature of the whole of the apparatus ought to be the same, and then the column exhibiting the pressure ought to be reduced to its equivalent at the mean temperature. The only observation where these circumstances would have place was that which appears to have been made when the thermometer was at 42°; then the column in the syphon was 29·7, and the barometer stood at

29·84 : the difference is the force of steam at 42°, and is 0·14 inches. By cooling down to 32° the force was not perceptibly different, and we know from later trials that this is nearly correct. Professor Robison, however, seems to have thought it was necessary to have the force 0 at 32°.*

Robison's Experiments on the Force of Steam.

Temperature of the steam.	Force of steam in inches of mercury.		Temperature of the steam.	Force of steam in inches of mercury.	
	By Dr. Robison's experiments.	By the formula.		By Dr. Robison's experiments.	By the formula.
32°	0·0	0·172	160°	8·65	10·05
40	0·1	0·245	170	11·05	12·6
50	0·2	0·37	180	14 05	15·67
60	0·35	0·55	190	17·85	19·35
70	0·55	0 78	200	22·62	23·71
80	0·82	1·106	210	28·68	28·86
90	1·18	1·53	220	35·8	34·92
100	1·6	2·08	230	44·5	42·0
110	2·25	2·79	240	54·9	50·24
120	3 0	3·68	250	66·8	59·79
130	3·95	4·81	260	80·3	70·8
140	5·15	6·21	270	94·1	83·45
150	6·72	7·94	280	105·9	97·92

If the elastic force ·14 from which Robison began to register had been added to all the experiments below 212°, as it ought to have been, they would have agreed extremely near with the results of later experiments. The experiments made by Achard seldom vary more than a degree or two from those in the above table.

96.—Mr. Dalton's inquiries were conducted by a different method. He took a barometer tube, made perfectly dry, and filled it with mercury just boiled, marking the place where it was stationary; then graduated the tube into inches and tenths by means

* Mechan. Phil. Vol. II. p. 36.

K

of a file; into this tube he poured a little water (or any other liquid the subject of experiment,) so as to moisten the whole inside; after this he again poured in mercury, carefully inverting the tube, to exclude all air. The barometer, by standing, sometimes exhibited a portion of water, &c. of one eighth or one tenth of an inch, upon the top of the mercurial column, because being lighter it ascends by the side of the tube; which may now be inclined and the mercury will rise to the top, manifesting a perfect vacuum from air. He then took a cylindrical glass tube open at both ends, of two inches diameter, and fourteen inches in length; to each end of which a cork was adapted, perforated in the middle so as to admit the barometer tube, to be pushed through and to be held fast by them; the upper cork was fixed two or three inches below the top of the tube, and half cut away so as to admit water, &c. to pass by; its service being merely to keep the tube steady. Things being thus circumstanced, water of any temperature may be poured into the wide tube, and made to surround the upper part of the vacuum of the barometer, and the effect of temperature in the production of vapour within can be observed from the depression of the mercurial column. In this way, he says, he had water as high as 155° surrounding the vacuum, but as the high temperature might endanger a glass apparatus, instead of it he used the following one for higher temperatures.

Having procured a tin tube of four inches in diameter, and two feet long, with a circular plate of the same soldered to one end, having a round tube in the centre, like the tube of a reflecting telescope; he got another smaller tube of the same length soldered into the larger, so as to be in the axis or centre of it; the small tube was open at both ends; and on this construction water could be poured into the larger vessel to fill it, whilst the central tube was exposed to its temperature. Into this central tube he could insert the upper half of a syphon barometer, and fix it by a cork, the top of the narrow tube also being corked; thus the effect of any temperature under 212° could be ascertained, the depression of the mercurial column being known by the ascent in the exterior leg of the syphon. Mr. Dalton also remarks, that the force of vapour from water between 80° and 212° may be determined by means of an air pump; and the results exactly agree with those determined as above. Take a florence flask half filled with hot water, into which insert the bulb of a thermometer; then cover the whole with a receiver on one of the pump plates, and place a barometer gauge on the other: the air being slowly exhausted, mark both the thermometer and barometer at the moment ebullition commences, and the height of the barometer gauge will denote the force of vapour from water of the observed temperature. This method may also be used for other liquids. It may be proper to observe, that the various thermometers used in these experiments were duly adjusted to a good standard one.

After repeated experiments by all these methods, and a careful comparison of the results, he was enabled to digest a table of the force of steam from water of all the temperatures from 32° to 212.°* The only experimental results were the following ones, which are compared with our formula.

Dalton's Experiments on the Force of Steam.

Temperature of steam.	Force in inches of mercury.		Temperature of steam.	Force in inches of mercury.	
	Dalton's observations.	By the formula.		Dalton's observations.	By the formula.
32°	0·2	0·172	133¼	4·76	5·24
43¼	0·297	0·281	144½	6·45	6·95
54½	0·435	0·442	155¾	8·55	9·10
65¾	0·63	0·675	167	11·25	11·7
77	0·91	1·00	178¼	14·6	15·1
88¼	1·29	1·447	189½	18·8	19·15
99½	1·82	2·05	200¾	24·0	24·07
110¾	2·54	2·85	212	30·0	30·0
122	3·5	3·89			

From these results he determined the ratio belonging to each interval, and filled in the intermediate degrees by interpolation, considering the forces to increase in a geometrical progression. Above 212° he made no trials at that period, though the table was extended to 325°, and has since been found to be erroneous for the temperatures above 212°.

97.—Mr. Dalton afterwards re-examined the subject, and considers from various trials, that the force of steam at 32° cannot be less than 0·2 of an inch; and is most probably 0·25. But with the advantage of having seen the results of Dr. Ure's, and Mr. Southern's experiments, and having made new experiments himself for the temperatures between 212° and 300°, he gives the following table formed from what he considers the most correct experiments on the subject.†

* Nich. Phil. Journal, Vol. VI. p. 263, 8vo.

† Annals of Philosophy, Vol. XV. p. 130. for 1820.

Temperature of steam.	Force of steam in inches of mercury.		Temperature of steam.	Force of steam in inches of mercury.	
	Dalton's numbers.	By the formula.		Dalton's numbers.	By the formula.
36°	0·29	0·201	173°	13·18	13·46
64	0·75	0·633	220	34·20	34·92
96	1·95	1·84	272	88·9	86·2
132	5·07	5·07	340	231·0	236·

It will appear from this that there is a greater difference between the results of different trials than between the numbers found by our rule and those results, and hence it may be presumed to be nearly true.

98.—For the further satisfaction of the reader the principal results of Dr. Ure's experiments shall be given, and his simple and elegant mode of making the experiments described, as in the event of any other species of fluid being found better adapted than water for furnishing vapour the same mode might be usefully adopted to try its force.*

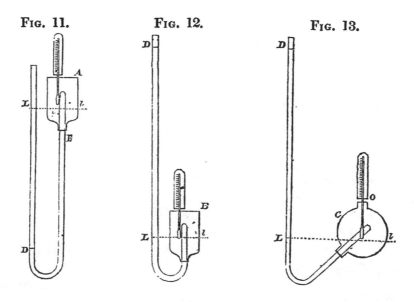

FIG. 11.   FIG. 12.   FIG. 13.

* Phil. Trans for 1818.

The preceding figure (Fig. 11,) represents the construction employed for temperatures under and a little above the boiling point. Fig. 12 and 13 were used for higher temperatures; the last is the more convenient of the two. It was suspended from a lofty window ceiling, and placed with the tube L D in a truly vertical position by means of a plumbline. One simple principle pervades the whole train of experiments, which is, that the progressive increase of elastic force developed by heat from the liquid, incumbent on the mercury at l, is measured by the length of column which must be added over L, in order to restore the quicksilver to its primitive level at l. These two stations, or points of departure, are nicely defined by a ring of fine platina wire twisted firmly round the tube.

At the commencement of the experiment, after the liquid, well freed from air, has been let up, the quicksilver is made to coincide with the edge of the ring l, by cautiously pouring mercury in a slender stream into the open leg of the syphon D. The level ring at L is then carefully adjusted.

From the mode of conducting the experiments, there remained always a quantity of liquid in contact with the vapour, a circumstance essential to accuracy in this research. Suppose the temperature of the water, or the oil in A (Fig. 11.) to be 32°, as denoted by a delicate thermometer, or by the liquefaction of ice; and let L D be a column equal to the atmospheric pressure; communicate heat to the cylinder A, by means of two Argand flames, playing gently against its shoulder at each side.

When the thermometer indicates 42°, modify the flames, or remove them, so as to maintain an uniform temperature for a few minutes. Then the elasticity will be faithfully represented and measured, by the mercurial column which we must add over D, in order to return the quicksilver to the line l its zero or initial level.

At E a piece of cork is fixed, between the parallel legs of the syphon, to sustain it, and to serve as a point by which the whole is steadily suspended.

For temperatures above the boiling point, the part of the syphon under E is evidently superfluous, merely containing in its two legs a useless weight of equipoised mercury. Accordingly for high heats, either the apparatus (Fig. 12) or (13,) is employed, and the same method of procedure is adopted. The aperture at O, (Fig. 13,) admits the bulb of the thermometer, which rests against the tube. The recurved part of the tube is filled with mercury, and then a little liquid is passed through it to the sealed end. Heat is applied by an Argand flame to the bottom of c, which is filled with oil or water, and the temperature is kept steadily at 212° for some minutes. Then a few drops of quicksilver may require to be added at D, till L and l be in the same horizontal plane. The further conduct of the experiment differs in no respect from what has been already described. The liquid in c is progressively heated, and at each stage mercury is progressively added over L to restore the initial level, or volume at l, by equipoising the progressive elasticity. The column above L being the accession of elastic force. When this column is wished to extend very high, the vertical tube requires to be placed for support in the groove of a long wooden prism.

Ure's Experiments on the Force of Steam.

Temperature of steam.	Force in inches of mercury.		Temperature of steam.	Force in inches of mercury.	
	Ure's observations.	Formula.		Ure's observations.	Formula.
24°	0·170	0·118	190°	19·000	19·35
32	0·200	0·172	200	23·600	23·71
40	0·250	0·245	210	28·880	28·86
50	0·360	0·37	212	30·000	30·00
55	0·416	0·45	220	35·540	34·92
60	0·516	0·55	225	39·110	38·32
70	0·726	0·78	230	43·100	42·00
80	1·010	1·106	240	51·700	50·24
90	1·360	1·53	250	61·900	59·79
100	1·860	2·08	260	72·300	70·8
110	2·456	2·79	270	86·300	82·45
120	3·300	3·68	280	101·900	97·92
130	4·366	4·81	290	120·150	114·4
140	5·770	6·21	295	129·000	123·5
150	7·53	7·94	300	139·700	133·2
160	9·600	10·05	310	161.300	154·5
170	12·050	12·6	312	167·000	159·
180	15·160	15·67	312	165·5	

If a nice agreement with a particular set of observations had been attempted, the formula could have easily been arranged to represent these better, but by so doing it appears to me that the elastic forces would have increased in a higher ratio than we are warranted in expecting from other experiments, and the later inquiries of Mr. Dalton justify the numbers being higher at or about 150° than Dr. Ure's.

99.—Mr. Southern's experiments on high pressure steam were made with a digester, with a thermometer fitted to a metallic tube, so that the stem of the thermometer might be immersed as far as it contained mercury. Also, instead of measuring the force of the steam by a loaded valve, a nicely bored cylinder was used, with a piston fitting it so as to have very little friction, to the rod of which a lever was applied, constructed to work on edges like those of a scale beam; and that no error might arise from this construction a column of mercury was substituted and the correspondence was within $\frac{1}{100}$ of an inch.

The observations at each of the points of temperature and pressure, were continued some minutes, the temperature being alternately raised, and lowered, so as to make the pressure in excess and defect, and a mean temperature was taken for the result. This method seems to me entitled to great confidence, and hence I have made the results the principal data for my formula. (See art. 86.)

The experiments below 212° were conducted nearly as Dr. Robison's, and those below 62° were made by Mr. W. Creighton. These low pressure experiments do not seem to be of equal value with the four high pressure ones.

*Southern's Experiments on the Force of Steam.**

Temperature of the steam.	Force in inches of mercury.		Temperature of the steam.	Force in inches of mercury.	
	Observed force.	Formula.		Observed force.	Formula.
32°	0·16	0·172	132°	4·71	5·07
42	0·23	0·266	142	6·10	6·53
52	0·35	0·401	152	7·90	8·33
62	0·52	0·587	162	10·05	10·52
72	0·73	0·842	172	12·72	13·17
82	1·02	1·182	182	16·01	16·35
92	1·42	1·629	212	30·00	30·00
102	1·96	2·21	250·3	60·00	60·00
112	2·66	2·95	293·4	120·00	120·50
122	3·58	3·89	343·6	240·00	247·80

100.—A scale of the elastic force of steam at high temperatures was published in

* Robison's Mechanical Philos. Vol. II. p. 173.

1822, by Mr. Philip Taylor,* which was formed by means of an apparatus not described, but it appears to correspond with the best experiments, and is likely to be near the effect in practice where we may expect some loss of elastic force, compared with the temperature.

Taylor's Scale of the Force of Steam.

Temperature of the steam.	Force in inches of mercury.	
	Taylor's observations.	Formula.
212°	30·00	30·00
220	34·95	34·92
230	41·51	42·00
240	50·00	50·24
250	59·12	59·79
260	70·10	70·8
270	82·50	83·45
280	97·75	97·92
290	114·50	114·4
300	133·75	133·2
320	179·40	178·5

101.—The experiments of Schmidt present a surprising accordance with the rule at the temperatures from 60° to 230°; at these points they slowly separate, the rule being in defect in the highest temperature 290° by eleven inches, and in excess at 43·¼° by 0·163.†

102.—The force of steam at high temperatures is still wanting to complete the experimental part of the inquiry. A few experiments have been made which appear to be entitled to some confidence by Professor Arsberger, of Vienna.‡

* Philosophical Mag. Vol. IX. p. 452.
† Dr. Young's Nat. Phil. Vol. II.
‡ Bulletin des Sciences Tech. Vol. I. p. 294.

Arsberger's Experiments on high Pressure Steam.

Temperature of the steam.	Force in inches of mercury.	
	By experiment.	By our rule.
232°	44·4	43·56
249	59·1	58·7
274	88·8	89·0
322	176·0	183·6
372	325·0	362·
432	620·0	737·

Here the rule is in excess at 432°, by more than one sixth, but in an experiment reported by M. Clement, to M. Poisson,* the force of steam at 419° is said to be 35 atmospheres, or 1050 inches of mercury, while our rule gives only 635 inches. I doubt the accuracy of the statement.

103.—M. Cagniard de la Tour,† made some essays to ascertain the space and temperature in which a given quantity of water became wholly steam, but from the frequent rupture of the glass tubes, and their loss of transparency, it was difficult to obtain a result. He states, however, that at a temperature but little removed from the melting point of zinc, water could be converted into vapour in a space nearly four times its volume. If this could have been really ascertained with accuracy it would have given an important datum, but on the above rude approximation no reliance can be placed.‡

* Philosophical Magazine, Vol. LXI. p. 60.

+ Philosophical Magazine, Vol. LXI. p. 58.

‡ In a paper on the elastic force of steam which has just been published by Mr. Ivory, in the Philosophical Magazine, a completely different process is followed for calculating the force of steam from that I have given; it does not however afford results more near to the experiments it is founded on, than those by my formula, while it is somewhat more difficult to apply, and becomes erroneous in high temperatures.

Counting t the temperature from 212°, and f the elastic force, his formula is

$$\log. \frac{f}{30} = \cdot0087466\, t - \cdot000015178\, t^2 + \cdot000000024825\, t^3.$$

It is derived from a comparison of Dr. Ure's experiments; and the following table shews the results by experiment, by Mr. Ivory's formula, by various experiments, and by my formula.

In the absence therefore of proper experiments to ascertain the force of steam, it is difficult to determine a rule that can be depended upon for high temperatures, and we must

Temperature of the steam.	Elastic force of steam in inches of mercury.			
	Dr. Ure's experiments.	Mr. Ivory's formula.	Various experiments.	My formula.
32°	0·2	0·185	0·16 Creighton	0·172
50°	0·36	0·36		0·37
70°	0·726	0·721		0·78
90°	1·360	1·378		1·53
110°	2·456	2·634		2·79
130°	4·336	4·408		4·81
150°	7·530	7·424		7·94
170°	12·05	12·05		12·60
190	19·00	18·93		19·35
210°	28·88	28·81		28·86
230°	43·10	42·63	41·51 Taylor	42·00
250°	61·90	61·50	60·0 Southern	59·79
270°	86·30	86 70	82·5 Taylor	83·45
290°	120·15	119 9	114·5 Taylor	114·40
310°	161·30	162·8		154·5
337°		240	234 Christian*	226·5
343·6°		264	240 Southern	247·8
419°		714	1050 Clement	635·0
432°		1852	620 Arsberger	737·0

At a temperature of about 770°, Mr. Ivory s formula gives an elastic force equal to the modulus of elasticity of water, the steam would if this were correct be more dense than water; while La Tour found it required a space four times its volume to become steam at about the same heat.   Arsberger's experiments had not been seen by Mr. Ivory or he would have had a reason for doubting the accuracy of M. Clement's observation, but as it is quite unsupported by either a description of the process, or any other observations at other temperatures, its deviation in excess both from formulæ founded on a considerable range of experiments, and also from other results, is to be regarded as a motive for doubt rather than for altering our formulæ.   Mr. Ivory most justly remarks, that this furnishes "another instance of the great difficulty of detecting general properties or laws by means of a comparison of particular results," and it is a difficulty which ought to direct mathematicians possessed of such great powers as Mr. Ivory certainly is, to endeavour to develope the first principles, rather than grope out analogies from experiments alone.

* Mecanique Industrielle Vol. II. p. 232.

now try to discover if the force of other vapours will afford any further insight into the subject.

Of the elastic Force of the Vapour of Alcohol.

104.—The elastic force of the vapour of alcohol, or spirit of wine, has been tried by several philosophers. The greater part of the experiments were made in the lower ranges of temperature, and in the same manner as those on the force of the steam of water; but in describing them it will be some advantage to begin with the experiments of Cagniard de la Tour on the space alcohol occupies when converted wholly into vapour. To ascertain this point, alcohol of the specific gravity ·837, was introduced into small tubes of glass, and hermetically sealed, with a handle of glass attached to each tube   A tube was two-fifths filled with alcohol, and then slowly, and carefully heated; as the fluid dilated its mobility increased, and, when its volume was nearly doubled, it completely disappeared, and became a vapour so transparent, that the tube appeared quite empty. On leaving it to cool for a moment, a very thick cloud formed in its interior, and the liquor returned to its first state. A second tube, nearly half occupied by the same fluid, gave a similar result; but a third, containing rather more than half, burst.

A process was next adopted to ascertain the pressure. It consisted in bending a tube into a syphon, one leg to hold the liquid to be tried, and the other leg containing air kept at a constant temperature of 73° by a cooling apparatus, and separated from the fluid by mercury; both legs being sealed, the end containing the liquid was heated, and when the liquid became vapour the diminution in the bulk of the air was marked.

Alcohol of the specific gravity ·837 was reduced into vapour at a temperature of 497° in a space a little less than three times its original bulk; and 476 parts of air were reduced to 4; indicating a pressure, according to M. Cagniard de la Tour of 119 atmospheres, or 3570 inches of mercury.*

105.—The experiments on alcohol vapour at lower temperatures are collected in the following table.

---

* By the same process as was adopted in finding the constants for calculating the force of the steam of water (art. 86.) the formula for the vapour of alcohol of sufficient purity to boil at 173° is

$$f = \left(\frac{t + 100}{154\cdot8}\right)^{6\cdot};$$

or, in logarithms,

$$\log. f = 6\left(\log.(t + 100) - 2\cdot189976\right);$$

where t is the temperature of the vapour, and f its force in inches of mercury. By this rule the force in inches for a temperature of 497° is 3280 inches; the experiment of M. Cagniard de la Tour gives 3570 inches.

Experiments on the Force of the Vapour of Alcohol.

Temperature of vapour.	Force in inches of mercury.					
	Ure's experiments.	Watt's experiments.	Robison's experiments.	Dalton's experiments.	Bettancourt's experiments.	Formula.
32°	0·40		0·0		0·0	0·383
40	0·56	0·929	0·1			0·546
50	0·86					0·826
54·5					48	0·986
60	1·23		0·8	1·4		1·215
64				1·51		1·41
70	1·76					1·75
77					1·62	2·228
80	2·45		1·8			2·465
90	3·40					3·41
96				4 07		4·11
99·5					3·63	4·57
100	4·50		3·9			4·64
110	6·00	5·63				6·22
120	8·10	7·12	6·9			8·22
122					7·36	8·67
130	10·60					10·73
132		10·34		11·0		11·3
140	13·90		12·2			13·85
144·5					13·7	15·48
150	18·00					17·7
160	22·60	20·71	21·3			22·4
167		24·47				25·4 26·25
170	28·30				25·4	28·1
173	30·00			29·70		30·00
180	34·73		34·			34·92
189·5					42·0	42·66
190	43·20					43·11
200	53·00		52·4			52·83
210	65·00					64·3
212					68·	66·84
220	78·50		78·5	80·20		77·81
230	94·10					93·6
234·5					105·	101·5
240	111·24		115·			112·0
250	132·30					133·2
260	155·20					157·7
264	166·10					168·6

The specific gravity of the alcohol used by Dr. Ure, was ·813, and its boiling point 173°.* The properties of the alcohol employed by Mr. Watt are not given,† his experiments are very irregular. Dr. Robison's boiled at 173°;‡ and above 100° agree well with later observations. Mr. Dalton's appears to have boiled at 175°;§ Bettancourt's boiling point is not stated, but appears to have been 173°;|| and his results like Dr. Robison's are too small at low temperatures.

Dr. Ure's experiments are confirmed by those of Mr. Dalton, and may be relied on as approaching very near the truth. The formula, it will be observed, represents them with considerable accuracy.

Of the elastic Force of the Vapour of sulphuric Ether.

106.—M. Cagniard de la Tour made several experiments on ether in the same manner as those on alcohol. (art. 104.) The ether was converted into vapour in a space less than twice its original volume by a temperature of 392°. This experiment was thrice repeated, with the same result, and 528 parts of air were compressed to fourteen, giving an elastic force of 37·5 atmospheres.¶

107.—Other trials were made the results of which are shewn in the following table.

---

* Phil. Trans. 1818.   †Robison's Mech. Phil. Vol. II. p. 33.
‡ Robison's Mech. Phil. Vol. II. p. 35.   § Annals of Philo. 1820, Vol. XV. p. 130.
|| Prony's Architecture Hydraulique. Vol. II. p. 160.
¶ The experiments on sulphuric ether may be very nearly represented by the formula,

$$f = \left(\frac{t + 210}{178 \cdot 7}\right)^6; \text{ or log.} f = 6 \left(\log. (t + 210) - 2 \cdot 252124\right)$$

when the ether boils at 104° or 105°; but for ether boiling at 98°, the constant logarithm should be 2·239534.

In the above experiment the formula for ether boiling at 105° gives forty-eight atmospheres for its elastic force at 392°; but the correspondence with the tabular experiments is nearer.

M. Cagniard de la Tour's Experiments on Ether.

Temperature by Fahrenheit.	Volume in the liquid state 7 parts. Volume in the state of vapour 20 parts.		Force as expanding gas by formula (art. 119.) in atmospheres.	Volume in the liquid state 3½ parts. Volume in the state of vapour 20 parts.		Force of vapour by formula(art. 106, note) in atmospheres.
	Force of vapour in atmospheres.	Differences.		Force of vapour in atmospheres.	Differences.	
212°	5·6					5·78
234·5	7·9	2·3				7·9
257	10·6	2·7		14·0		10·63
279·5	12·9	2·3		17·5	3·5	14·1
302	18·0	5·1		22·5	5·0	18·4
324·5	22·2	4·2		28·5	6·0	23·8
347	28·3 {State of vapour.	6·1		35·0 {State of vapour.	6·5	30·6
369·5	37·5	9·2		42·0	7·0	38·7
392	48·5	11·0		50·5	8·5	48·
414·5	59·7	11·2		58·0	7·5	60·7
447	68·8	9·1	68·8	63·5	5·5	82·3
469·5	78·0	9·2	70·5	66·0	2·5	100·7
492	86·3	8·3	72·2	70·5	4·5	
514·5	92·3	6·0	73·9	74·0	3·5	
537	104·1	11·8	75·6	78·0	4·0	
559·5	112·7	8·6	77·4	81·0	3·0	
572	119·4	6·7	78·3	85·0	4·0	
594·5	123·7	4·3	80·0	89·0	4·0	
617	130·9	7·2	81·8	94·0	5·0	

On comparing the two series it will be observed, that the pressure up to the point where the liquids change wholly into vapour is greater in the tube containing the least proportion of liquid; but this I expect is entirely owing to the mode of trial not being susceptible of much accuracy. Up to the point where the change to vapour takes place the formula derived from Dr. Ure's experiments applies with admirable precision, a new formula is necessary after the change. The formation of vapour from the mercury in the apparatus most probably affects the results in high temperatures.

108.   *Ure's and Dalton's Experiments on Ether.*

Temperature of vapour.	Force in inches of mercury.			Temperature of vapour.	Force in inches of mercury.		
	Ure's experiments.	Dalton's experiments.	Formula.		Ure's experiments.	Dalton's experiments.	Formula.
34°	6·20		6·48	140°	56·90		56·4
36		7·5	6·8				
44	8·10		8·25				
54	10·30		10·4	150	67·60		66·9
64	13·00	15·0	13·0				
74	16·10		16·1	160	80·30		78·8
84	20·00		19·83				
94	24·70		24·2	170	92·80		92·5
96		30·00	25·2	173		120·0	96·9
104	30·00		30 00				
Second kind of ether							
105	30·00		30·00	180	108·30		108·1
110	32·54		33·00				
115	35·90		36·2	190	124·80		125·8
120	39·47		39·7				
125	43·24		43·4	200	142·80		146·
130	47·14		47·4				
132		60·0	49·1				
135	51·90		51·8	210	166·00		168·5
				220		240	194·

The ether employed by Mr. Dalton boiled in a tube at 96°*, and will be very nearly represented by increasing by one-fifth the calculated quantity for the temperature.   Thus for 132°, we have

$$49·1 + \frac{49·1}{5} = 58·92,$$

and for 220°, we have

$$194 + \frac{194}{5} = 232·8.$$

Dr. Ure's ether boiled at 104° or 105°, and his experiments are very regular.†

* Thomson's Annals of Philosophy, Vol. XV. p. 130.   † Dict. of Chemistry.

Of the elastic Force of the Vapour of Sulphuret of Carbon.

109.—There is a remarkable compound of sulphur with carbon which is usually distinguished by the name sulphuret of carbon, but is sometimes called carburet of sulphur. It is liquid and as transparent and colourless as water. It has an acrid and pungent taste, somewhat aromatic; its smell is nauseous and peculiar, its specific gravity is 1·272, and it boils briskly and distils at from 110° to 116° depending on its purity. When heated to about 680° or 700° in the air it takes fire and burns with a blue flame. It is scarcely soluble in water. It appears to be a compound of

$$\begin{aligned}
\text{Sulphur} &\quad - \quad - \quad - \quad - \quad - \quad 84{\cdot}21 \\
\text{Carbon} &\quad - \quad - \quad - \quad - \quad - \quad 15{\cdot}79 \\
\hline
&\quad\quad\quad\quad\quad\quad\quad 100{\cdot}00
\end{aligned}$$

It may be prepared by mixing about ten parts of well calcined charcoal in powder with fifty parts of pulverized native pyrites, and distilling the mixture from a retort into a tubulated receiver surrounded by ice; somewhat more than one part of sulphuret of carbon may be obtained from the above quantities.

110.—It appears to me that it might be used in a steam engine with some advantage, provided it does not act too much on the metallic parts, nor undergo a change by the continued transition from heat to cold. For it has a high elastic force at a low temperature, being equal to about four atmospheres at 212°, and therefore the advantage of a high pressure engine may be obtained without the inconvenience of a high temperature.

Experiments on the elastic Force of Vapour from Sulphuret of Carbon.

Temperature.	Forces in inches of mercury.	
	Observed.	Calculated by formula.
53·5°	7·4	11·73
72·5	12.55	16·35
110	30·00	30·00

These two experiments I have not attempted to represent by calculation; as the rule by which the numbers were calculated was formed from the experiments in the following table, they serve here to indicate that the observed numbers are probably too low for the true ones.*

111.   *Experiments on the Force of the Vapour of Sulphuret of Carbon by M. Cagniard de la Tour.*

Degrees of heat by Fahrenheit	Volume in liquid state 8 parts. Volume in the state of vapour 20 parts.		Force of vapour by the formula in atmospheres.
	Force in the atmospheres.	Difference.	
		Atmospheres.	
212°	4·2		4·03
234·5	5·5	1·3	5·3
257	7·9	2·4	6·8
279·5	10·0	2·1	8·7
302	13·0	3·0	11·0
324·5	16·5	3·5	13·8
347	20·2	3·7	17·3
369·5	24·2	4·0	21·3
392	28·8	4·6	26·2
414·5	33·6	4.8	31·9
447	40·2	6·6	42·0
469·5	47·5	7·3	50·3
492	57·2	9·7	60·2
514·5	66·5 ⎱ State of	9·3	71·4
537	77·8 ⎰ vapour.	11·3	84·5
559·5	89·2	11·4	99·5
572	98·9	9·7	
594·5	114·3	15·4	
617	129·6	15·3	
628·25	133·5	3·9	

* The rule in logarithms for sulphuret of carbon by which the calculated numbers in these tables were found is

$$\log. f = 6 \left(\log. (t + 280) - 2\cdot344878 \right)$$

to the point where the liquid becomes wholly vapour.

The irregularities in all M. Cagniard de la Tour's experiments, would be in part occasioned by the expansion of the tubes under such high pressures and temperatures; hence, to attempt a minute comparison would only shew a want of attention to physical effects too common in such inquiries. The usual practice of attempting to supply want of observation by minute calculations, is one of the great defects of the present mode of scientific inquiry as applied to improve the practice of the scientific arts.

112.—The forces of various other substances have been tried, but not with much attention to the selection of such as are adopted for the acting vapours in an engine, as for that purpose one should be chosen which affords the highest power with the least range of temperature above the one convenient for condensation; to a vapour of this kind, heat may be applied without requiring so extensive a surface for the fire to act on as when water is used.

On the other part a fluid which has a low elastic force at a high temperature may sometimes be conveniently, and safely applied to afford a regular heat to the acting vapour; hence, it becomes difficult to say to what objects it is improper to extend our inquiries.

Mr. Dalton made some experiments on the vapour of ammonia. The ammonia he used boiled at 140°; and its specific gravity was ·9474. It had a force of 4·3 inches at 60°, but on increasing the temperature, the volatile parts separated first, and left the rest with a greater proportion of water, requiring a still higher temperature to convert them into steam; this fluid is therefore inapplicable.

113.—The force of the vapours of petroleum, and of oil of turpentine has been ascertained by Dr. Ure; the following tables contain his results.

Experiments on the Force of Vapour of Petroleum,* or Naptha.

Temperature.	Force in inches of mercury.		Temperature.	Force in inches of mercury.	
	Ure's experiments.	Formula.		Ure's experiments.	Formula.
316°	30·00	30·00	350°	46·86	48·1
320	31·70	31·8	355	50·20	
325	34·00	34·1	360	53·30	54·8
330	36·40	36·6	365	56·90	
335	38·90		370	60·70	62·4
340	41·60	42·	372	61·90	
345	44·10		375	64·00	66·5

Experiments on the Force of the Vapour of Oil of Turpentine.†

Temperature.	Force in inches of mercury.		Temperature.	Force in inches of mercury.	
	Ure's experiments.	Formula.		Ure's experiments.	Formula.
304°	30·00	30·00	340°	47·30	50.10
307·6	32·60	31·6	343	49·40	52·5
310	33·50	32·7	347	51·70	
315	35·20	35·3	350	53·80	57·3
320	37·06	38·0	354	56·60	
322	37·80	39·0	357	58·70	
326	40·20	41·1	360	60·80	65·4
330	42·10	43·6	362	62·40	
336	45·00				

* For the steam of petroleum, the boiling point being 316°, the rule in logarithms is

$$\log. f = 6 \left(\log. (t + 100) - 2·372906 \right).$$

† For the steam of oil of turpentine, which boils in a tube at 304°, the rule in logarithms is

$$\log. f = 6 \left(\log. (t + 100) - 2·360194 \right).$$

114.—There yet remains a substance which seems to possess the properties desirable in the acting vapour of an engine. It is called oil gas vapour, and is separated from oil gas by the compression used to render that gas portable. It has been examined by Mr. Faraday,[*] who found that it is insoluble in water except in very minute quantities. It boils at about 170° but remains liquid at common temperatures; it consists of a combination of fluids of different degrees of volatility, and by repeated distillations at different temperatures the volatile fluids may be separated; the most abundant separates between 170° and 200°.

At common temperatures the fluid which separates between 170° and 200° appears as a colourless transparent liquid, of the specific gravity 0·85 at 60°, having the general odour of oil gas. Below 42° it is a solid body which contracts much during its congelation. At zero it appears as a white or transparent substance, brittle, pulverulent, and of the hardness nearly of loaf sugar. It evaporates entirely in the air, and when its temperature is raised to 186° it boils, furnishing a vapour, which is 2·7 times the weight of the same bulk of common air. It appears, however, that at a higher temperature the vapour is decomposed, depositing carbon.

It is composed of six volumes of carbon, and three volumes of hydrogen, condensed into one.

115.—In a paper in the Philosophical Transactions on the application of liquids formed by the condensation of gases as mechanical agents, Sir H. Davy anticipates the probability of the application of the elastic force of compressed gases to the movement of machines.[†] He founds this anticipation upon the immense difference between the increase of elastic force in gases under high and low temperatures by similar increments of temperature. The force of carbonic acid was found to be equal to that of air compressed to $\frac{1}{13}$ at 12°, and of air compressed to $\frac{1}{13}$ at 32°, making an increase of pressure equal to the weight of thirteen atmospheres.

116.—I think, however, it will be found, that two other circumstances should be considered in estimating the fitness of compressed gases as mechanical agents. First, The distance through which the force will act; for if this distance of its action be less in the same proportion, as the force is increased by compression, no advantage will be gained: the power of a mechanical agent being jointly as the force, and the distance through which that force acts. Secondly, The quantity of heat required to produce the change of temperature is also to be considered. For if the mechanical power requires as great an expenditure of heat as common steam, no advantage worthy of notice would be gained. In fact the only prospect they afford of being useful is through lessening the extent of surface to be heated.

* Philosophical Transactions, 1826.   † Idem, for 1823.

The idea of employing very powerful pressures, acting through a short space, seems more valuable at first sight than it proves on examination. It is considered that an engine of high power can be got into a small place, and will be of less weight. But the real inconveniences, are, the large mass of fuel required to supply the engine a given time, and the immense surface that must be exposed to an intense heat to obtain a given quantity of heat in a given time. Besides, when we attempt to use high degrees of pressure, an accuracy of workmanship, and attention to the elasticity of materials, becomes necessary, which renders the work expensive, and of short duration.

The success of Mr. Faraday in reducing various gases into the liquid state is not however the less important. His method consisted in generating the substances in a bent tube of glass hermetically sealed at both ends. Then, by cooling one end of the bent tube and heating the other, when heat was necessary, the gas was condensed in a liquid state at the cold end of the tube.

117.—Carbonic acid required the greatest precautions to effect the condensation with safety. The liquid obtained is a limpid, colourless body, extremely fluid, and floated upon the contents of the tube, without mixing. It distils readily at the difference of temperature between 32° and 0°; its refractive power is much less than that of water, and its vapour exerts a pressure of thirty-six atmospheres at a temperature of 32°. In endeavouring to open the tubes which contained it, at one end, Mr. Faraday states, that they uniformly burst with powerful explosions.*

The gases reduced to a liquid state by Mr. Faraday, with their densities as far as known, are collected in the following table, with a column to shew the mechanical power compared with steam.†

* The ingenious Mr. Brunel is attempting to work an engine where the acting vapour is to be liquid carbonic acid. It is to be regretted that his great talent for mechanical combination should be employed where there is so little chance of success.

† The power is as the force and the space through which the gas passes in its reduction to the state of liquid, (See Sect. IV.) The space is found by comparing the density of the body in the liquid state with its density in the gaseous under the same pressure; and as the weight of air is to water as $1 : 828$; to find the mechanical power of equal volumes of the liquid, we have simply to multiply 828 by the specific gravity of the liquid, and divide the product by the specific gravity of the body in the state of gas. The force does not enter into the calculation, because the density of the gas must obviously be greater in the same proportion. The quantity of heat is most probably in the ratio of the power, and if this be the case, all substances will afford equal powers with equal quantities of heat.

Body.	Specific gravity of the gas, air being unity.	Specific gravity of liquid, water being unity.	Temperature.	Force in atmospheres.	Mechanical power of equal weights of the gases.
Carbonic acid gas	1·527		32°	36	
Sulphuric acid gas	2·777	1·42	45°	2	426
Sulphuretted hydrogen gas	1·192	·9	50°	17	630
Euchlorine gas	2·365				
Nitrous oxide	1·527		45°	50	
Cyanogen	1·818	·9	45°	3·6	395
Ammonia	·5962	·76	50°	6·5	1057
Muriatic acid gas	1·285		50°	40·	
Chlorine	2·496	1·33	50°	4·	440
Steam of water	·48	1·000	212°	1·	1711

These are the principal researches that have been made on the force of vapours at different temperatures, when in contact with liquids; but in order to render the subject more complete, we must consider the force when not in contact with the liquids which generate them, and their density and volume.

Of the elastic Force of Vapour separated from the Liquids from which they were generated.

118.—It has been remarked, that the elastic force of steam or vapour produced by increase of temperature, ceases to follow the same law where it is not in contact with the liquid from which it was formed, (art. 87.) The density of the steam no longer increases, the force being solely that which prevents it expanding, and is measured from the quantity it would expand if unconfined. The expansion by the same increase of temperature having been found to be the same in all gases and vapours, and the density as the compressing force, as far at least as 60 atmospheres, it becomes an easy task to compute this species of force within that range of compressive force.

This will also be further useful in determining the volume steam of a given density and temperature occupies as far as about 60 atmospheres; higher we need not attempt to go for useful purposes; and if we did our rules would fail, for there is not even a probable chance of the law of the density being as the force extending to very high degrees of compression.

119.—The quantity a gas or vapour expands is found by the following rule.

RULE. To the temperature before and after expansion, add 459. Then divide the greater sum by the less, and the quotient multiplied by the volume at the lower temperature, will give the volume at the higher temperature.

Or let t be the temperature with the volume v, and t' any other temperature, then

$$\left(\frac{459 + t'}{459 + t}\right) v =$$

the volume at the temperature t'.

As the volume the vapour occupies at the lower temperature is to the volume it would become by expansion, so is the elastic force at the lower temperature to that at the higher one.

$$\text{Or } v : \left(\frac{459 + t'}{459 + t}\right) v :: f : f\left(\frac{459 + t'}{459 + t}\right).$$

Taking as an example M. Cagniard de la Tour's experiments on ether, it is stated, that it was completely in a state of vapour at a lower degree, but the differences do not indicate this to have taken place till it was 447°, and its force was 68·8 atmospheres; required its force at 617.°　In this case

$$\frac{459 + 617}{459 + 447} \times 68·8 = 81·8 \text{ atmospheres.}$$

In the experiment he states it as 94 atmospheres, and undoubtedly in consequence of the vapour of mercury forming in the apparatus, (art. 107.) and a like remark applies to all his experiments; for our rule for the expansion rather exceeds the truth than otherwise.

120.—By reversing the process, we may find the volume steam will occupy under any compressive force not exceeding 60 atmospheres, when its volume is known for a given temperature and pressure. For example at 60° its force being thirty inches of mercury, its volume is 1324 times its volume in water.*　Now by increasing its temperature

---

* The volume of any vapour or gas at 60° and 30 in. is easily found from chemical tables containing their specific gravity, compared with air at that temperature and pressure; for air is 828 times the volume of an equal weight of water: consequently, the number 828 being multiplied by the specific gravity of the liquid, and divided by the specific gravity of the vapour in question, gives its proportion of volume to an unit of volume of the

to the degree t' its volume would be

$$1324 \times \frac{459 + t'}{459 + 60} = 2.55 \left(459 + t'\right).$$

And $f : 30 :: 2.55 \left(459 + t'\right) : \dfrac{30 \times 2.55 (489 + t')}{f} = \dfrac{76.5 (459 + t')}{f} =$

the volume at the force f and temperature t'.

121.—Hence, we have this convenient rule for finding the volume or space the steam of a cubic foot of water occupies, when the steam is of any given elastic force and temperature.

RULE. To 459 add the temperature in degrees, and multiply the sum by 76·5. Divide the product by the force of the steam in inches of mercury, and the result will be the space in feet the steam of a cubic foot of water will occupy.

Example. If the force of the steam be four atmospheres, or 120 inches of mercury, the temperature to that force being according to Mr. Southern's experiments 295°, (art. 77.) then 459 + 295 = 754; and

$$\frac{754 \times 76.5}{120} = \frac{57681}{120} = 480.7.$$

Its volume found by experiment was 404; and considering the difficulty of ascertain-

fluid. Thus steam is of the specific gravity ·625 ; and

$$\frac{828}{·625} = 1324.$$

Substance.	Specific gravity in liquid state, water being unity.	Specific gravity in vapour, air being unity	Volume of vapour for one of liquid at 60° and 30 in.	Constant number for formula.	Volume at the boiling point of the liquid.	Boiling point.
Water	1·000	0·625	1324	76·5	1711	212°
Alcohol	·825	1·6133	423	24·5	476	173°
Sulphuric ether	·632	2·586	203	11·7	220	104°
Sulphuret of carbon	1·272	2·6447	398	23	440	116°
Naptha	·758	2·833	224	13	280	186°
Oil of turpentine	·792	5·013	130	7·5	193	314°
Oil gas liquid	·85	2·7	260	15	337	186°

From this table it appears that one volume of water produces more vapour than an equal volume of any other substance in the list.

ing the volume, on account of the allowances to be made for escape of steam of such a high temperature, it agrees very well with the calculated result. According to Dr. Ure's experiments the force of steam at 295°, is 129 inches, which gives 446 for the times the volume is increased by converting into steam of that force and pressure.

Of the Mixture of Air and Steam.

122.—It is a well known fact that common water contains a considerable portion of air or other uncondensible gaseous matter, and when water is converted into steam, this air mixes with it, and when the steam is condensed remains in the gaseous state. If means were not taken to remove this gaseous matter from the condenser of an engine, it would collect so as to obstruct the motion of the piston. But even when means for removing it are employed, a certain quantity constantly remains in the condenser of an engine, and in order to determine its state, we must consider the effects produced by mixing air with steam, or vapour, at different temperatures and pressures.

Let us suppose that we have air and vapour of the same temperature t, and elastic force p; and that the volumes are v and v'. If they were now put one on the other in a closed vessel of the capacity $v + v'$, it is plain they could preserve an equilibrium; because, the temperature is the same, and the mutual pressures are equal; but this equilibrium would not be stable.

Experience proves that these gases would gradually mix together till they became completely intermixed. It further shews that during this operation heat is neither evolved nor absorbed; so that after a certain time the mixture is perfectly homogeneous, the two gases holding the same proportion in every part, and the temperature and pressure being t and p. From these facts, established by observation, we may deduce another equally well verified by experience.

123.—If two gases, or a gas and vapour, mixed together at the temperature t fill a volume v; and if p and f denote the pressures they would separately exert when separately occupying the same volume v, at the same temperature t, the pressure of the mixture will be $p + f$.

In effect, let us suppose that the two gases at first are distinct, and let f be greater than p; then dilating the gas under the pressure f, until f changes to p, its volume will become

$$\frac{v\,f}{p}$$

provided the same temperature t has been preserved. Placing the two gases now one on the other, their united volume is

$$v + \frac{vf}{p} \text{ or } \frac{v}{p}(p+f).$$

124.—These gases, according to what we have said above, will equally intermix without changing their temperature or common pressure p. Now according to the law of the volume being conversely as the pressure, which is as true of mixed as of simple gases, if we compress the mixture without changing its temperature until its volume

$$\frac{v}{p}(p \pm f)$$

becomes v, the pressure p will become $p+f$, the same as we had to prove. Equally good would the principle hold with three or more gases, or with a mixture of gases and vapour; in all cases the united pressure will be equal to the sum of all the pressures which the gases or vapours would singly exert, *when separately occupying* the same volume v at the same temperature t.

When a change of temperature takes place either after or during the mixture, and the first temperature being t; then

$$\left(\frac{459 + t'}{459 + t}\right) \times \frac{v(p+f)}{p} =$$

the volume when at the temperature t'.

125.—This is compared with General Roy's experiments in the following table, formed from the mean results which he obtained.* Commencing at zero, 1000 parts of air, in contact with water, and under a pressure of 32·18 inches, increased in volume by the formation of vapour, and increase of temperature as shewn in the second column of the table; while the third is the force of vapour at these temperatures by our rule; the fourth is computed by the rule in the preceding article.†

* Philosophical Transactions, Vol. LXVII. p. 653.

† An erroneous formula for this purpose has been copied into several works; it is

$$\frac{v\,p}{p-f} =$$

the volume; and does not at all agree with the experiments. I gave an analysis of the correct rule in my work on warming and ventilating. p. 291. It has also been investigated by M. Poisson, whose mode of illustration have followed in the above.

Temperature.	Volume of air and vapour by experiment.	Force of vapour.	Volume of air and vapour by calculation.
0°	1000·00	0·032	1000
32	1071·29	0·172	1076
52	1123·05	0·401	1132
72	1182·50	0·842	1190
92	1255·14	1·629	1260
112	1353·75	2·95	1360
132	1491·06	5·07	1500
152	1688·96	8·33	1680
172	1929·78	13·17	1930
192	2287·44	20·16	2300
212	2671·94	30·00	2850

The agreement with experiment is in this case very near, and it adds further confirmation of the accuracy of the formula for the force of steam below the boiling point.

126.—In the condenser of a steam engine the vapour will be of the elastic force corresponding to its temperature, and that temperature is determined by that of the fluids which condense it.

It will also always become, after a few strokes of the engine, mixed with as much air as it will saturate at the given temperature and pressure; and by the preceding inquiry it appears, that this saturation will take place when there is an equal mixture of air and vapour in the condenser; consequently, only half the quantity drawn out by the air pump at one stroke will be air, the rest will be uncondensed vapour; and the quantity of air drawn out at each stroke must be at least equal to all the air which enters both from the boiler, from the injection water, and from leakage at the joints in the time between stroke and stroke; a slight variation on either side, however, will not, it may easily be proved, have much effect in retarding the engine.

As the volume the air and vapour occupies determines the air pump to be of a large size, and consequently expensive both in construction and power, in order to lessen its bulk, a second injection might be made within the air pump. But the utmost that could be gained by this method would be very little more than the difference of volume due to temperature, not perhaps one-tenth of the volume of the pump in any case.

It is important to remark, that in steam from salt water, the same quantity of air will occupy more space, on account of the steam being of less elastic force at the same temperature; but perhaps this is much more than compensated for, by salt water containing less air.

Of the Motion of elastic Fluids and Vapours.

127.—A knowledge of the principles and circumstances which affect the motion of elastic fluids is of considerable importance in assigning the relative proportion of the parts of a steam engine. It is a subject that has been very little studied in discussing the theory of this invaluable machine, and therefore, it is one which will engage a considerable share of our attention in this work. Steam is in motion during its action; it must move through passages to perform its office, and be forced through others as it retires; and the effect of disproportion it is difficult to determine from practice alone, because the result depends on so many contingent circumstances.

The best method, therefore, must be to separate the effects, and study each independently; there is then reason to hope that they may be united into a perfect system; and at least it shall be our endeavour to forward this desirable end to the extent of our power.

128.—The condition of free elastic fluids, has been shewn to be regulated by the pressure and temperature of the atmosphere. And, when an elastic fluid is confined in a close vessel, its condition as to temperature and pressure must be similar to that it would be in, if in an atmosphere of the same fluid capable of producing the same pressure upon it.

129.—The most convenient method of investigating the motion of an elastic fluid is, to find the height of a homogeneous column of the same fluid, capable of producing the same pressure as that to which the fluid is subjected. For then the fluid would rush into a perfect vacuum with the velocity a heavy body would acquire by falling through the height of the homogeneous column, when a proper reduction is made for the contraction of the aperture.

130.—If a pipe of communication be opened between two vessels containing elastic fluids of different elastic forces, the velocity of the efflux through the pipe at the first instant, will be that which a heavy body would acquire by falling through the difference between the heights of homogeneous columns, of the fluid of greatest elastic force equivalent to the pressures. And it would be as the velocity acquired by falling through the difference between the heights of the columns equivalent to the pressure at any other instant; the height to be ascertained for the instant at which the velocity is required. After a certain time the pressures or elastic forces would become equal, and the velocity of course would be nothing.

131.—The consideration of chimneys is another case of the motion of elastic fluids, where by increase of temperature, a part of an atmospheric column is rendered of a different density. Some mistakes have been committed in treating this case, but we must proceed to treat of the motions which take place in engines; and first of the allowances to be made for contractions.

132.—In the motion of elastic fluids, it appears from experiments, that oblique action produces nearly the same effect as in the motion of water, in the passage of apertures, and that eddies take place under the same circumstances, tending to retard the motions in a considerable degree.

133.—The velocity of motion that would result from the direct unretarded action

of the column of the fluid which produces it being unity - - - - - - - - - -	1·000	or	8
The velocity through an aperture in a thin plate by the same pressure is - - - -	·625*	or	5
Through a tube from two to three diameters in length projecting outward - - - -	·813	or	6·5
Through a tube of the same length projecting inwards - - - - - - - - - -	·681	or	5·45
Through a conical tube, or mouth piece, of the form of the contracted vein　 - - -	·983	or	7·9

134.—Every enlargement of a pipe which is succeeded by a contraction, reduces the velocity of the motion, and in proportion to the nature of the contraction, and every bend and angle in a pipe, is attended with a diminution of velocity. Hence, as far as convenience will admit, these causes of loss should be avoided; and where they must be introduced, such forms should be given as will lessen the defect as much as possible.

Of the Motion of Steam in an Engine.

135.—We have stated (art. 129.) that the most convenient mode of determining the motion of steam is, by finding the height of a column of the same fluid which would produce the same pressure upon a base of equal area; the manner of determining this column is, therefore, the first point to be considered. The force of steam is sometimes expressed by the pounds on a square inch; sometimes by the inches in height of a column of mercury, and not unfrequently by the number of atmospheres; it will therefore be an advantage to find the height of a column of water equivalent to each of these measures, and then,

---

* According to experiments on air made by Mr. Banks, 0·634.　See Power of Machines, p. 13.

that being multiplied by the relative bulk and pressure of the steam, the height of the column of steam will be found.

The height of a column of water at 60° equivalent to a pressure of 1 lb. per square inch is 2·31 feet.
—————————————————————————————————— of 1 lb. per circular inch is 2·94 feet.
—————————————————————————————————— of 1 inch of mercury is   1·133 feet.
—————————————————————————————————— of the atmosphere is   34·0   feet.

The water is supposed to be of the temperature 60°, and the atmosphere equal to a pressure of thirty inches of mercury; the bulk of the steam will depend on the pressure and temperature, and will be given for the range of practice in a table at the end of the volume, or may be found by (art. 121.)   For example, the volume of steam at 212° being 1711 times the bulk of the square quantity of water at 60°, and the pressure being thirty inches, we have 1711 × 34 = 58,174 feet, the height of an atmosphere of steam at 212°

136.—If an aperture were formed so that there would be no oblique action in passing it, a gaseous fluid or vapour would rush through it into a perfect vacuum, with the velocity a heavy body would acquire in falling through the height of the column of the same fluid equivalent to the pressure.

And this velocity in feet per second is equal to eight times the square root of the height of the column.*   But through pipes and other apertures the velocity will be only 5, or 6½, or other number of times the square root of the height of the column, as shewn in the table, (art. 133.) for each kind of aperture.

137.—If the height of a column of steam equivalent to the pressure of steam in a boiler be determined, and also the height of a column of the same steam equivalent to the pressure on the piston of a steam cylinder, then the velocity will be equal to 6·5 times the

* In algebra c notation : let f be the inches of mercury equal to the force of the steam or the pressure on the fluid, b the bulk of the fluid when the same weight of water is 1, and $h =$ the height of an atmosphere of the fluid of uniform density.   Then, $1·13 f\ b = h$; and

$$8 \sqrt{h} = v = 8 \sqrt{1·13 f b,}$$

when the fluid flows into a perfect vacuum without contraction at the aperture.   In the best-formed pipes it is

$$79 \sqrt{h} = v,$$

in common formed ones

$$6·5 \sqrt{h} = v. \quad \text{But } \dot{b} = \frac{76·5\ (459 + t)}{f}$$

(art. 121,) hence

$$v = 6·5 \sqrt{86·5\ (459 + t)} ;$$

when t' is the temperature of the steam

square root of the difference between the heights of the two columns. This result is the velocity in feet per second through a straight pipe.

138.—The quantity of steam generated may be considered to be equal to the quantity consumed in the same time, or that the boiler is of sufficient capacity to admit of its being taken at intervals without a sensible loss of elastic force; and as these conditions are essential to a good engine we shall consider them to be fulfilled. (See Sect. III. art. 210.) for the proportion of space in boilers.

139.—The volume of steam required in a second is equal to the area of the piston multiplied by its velocity in feet per second. And its density or elastic force must be as much less than that of the steam in the boiler, as to allow the same weight of steam to pass in a second through the steam passages. For if it passed through the steam passages with no greater velocity than that of the piston, those passages must be of the same area as the cylinder; but as they are less than the cylinder, the excess of velocity must be produced by a corresponding excess of force in the boiler.

140.—The steam, till it has passed the narrowest part of the passages, will have the same density as in the boiler, but in the cylinder it must expand till its density be so reduced as to cause the difference of pressure producing the velocity through the contracted passages; and, as the density is as the elastic force, the force of the steam in the boiler multiplied by the velocity, and the area of the passage, must be equal to the elastic force on the piston multiplied by its area and velocity.

That is $f\,a\,v = p\,\mathrm{A}\,\mathrm{V}$, when f is the force of the steam in the boiler in inches of mercury; a, the area of the steam passages; v the velocity; p the force on the piston in inches, A its area and V its velocity.

From this we have

$$v = \frac{p\,\mathrm{A}\,\mathrm{V}}{f\,a}.$$

But by the rule (art. 137.)

$$v = 6\cdot5\,\sqrt{1\cdot13\,b\,(f-p)}; \text{ therefore}$$

$$\frac{p\,\mathrm{A}\,\mathrm{V}}{f\,a} = 6\cdot5\,\sqrt{1\cdot13\,b\,(f-p)}.$$

If the area of the steam passage be required, we have

$$\frac{p\,\mathrm{A}\,\mathrm{V}}{6\cdot5\,f\,\sqrt{1\cdot13\,b\,(f-p)}} = a. \text{ Or } \frac{p\,\mathrm{A}\,\mathrm{V}}{6\cdot5\,f\,\sqrt{86\cdot5\,n\,(459 + t')}} = a;$$

when $n\,f = (f-p) = $ the loss of force; which should not be exceeded.

141.—In practice for low pressure engines it is usual to make the diameter of the passage about one-fifth of the diameter of the cylinder, and then its area is $\frac{1}{25}$ of the area of the cylinder. And as this proportion is grounded upon the experience of the difficul-

ties involved by making the passages larger, it ought not to be varied from without a sufficiently obvious advantage.

142.—This formula applies only to the case of a pipe without obstructions, and we
have no experiments by which the effect of these causes of diminution can be estimated
with accuracy, but we may endeavour to allow for them on the principles which operate
in similar circumstances. For this purpose let the part of the pipe from whence the
change of figure takes place be considered a vessel with an aperture of the kind nearest
resembling the figure of the branching pipe; and the loss of motion at the place equal to
that such an aperture would cause.

Thus when the angle is a right angle, the loss of velocity may be considered half that
which takes place when a pipe is inserted in the side of a vessel, as the diminutions in
the exterior half of the aperture will not be so great in this case, therefore, the loss
will be

$$\frac{1 \cdot 000 - \cdot 813}{2} = \cdot 094, \text{ nearly,}$$

and may be allowed for by diminishing the velocity *one-tenth*, for each *right angled
bend*.

The same allowance for loss should be made when one pipe branches at right angles
from another.

143.—In a pipe formed to a regular curve, or bent only to an obtuse angle, the reduction will not exceed that which happens with a conical mouth piece which is about
$\frac{1}{10}$ of the velocity.

If a pipe be terminated in a valve box the allowance of *two-tenths* should be made for
the loss of velocity in passing the valve.*

---

* When a series of obstructions of the same kind occur in a pipe, the reduction for the first being

$$\frac{1}{a},$$

the velocity will be reduced from V to

$$V \left(1 - \frac{1}{a} \right)^n$$

in passing *n* obstructions. For the loss of force at the first obstruction must be as

$$\frac{V}{a};$$

hence. it will be reduced to

$$V - \frac{V}{a} = V \left(1 - \frac{1}{a} \right);$$

144.—Few engines have less than three obstructions equivalent to passing so many different apertures, which together may be expected to reduce the velocity so as to require the number 6·5 to be reduced to 4·5, consequently the formula becomes.

$$\frac{p\,\text{A}\,\text{V}}{4\cdot4\,f\,\sqrt{86\cdot5\,n\,(459+t')}} = \frac{\text{A}\,\text{V}}{40\cdot5\,(n+1)\,\sqrt{n\,(459+t')}} = a;$$

which will enable us to compare with practice after considering the other causes of loss.

145.—*Loss of force by cooling.* But much of the force of the steam will also be lost in the passage through the steam pipe by cooling. The quantity of steam exposed during a second is as the area and velocity of the steam ; or

$$= \frac{a\,v}{144},$$

a being in inches, the rest in feet. The surface is as the length and diameter, or

$$= \frac{4\,l\,\sqrt{a}}{12}.$$

Hence the loss of heat being directly as the surface, and inversely as the velocity, we have for cooling in metals

$$\frac{2\cdot1\,(\text{T}-t'')\times 4\,l\,\sqrt{a}\times 144}{60\,a\,v\times 12} = t''',$$

the loss of heat in a second; or rather the loss of heat which the quantity passing in one second experiences.* By reducing the numbers to their lowest fractions it becomes

$$\frac{1\cdot7\,l\,(\text{T}-t'')}{d\,v} = t'''.$$

In this equation T is the temperature of the surface of the steam pipe, which will be

this quantity will be again reduced

$$\frac{\text{V}\,(1-\frac{1}{a})}{a},$$

at the second contraction, and the velocity will become

$$\text{V}\,(1-\frac{1}{a}) - \frac{\text{V}\,(1-\frac{1}{a})}{a} = \text{V}\,(1-\frac{1}{a})^2;$$

and so on. To calculate such a succession of diminutions, we have

$$\log.\text{V} + n\log.(1-\frac{1}{a}) =$$

the logarithm of the reduced velocity.

* Tredgold on Warming and Ventilating, art. 44.

o

about one-twentieth less than that of the steam; t is the temperature of the air, l the length of the pipe in feet, d its diameter in inches, and v the velocity in feet per second.

146.—In applying this equation to find the loss of heat, there are no other circumstances to be considered, but in its application to determine the loss of elastic force, there is a most important point to which I wish particularly to direct the attention of the manufacturers of engines. It is the degree to which the temperature of the steam is reduced by passing through the pipe. It is said to be frequently as much as would reduce its temperature below 212°; when this is the case we know that part of the steam must become water, and the rest of it become of the force equivalent to a temperature of 212°, and therefore all the excess of force which was generated in the boiler would be destroyed by the cooling in the passage to the engine.

147.—A knowledge of this cause of the reduction of the force of steam to atmospheric elastic force, and of the importance of not losing force where either economy of heat or of space is desirable, makes one feel a strong desire to know its amount, knowing that the most esteemed manufacturers of steam-boat engines, Boulton and Watt not excepted, cause the steam to pass round between the jacket and the cylinder, and as if intentionally to expose it as much as possible to the cooling effect of the atmosphere, to reduce its elastic force before it enters the cylinder to exert its power.

148.—The reduction of the temperature of steam reduces its elastic force to that of a lower temperature, and during this reduction a portion of the steam becomes water. If f denote the elastic force in the boiler, and f' that after the heat has been lost,

$$\frac{f - f'}{f}$$

will be the quantity reduced to water, and this multiplied by its heat of conversion into steam, must be equal to the heat the whole has lost by cooling; therefore

$$\frac{967\ (f - f')}{f} = t'''; \text{ or } f\left(1 - \frac{t'''}{967}\right) = f'.{}^{*}$$

And here it will be remarked, that when t''' is equal to the whole heat of conversion, f will be nothing; or the whole will be cooled into water as it is in an apparatus for warming buildings. We are now in a condition to give an answer to the question of what is the loss of force in any particular case. Let the temperature of the steam be 220, and its force thirty-five inches, the length of the steam pipe twelve feet, its diameter six inches, and the velocity of the steam in the pipe eighty feet per second, and the temperature of the air 60°. Then by (art. 145.) we have

---

* The number 967° is here taken as the heat of conversion into steam, but in general I use 1000° as more accurate. (See art. 82.)

$$\frac{1\cdot7\, l\, (\mathrm{T} - t'')}{d\, v} = \frac{1\cdot7 \times 12 \times (209 - 60)}{6 \times 80} = 6\cdot3 \text{ degrees};$$

and therefore by the equation above we have

$$f \left(1 - \frac{t'''}{967} \right) = 35 \left(1 - \frac{6\cdot3}{967} \right) = 34\cdot77;$$

consequently there is in this case a loss of force equivalent to 0·23 inches of mercury; or $\frac{1}{150}$ of the force; but this, it may be said, is one of the most favourable of the cases that usually occurs in practice. In steam boat engines where the steam has to pass round the cylinder, the force in the cylinder is said, from observation, not to exceed about twenty-eight inches, when the force in the boiler is about thirty-six inches.

149.—It is obvious that the higher the force and temperature, the greater will be the reduction by cooling, and therefore, the loss in engines of Woolf's method of construction, where the steam has to make its way round the cylinders, must be greater; and take away much from that increase of effect arising from the use of high pressure steam, to gain which so much risk at the boiler is encountered.

Of the Area of the Steam Passages.

150.—The formula for calculating the motion of steam in an engine, has no maximum value to assist us in the choice of a proportion for applying it in practice; but it shews that the larger we make the aperture the less we shall lose of the elastic force of the steam. On the other hand, we have shewn that the loss of force by loss of heat is greater, the less the velocity, and its variation increases in proportion to the increase of the diameter of the pipe. The proportions, however, which about render the loss by the two causes equal have been found most convenient in practice, and therefore claim the preference. There are two rules in use, and neither of these is exactly the same as the theoretical one.

151.—The one is to make the diameter of the steam-ways one-fifth of the diameter of the cylinder. This appears to be Boulton and Watt's proportion.

152.—The other is to make the area of the passage one superficial inch for each horse power.

153.—The obvious intention of these rules, is, that the steam should move with the same velocity, or require the same impelling force, in any sized engine. Either of them gives nearly the result, but neither of them gives it exactly. For the horse power in a small engine requires more steam than in a large one, and therefore the aperture should be greater in small engines or less in large ones than one inch area for each horse power.

Again, engines having a short stroke move slower than those with a long one, and therefore should have the steam passage of a different proportion of the diameter according to the velocity.

154.—To render the velocity very nearly the same in all cases we have this rule :*

Multiply the length of the stroke by the number of strokes per minute, and divide the product by 2400; the square root of the quotient multiplied by the diameter of the cylinder is the diameter of the pipe.

Example. To find the diameter for the steam pipe of an engine of which the diameter of the cylinder is two feet, the length of the stroke 2·5 feet, and the number of strokes per minute thirty-eight;

$$\frac{38 \times 2\cdot 5}{2400} = \frac{95}{2400} = \frac{1}{25}$$

and the square root of $\frac{1}{25}$ is one-fifth, hence, the diameter of the steam pipe in this case is one-fifth of that of the cylinder.

The same rule applies to both high pressure and low pressure steam engines, and both to the steam passages and the passages to the condenser; and the excess of force necessary to produce the velocity is very nearly one 144th part of the force of the steam.

---

* From the equation (art. 144,) we have, when $n = \cdot 00694$, supposing $\frac{1}{144}$ part of the force to be lost in producing the velocity,

$$\frac{A\ V}{3\cdot 357\ \sqrt{459 + t}} = a;$$

and when we use the length of the stroke l, and the number per minute m; $2\,l\,m = 60$ V, and

$$\frac{A\,l\,m}{90\ \sqrt{459 + t'}} = a.$$

When $t' = 220°$, it becomes

$$\frac{A\,l\,m}{2400} = a,$$

which is the same as the rule. If $t' = 320°$, then

$$\frac{A\,l\,m}{2520} = a.$$

showing that a rather smaller aperture will do for high pressure steam. A is the area of the cylinder in circular inches, and a the area of the pipe in circular inches.

Of the Loss of Force by the cooling of the Cylinder.

155.—The steam after it gets within the cylinder is liable to a loss of force by cooling. It is, in large engines, usually inclosed by a case called a jacket, and steam is introduced between this case and the cylinder to keep the latter hot, but the loss in fuel by this mode is the same as with a naked cylinder, and there is clearly no advantage in preserving the force of the steam by adding this case, unless it be supplied with steam by a separate pipe. (See art. 147.)

156.—The investigation for the loss of force in the steam pipe applies in the case of a naked cylinder with a very slight alteration. The steam in this case is progressively exposed to the sides of the cylinder; hence, the loss will be some little, but not materially less, than that which would take place were it kept constantly exposed to the sides. But to the convex surface the ends of the cylinder have to be added.

With the addition of the ends to the surface, the quantity of cooling in degrees per second, from (art. 145.) becomes

$$\frac{\cdot 07\,(24\,l + d).\,(T - t'')}{d\,v} = t'''.$$

Where l is the length of the cylinder in feet, d its diameter in inches, $v =$ the velocity of the piston in feet per second; $T =$ the temperature of the steam less $\frac{1}{74}$ part, and $t'' =$ the temperature of the air. The force is reduced to

$$f\left(1 - \frac{t'''}{967}\right) =$$

the force on the piston.

157.—When low pressure steam is employed the temperature T will be 212°, and putting $t'' = 60°$, and

$$l = \frac{2\,d}{12}, \text{and } v = 3\cdot5,$$

we shall have

$$\frac{\cdot07\,(24\,l + d).\,(T - t'')}{d\,v} = 15\cdot2 \text{ degrees; and}$$

$$\frac{9\cdot1}{967} = \frac{1}{106}.$$

Therefore in low pressure engines there is a constant loss for all sized engines of $\frac{1}{106}$ of the power. When a casing is used and kept constantly filled with steam, the loss of heat and consequently of power from the same fuel will be greater; because the surface will be constantly kept at the temperature of the steam.   I hope this will be sufficient to establish the truth that the steam case is an useless addition to the expense of an engine.

158.—In a high pressure engine working at 300°, the loss by a naked cylinder is only about $\frac{1}{64}$ part of the force.

159.—The best mode of preventing loss is to put a case with an air-tight cavity between it and the cylinder, instead of filling this case with steam, and besides the advantage of saving fuel the engine rooms will not be heated so much.

160.—The single engine will lose more heat but not quite double the quantity of the double engine; hence, we shall be about its amount in stating the cylinder at $\frac{1}{41}$ of the power.   It will also lose double the quantity by the passage of the steam from the boiler to the cylinder.

161.—In atmospheric engines the loss of force by cooling in the cylinder, when a separate vessel is used for a condenser, is an interesting inquiry.   Assuming that the piston is kept steam-tight without the use of water, the loss must be greater than in the single steam engines by the amount lost in cooling the inside of the cylinder half the time; hence, the value of l the length of the cylinder must be increased one half, besides doubling the area exposed in a given time.   This will render the equation for the loss of temperature, (art. 156.)

$$\frac{\cdot 14\,(\,36\,l\,+\,d\,).\quad(\,\mathrm{T}\,-\,t''\,)}{d\,v} = t'''.$$

With the proportions and temperatures of the example, (art. 157.) the loss by cooling is about $\frac{1}{77}$ of the power, therefore it is not this species of loss which should prevent this simple kind of engine being employed for mines.

If water be applied to keep the cylinder tight, the additional loss from converting this water into vapour will be considerable.   If the mean temperature of this water be 180°, the effect of each foot of area will be to abstract, or to destroy a cubic foot of steam per minute, this being the quantity of evaporation from a foot of surface of water sustained at that temperature.   Therefore in an engine working at the rate of 170 feet per minute, that is expending eighty-five cubic feet of steam of atmospheric density per minute, for each foot in area of the cylinder the loss will be $\frac{1}{85} = \frac{1}{47}$ of its power; hence, adding this to the cooling effect, we have $\frac{1}{77} + \frac{1}{47} = $ about $\frac{1}{25}$th of the power

162.—In the common atmospheric engine where the injection is made within the cylinder, the only person who had attempted to calculate the loss of force was Smeaton; of which some account has been given by Mr. Farey, in Rees's Cyclopædia.

The mode of calculation is not very clearly given, and it was formed at a time when the properties of heat were less known.

163.—Cylinders are usually made of the same thickness, or so nearly so as to render the variation not worthy of notice; hence, we will assume them to be of the same thickness. The quantity of matter in them is cooled by the injection from 212°, to about 150°, rarely lower, and in good engines not lower than 170°, or 180°; the mean 160° may be taken for the effect. The specific heat of iron is about 200 times that of steam, and calculating the mass of iron which must have its temperature raised from 160 to between 160 and 212 degrees, by each cylinder full of steam, we have the quantity which that of the steam must be lowered.

The surface of a cylinder is equal to its length, increased by half its diameter, multiplied by its circumference $= (l + \frac{1}{2} d) d p;$ and the thickness, with an allowance equivalent to the escape of heat from the external surface, is one inch and a half $=$ one-eighth of a foot; and the mass of metal equivalent to the absorption of heat is

$$\frac{(l + \frac{1}{2} d) \, d p}{8};$$

its specific heat, allowing for the time the exposure is reduced by the exposed side of the cylinder decreasing, is equal to that of

$$\frac{200 \, (l + d) \, d p}{16}$$

cubic feet of steam heated one degree; but the temperature will rise to the mean between the condensing and boiling points, or to

$$\frac{160 + 212}{2} = 186°,$$

or the addition of heat will be 26 degrees. The whole quantity of heat consumed will therefore be

$$\frac{200 \times 26 \, (l + d) \, d \, p.}{16}$$

This divided by the capacity of the cylinder, or

$$\frac{l \, d^2 \, p}{4} \text{ gives } \frac{50 \times 26 \, (l + d)}{l \, d} = t''',$$

the loss it would sustain in temperature, or

$$\frac{1300 \, (l + d)}{l \, d} = t'''.$$

When the length of the cylinder is twice its diameter, or $2 \, d = l$ the loss becomes

$$\frac{1950}{d} = t'''$$

Now one-fifth of the power is lost by imperfect condensation, more than in engines with a separate condenser, which is equal

$$\frac{1127}{5} = 225$$

degrees of heat; and by the condensation and cooling in the cylinder

$$\frac{1950}{d}$$

Hence, the total heat required over and above what is required in other engines is equivalent to converting into steam,

$$\frac{225 + \frac{1950}{d}}{1500}$$

times the water necessary for the steam engine, with a condenser and steam pressure.

With a cylinder 1·5 feet diameter double the fuel is required, but for a six feet cylinder only one-third more than in the single engine of Watt's construction.

164.—This enables us to illustrate the fact observed by Mr. Watt, when he repaired a working model of a steam engine for the university of Glasgow, in 1763. The cylinder of the model was 0·5 feet stroke, and two inches, or one-sixth of a foot diameter. He "was surprised to find that its boiler would not supply it with steam though apparently quite large enough." By blowing the fire it was made to make a few strokes, but required an enormous quantity of injection water, though it was very lightly loaded by the column of water in the pump. It soon occured that this was caused by the little cylinder exposing a greater surface to condense the steam, than the cylinders of larger engines did in proportion to their respective contents.*

There is no doubt this difficulty was the cause of Mr. Watt turning his thoughts to improve the steam engine. Our rule being applied to this case, $l = \cdot 5$, $d =$ one-sixth,

$$\frac{1300\,(\,l + d\,)}{l\,d} = \frac{1300\,(\cdot 5 + \frac{1}{6})}{\frac{1}{6} \times \cdot 5} = 2600; \text{ and } \frac{2600 + 967}{967} = 3\cdot 8$$

times the volume of steam which would fill the cylinder would be consumed to condense at 160°. By lessening the load lifted, and consequently not condensing the steam to so low a temperature, Mr. Watt made the engine work.

165.—Now in our formula it will be observed, that 26° is half the degrees the temperature of the steam falls by condensation, and that if we lessen this the quantity of heat

---

* Robison's Mechan. Phil. Vol. II. p. 114. Note by Mr. Watt.

lost will lessen in the same proportion, but the loss by uncondensed steam will be greater. The effect of the engine will be greatest when the sum of these losses is a minimum, and its load should be arranged accordingly.

The loss by cooling the cylinder is

$$\frac{25\,(\,212 - t\,)\,(\,l + d\,)}{l\,d};$$

when t is the temperature of condensation.

The loss by imperfect condensation is

$$\frac{f'\,1127}{30},$$

but by our formula (art. 86.)

$$f' = \left(\frac{t + 100}{177}\right)^{6}$$

Hence,

$$\frac{1127\,(\,t + 100\,)^{6}}{30 \times 177^{6}} + \frac{25\,(\,212 - t\,)\,(\,l + d\,)}{l\,d} = \text{a minimum.}$$

Its fluxion must therefore be $= 0$, or

$$\frac{1127 \times 6\,(\,t + 100\,)^{5}\,\dot{t}}{30 \times 177^{6}} - \frac{25\,(\,l + d\,)\,\dot{t}}{l\,d} = 0,$$

whence

$$t = \left(\frac{30 \times 177^{6} \times 25\,(\,l + d\,)}{1127 \times 6\,l\,d}\right)^{\frac{1}{5}} - 100.$$

Which reduce to

$$t = 321\left(\frac{l + d}{l\,d}\right)^{\frac{1}{5}} - 100.$$

166.—When $l = 2\,d$, or the length of stroke is double the diameter,

$$t = \frac{348\cdot3}{d^{\frac{1}{5}}} - 100;$$

or in logarithms, log. $(\,t + 100\,) = 2\cdot541931 -$ one-fifth log. d.

167.—Hence, it appears, that when the length of the cylinder is double its diameter, the temperature of condensation, which gives the maximum, varies inversely as the fifth

P

root of the diameter of the cylinder. When the diameter $d =$ six feet, the temperature of condensation $t = 143\cdot4$; when $d =$ three feet, the temperature of condensation should be $179\cdot6$; and by using a table of logarithms the best temperature for condensation for any other diameter may be easily found by the rule.

Of the Ascent of Smoke in Chimnies

168.—If a bent tube, of uniform diameter, A C B, were continued to the surface of the atmosphere, the lowest point of the curve being at C, the centre of the aperture of a chimney, and the tube of the same size as a chimney, then the temperatures being equal at the same height in the two branches, the whole would be in equilibrio. But if a part, C D, be of a more elevated temperature than the corresponding part of the other branch of the tube, that air being of less density than cold air, the balance will be destroyed, and motion will take place, the moving force being the difference between the weights of the columns of air. Now a chimney may be considered part of a bent tube, for though in a chimney, the column of air is confined only as far as the short canal or tube of the chimney extends, the actual pressures which occur in the atmosphere are equivalent to the pressures in the bent tube; and must be measured in the same manner.

169.—When C A is the height of a uniform atmosphere, C D the height of the chimney, and D E the quantity the air expands by the heat it receives in passing through the fire, the height E D, or F G its equal, represents the height of the column of air which produces the motion, and the velocity will be that a heavy body would acquire by falling through the height F G. If the whole of C A were empty, then A H, the height of the atmosphere, would be the height through which the body must fall to acquire the velocity with which the air would move into the tube, provided it suffered no contraction at the entrance, but such a contraction is well known to take place in air as well as in water.

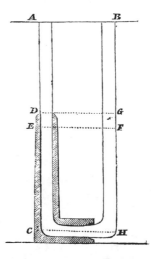

FIG. 14.

170.—When this is applied to a chimney, the smoke being sometimes of a density different from common air at the same pressure and temperature, the same excess of temperature will produce a greater or less effect in proportion as it is of less or greater density than common air. This will be found by subtracting from the expansion, the specific gravity of the smoke or vapour; that of air of the same temperature and pressure being unity. Or it may be done by an allowance of a portion of the temperature for the difference of density; either method gives the same result when properly calculated. In this case I intend to adopt the former method. The latter is followed in my book on Warming and Ventilating Buildings.*

171.—Let h be the height in feet from the place where the flue enters to the top of the chimney, $\varepsilon =$ the bulk to which one foot of air increases by the change of temperature; v the velocity; s the specific gravity of the smoke, air being 1; and $a =$ the area of the chimney in inches.

Then $h = $ C D, and the expansion being as the height in feet, $h(\varepsilon - 1) = $ D E $ = $ F G. But the velocity is that which a heavy body would acquire by falling the height F G; hence, $v = 8 \sqrt{\text{F G}} = 8 \sqrt{h(\varepsilon - s.)}$ When F G is equal to B H, the line A B representing the upper line of a uniform atmosphere, it becomes $v = 8 \sqrt{\text{B H}}$; and in all other cases it is as the difference, D C $-$ C E $=$ E D, when E C is reduced to the same density as B H.

If B be the volume of air before it be heated, then in its heated state it is B ε; therefore,

$$\frac{v\,a}{144} = \frac{8\,a}{144} \sqrt{h(\varepsilon - s)} = \text{B } \varepsilon; \; - \text{ or } \frac{\varepsilon\,a}{144\,\varepsilon} \sqrt{h(\varepsilon - s)} = \text{B };$$

also

$$\frac{a}{144} = \frac{\text{B } \varepsilon}{8} \sqrt{\frac{1}{h(\varepsilon - s)}}.$$

The expansion ε may be found from the table, (Sect. XI.)

But the bulk of a gaseous body at the temperature t' is

$$\frac{459 + t'}{459 + t} = \varepsilon,$$

---

* The principles of calculation followed, both in this and in the work referred to, are perfectly identical with those employed by Mr. Gilbert, in an excellent paper on the subject in the Quarterly Journal of Science, Vol. XIII. p. 113. but the notation and methods of managing the processes are different; and Mr. Gilbert's mode of calculating the expansion does not afford quite satisfactory results; besides, he makes no allowances for the contractions and loss of force in curvilinear motion. I mention the circumstance, because some people compare and criticise, and imagine those things to be different which are in reality identical, as may easily be shewn by putting both in the same notation, and reducing by the rules of algebra. The great object of a practical analyst is to render the final equation as easy of application as possible. As to those who question principles, it is rather unfortunate for them to question those established principles of pneumatics which are confirmed by experiment. It is only when theory and experiment do not agree that the principles can be called in question.

when the bulk at the temperature t is one; hence,

$$\frac{459 + t'}{459 + t} - s = 1 - s.$$

Substituting this expression for s in the equation, we have

$$\frac{B s}{s \sqrt{h(1-s)}} = \frac{B(459 + t')}{8(459 + t)} \sqrt{\frac{(459 + t)}{h(t' - ts - 459(s-1))}} = \frac{a}{144}.$$

172.—The divisor, 8, should be changed according to the species of aperture, (see art. 133.) but that which generally applies is 5; and t will be the mean temperature of 52°; and in this case B being the quantity per hour,

$$\frac{B(459 + t')}{5 \times 9\cdot4 \times 60} \times \frac{1}{\sqrt{h(t' - 52 s - 459(s-1))}} = a.$$

For coal smoke $s = 1\cdot05$, and the formula

$$\frac{B(459 + t')}{28\cdot20 \sqrt{h(t' - 32)}} = a.$$

For low pressure steam

$$\frac{B}{56 \sqrt{h}} = a,$$

the area in square inches.

The application of the formula, with some simple rules derived from it for engines of different powers, will be given with the proportions of fire places: the investigation being given here to separate in some measure scientific inquiry from practical details.

Of the Escape of Steam at Safety Valves.

173.—This is a subject which has been little studied. If we suppose the steam to be of the same density as atmospheric air, its elastic force is twice as great, and it would rush into the atmosphere with the same velocity that atmospheric air rushes into a vacuum. Also, in any case, whether the elastic force of the steam be greater or less than this, if n be the number of times its specific gravity exceeds that of atmospheric steam, when that of air at the same pressure is 1, we shall have $8 \sqrt{(n-1)h} = v$, when h is the height of a uniform atmosphere of air. A uniform atmosphere equivalent to thirty inches of mercury, is 28,000 feet; hence, $1340 \sqrt{n-1} = v$.

174.—In certain cases this will be aided by the buoyancy of the escaping steam, and in very dense steam it may be slightly retarded by the same cause, but these effects are not so great as to need to be introduced in the calculation. They may, however, be sensibly observed by turning the aperture up or down, a light fluid escaping with the greatest velocity when the aperture is turned up, a heavy one when it is downwards.

175.—Let a be the area of the aperture in inches; then, reducing the velocity for the contraction in passing the aperture, we have $300\, a \sqrt{n-1} =$ the quantity of steam escaping per minute in cubic feet; n being the number of times the density is greater than that of atmospheric steam : or making $c =$ the number of cubic feet of steam generated in an hour, then

$$\frac{c}{10\, n \sqrt{n-1}} = a =$$

the area in inches.

This quantity should obviously be the greatest which the fire could, under any possible circumstances, produce.

When n is less than 1, or the density less than the density of atmospheric steam, that is, steam at 212° and thirty inches pressure, the equation becomes negative, and steam rises only by the difference between this negative quantity and the buoyancy; leading us to the beautiful theory of evaporation.

An equivalent rule is given in my book on Warming and Ventilating, but though it is from the same principles, it was not so directly nor so generally derived.*

* Treatise on Warming. &c. p. 148.

SECTION III.

OF THE GENERATION AND CONDENSATION OF STEAM,
AND THE APPARATUS FOR THOSE PURPOSES.

ART. 176.—STEAM is generated by the application of heat, and it is condensed by cold, (art. 71.) and we have now to consider the best sources for obtaining the heat for its generation with economy, and the means of applying it, so as to obtain the most effect. Our section therefore naturally divides into an inquiry concerning combustion and fuel; the effect of and application of fuel; the structure of boilers, and fire places; the principles of condensation, and the apparatus.

Of Combustion and Combustibles.

177.—There are various substances which, when heated to a certain temperature depending on their nature, begin to give out heat, and continue to do so till the whole of the substance be completely changed into new products, most of them gaseous, which in ordinary cases are dissipated in the atmosphere. A substance which undergoes this change is termed a *combustible*, or burning body, and if it be commonly used for producing heat, it is called *fuel*.

178.—The quantity of heat given out during combustion is the difference between that which the matter operated upon contained before, and that which it contains after combustion. This is an invariable quantity when the same quantity of matter is operated upon, and proportional simply to the quantity of fuel used; unless indeed the process be imperfectly managed, or that we could render the products of such a nature, that they would contain a less quantity of heat than those usually produced. The latter would perhaps be a fruitless research, but chemistry is making rapid advances in the means of fully establishing this point. It is, however, of the utmost importance, since the application of steam to navigation, to determine the effect of the mixture of combustible bodies, both with a view to fix on those which contain most heat to a given quantity of matter, and to render the products of less capacity for heat so as to gain the greatest effect; using capacity to signify the whole heat these products contain, as defined (art. 69.)

179.—There is no question that a solid contains less heat than the same substance when liquid, and the substance in the liquid less than in the gaseous state; provided it remain the same chemical compound. But if to a solid, which is a mixture of different simple substances, heat of a certain degree of intensity be applied, the parts of the mixture act on each other, and gaseous products are obtained which contain less heat than the mixture. This is the case with gunpowder, which is a mixture of charcoal, nitre, and sulphur; and it seems necessary in this species of combustion that one of the substances composing the mixture should be easily fusible. It is a completely mistaken notion to imagine that the presence of any particular substance is essential to combustion; for it must take place in any mixture of bodies which act chemically on one another at a certain temperature, so as to form new products containing less heat than the ones mixed.

180.—That which takes place in a mixture of bodies will also take place if either a simple body or a chemical compound be exposed to the action of another body, with which it always forms a new chemical combination if they be brought in contact at a high temperature. Thus charcoal, heated to about 700°, in contact with oxygen burns; and the new product formed is carbonic acid gas; consisting of the charcoal united to oxygen. At about 800°, charcoal abstracts the oxygen freely from the atmospheric air, and therefore burns. Now as the oxygen gas changes neither its volume, nor its elastic force, it may be inferred that the whole of the heat contained in the charcoal is liberated, besides some portion of that previously in the oxygen.

181.—It is important in this inquiry to know in what state the elements of bodies exist, because this must greatly affect the quantity of heat. If hydrogen in solid compounds be itself in its solid state, then it ought to give out less heat than gaseous hydrogen. But I am of opinion that hydrogen, carbon, and other permanent gases, exist in combination in the state of highly compressed gases, and not in their solid state. The experiments I have to compare on combustion will be found to support this opinion; rendering it tolerably certain, that in the range of temperature we can command, these elementary bodies are never even in the liqnid state in combination; and we are thus freed from what I had regarded as the greatest difficulty in rendering the theory of combustion applicable to useful objects.* That this theory has been so neglected, since Count Rumford

---

* It has been assumed by the few who have considered this subject, that the combination of oxygen and hydrogen de-

paid attention to it, is wonderful considering that it becomes every day more important. It is universally admitted, that steam navigation loudly calls for some inquiry. The immense weight of a supply of ordinary fuel renders long voyages nearly impracticable; and while the possibility of making fuel more effective, or of selecting one fuel more effective than another, remains probable, it is worthy of inquiry.

182.—The first and most difficult point is, to determine the heat afforded when two simple or elementary bodies unite and form a compound body; but when the heat is determined for each of the binary combinations, that afforded by any other combinations of them may be calculated.

183.—The measure of the effect of a combustible is, the number of degrees the heat developed in its combustion will raise the temperature of the same weight of water. Or the weight of water that would be heated one degree, the weight of the substance being unity.

184.—The heat afforded by carbon when it combines with oxygen is variously stated, and the results are in some measure dependant on the method employed for taking the quantity of heat, and in others the difference is owing to the quality of the charcoal.* It combines with two-thirds of its weight of oxygen.

According to Dr. Crawford 1 lb. of carbon raises	10369 lbs. of water 1 degree.
Lavoisier - - - - -	13370
Count Rumford - - -	9720
Clement and Desormes -	13300
Hassenfratz - - - -	12880
Dalton - - - - - -	5600

$$6\,)\,65239$$

Mean.   10873

The greatest discrepancy is in Dalton's experiments, which it appears was owing to the method he employed; and in taking 10800 lbs. of water raised one degree as the measure of the effect of carbon we shall be nearly correct.

185.—The heat afforded by hydrogen when it combines with oxygen is also differ-

veloped the same quantity of heat whether the hydrogen was in a gaseous, a liquid, or a solid compound; but this could not happen if in a solid compound the hydrogen were united particle to particle; and hence, I conclude it exists only as a highly compressed gas in solids containing it, for it appears that it does afford the same quantity of heat in all states.

* Phil. Mag. Vol. XLI. p. 295; Thomson's System of Chemistry, Vol. I. p. 148.

ently stated. It combines with eight times its weight of oxygen, and according to the experiments of

$$
\begin{array}{rr}
\text{Dr. Crawford} \ - \ - \ - \ - & 67200 \\
\text{Lavoisier} \ - \ - \ - \ - & 41440 \\
\text{Dalton} \ - \ - \ - \ - \ - & 44800 \\
\hline
3)153440 \\
\hline
\text{Mean.} \qquad 51146 \text{ lbs. of}
\end{array}
$$

water is raised one degree by one pound of hydrogen.

The number 50,000 represents the mean effect of hydrogen very nearly, and as far as we have compared it with other experience seems to be about the true effect.

186.—By the experiments of

Lavoisier one lb. of phosphorus combining with oxygen,

$$
\begin{array}{ll}
\text{raises} \ - \ - \ - \ - \ - \ - \ - \ - \ - \ - \ - \ - \ - \ - & 15400 \text{ lbs. of water } 1^{\circ}. \\
\text{Dalton} \ - \ - \ - \ - \ - \ - \ - \ - \ - \ - \ - \ - \ - & 8400 \ \text{——————} \\
\hline
& 2)23800 \\
\hline
\text{Mean.} & 11,900
\end{array}
$$

Sulphur combining with oxygen, according to Dalton, heats 2800 times its weight of water one degree.

187.—From these we may proceed to compare the several compound bodies on which experiments have been made; and also to shew the proportions of oxygen they consume. The first column of the following table contains the name of the substance and the author of the analysis; the second column its composition in decimal parts of its weight; the third shews the quantity of oxygen each of its components requires, and the sum, or that required for the substance; the fourth column shews the heat each component affords, and their sum is the whole the substance yields; the fifth and last column contains the whole heat afforded by the substance by experiment.

Names.	Composition of combustible portion.		Weight of oxygen to support combustion, the weight of the combustible being unity.	Number of pounds weight of water heated 1° by one pound of the combustible.	
				By calculation.	By experiment.
GASES.					
Carbureted hydrogen	Hydrogen	·25	2·	12500	
	Carbon	·75	2·	8100	
			4·	20600	11900 Da.
Olefiant gas	Hydrogen	$\frac{1}{7}$	1·14	7143	
	Carbon	$\frac{6}{7}$	2·3	9257	
			3·44	16400	12300 Da.
Carbonic oxide	Carbon	·43	·57	4744	3500 Da.
LIQUIDS.					
Alcohol, specific gravity ·812 Ure's analysis	Hydrogen	·1224	·98	6120	
	Carbon	·4785	1·27	5167	
			2·25	11287	8120 Da. 11150 Ru.
Sulphuric ether, spe. grav. ·7 Ure's analysis	Hydrogen	·133	1·06	6650	
	Carbon	·596	1·59	6437	
			2·65	13087	8680 Da. 14454 Ru.
Oil of turpentine Ure's analysis	Hydrogen	·0962	·77	4810	
	Carbon	·825	2·2	8910	
			2·97	13720	8400 Da.
Naptha Ure's analysis	Hydrogen	·123	·98	6150	
	Carbon	·83	2·22	8964	
			3·20	15114	13200 Ru.

Table continued.

Names.	Composition of, &c.		Weight of, &c.	Number of pounds weight of, &c.	
				By calculation.	By experiment.
Olive oil	{ Hydrogen	·1336	1·07	6680	
	{ Carbon	·772	2·06	8337	⌈20720 La.
					| 12460 Cr.
			3.13	15017	{ 16300 Ru.
					⌊14560 Da.
Rape oil, or oil of colza					16750 Ru.
SOLIDS.					
Bees' wax, yellow	{ Hydrogen	·1137	·91	5685	
	{ Carbon	·8069	2·15	8710	{ 18620 La.
					{ 13580 Cr.
			3·06	14395	
Bees' wax, white					{ 17673 Ru.
					{ 14560 Da.
Tallow					{ 15064 Ru.
					{ 14560 Da.
Oak wood, dry & pure woody fibre	{ Hydrogen	·0569	·455	2845	
Gay Lussac & Thenard's expts.	{ Carbon	·5253	1·4	5673	
			1·855	8518	
Oak wood	Allowing 20 per cent for water, mucilage, &c. }		1·484	6825	5662 Ru.
Caking coal from Newcastle	{ Hydrogen	·0416	·334	2080	
Thomson's analysis	{ Carbon	·7516	2·000	8100	{ 9230 Black
			2·334	10180	{ 8675 Watt
Cherry coal from Glasgow	{ Hydrogen	·100	·8	5000	
Thomson's analysis	{ Carbon	·666	1·78	7192	
			2·58	12192	

Table continued.

Names.	Composition of, &c.		Weight of, &c.	Number of pounds weight of, &c.	
				By calculation.	By experiment.*
Splint coal from Glasgow Thomson's analysis	Hydrogen	·044	·35	2200	
	Carbon	·568	1·52	6134	
			1·87	8334	
Splint coal, earthy matters not stated Ure's analysis	Hydrogen	·043	·345	2150	
	Carbon	·709	1·89	7657	
			2.235	9807	
	Allowing 10 per cent for ashes			8826	
Cannel coal from near Coventry Thomson's analysis	Hydrogen	·2	1·6	10000	
	Carbon	·626	1·67	6760	
			3·27	16760	
Cannel coal from Woodhall near Glasgow Ure's analysis	Hydrogen	·0393	·315	1965	
	Carbon	·722	1·93	7799	
			2·245	9764	
	Allowing 10 per cent for ashes			8788	
Charred peat Klaproth's analysis	Carbon	·525	1·4	5670	
Coke prepared in a close vessel Mean of Dr. Thomson's exp.	Carbon	·84	2·27	9070	9128†

* The experiments marked Da were made by Dalton, Ru. Rumford, and Cr. Crawford. The analyses of the substances are referred to their authors; Dr. Ure's will be found in his Chemical Dictionary, and Dr. Thomson's in his Annals of Philosophy.

† This is the result collected from a comparison of Lavoisier's experiments. Treatise on Warming Buildings, art. 31.

188.—These are the results as far as the latest researches in chemistry enable us to carry the comparison; it is sufficient to shew that the correspondence is very close, and that the numbers we have selected for the binary compounds are very nearly the true ones. It will be particularly remarked that in wax, oil, and those substances which are likely to afford the most accurate results, theory and experiment agree

There is, however, still another mode of prosecuting our inquiries, and perhaps with equally satisfactory results.

189.—In gas works, from the quantity of gas and coke afforded by a given quantity of coal, or other matter, we have an approximate means of measuring its effect as fuel, but from the want of a correct knowledge of the density of the gas, in each case, we are obliged to assume it to be the same as carbureted hydrogen.

Kind of fuel.	Composition.		Oxygen.	Heat.
Wigan cannel coal*	{	·134 gas ·635 coke	0·54 1·45	2750 5759
			1·99	8509
Staffordshire coal, inferior†	{	·123 gas ·61 coke	·49 1·39	2460 5534
			1·88	7994
Peat. Klaproth's analysis‡	{	·100 gas ·200 carbon	·4 ·535	2060 2160
			·935	4220

190.—When any of these species of fuel are used for generating steam, there must be a loss of effect equivalent to the quantity of vapour formed from the hydrogen, and from the water in the fuel; one pound of hydrogen will form nine pounds of steam, and in

* Murdoch. Phil. Trans. 1808.   † Art. Gas Lights, Napier's Supp. to Ency. Brit.
‡ Phil. Mag. Vol. XVII. p. 312.

practice the loss of effect will be one-fifth of the power of the hydrogen. This proportion being deducted from the whole effect, and also 1170 for each pound of water the fuel contained, we have the following for the power of the most important species of fuel.*

Species of fuel.	Effect in pounds of water heated one degree by one lb. of fuel.	Effect in pounds of water converted into steam of 220°.	Quantity to convert a cubic foot of water into low pressure steam.	Quantity to convert a cubic foot of water into steam, allowing 10 per cent for loss.
Olive oil	13700 lbs.	11·7 lbs.	5·3 lbs.	5·89 lbs.
Caking coal	9800 —	8·4 —	7·45 —	8·22 —
Coke prepared in close vessels	9000 —	7·7 —	8·1 —	9·00 —
Splint coal	7900 —	6 75 —	9·25 —	10·28 —
Staffordshire coal	7500 —	6·4 —	9·75 —	10·83 —
Oak wood, dry	6000 —	5·13 —	12·2 —	13·6 —
Charred peat, charred in close vessels	5670 —	4·85 —	12·9 —	14·3 —
Peat compact and dry	3900 —	3·35 —	18·7 —	20·8 —
Ordinary oak	3600 —	3·07 —	20·31 —	22·6 —
Peat compact in the ordinary state of dryness	3250 —	2·8 —	22·5 —	25·0 —

These quantities, derived entirely from theoretical considerations, are so near to the actual effects obtained in practice, that they shew us we have little to expect in the form of improvement; and with the addition of one-tenth for various causes tending to decrease the effect, they may be adopted as the measure of effect in those computations we have to make; and the table affords an easy means of comparing the expense of different kinds of fuel.

191.—The trials of the quantity of steam a given quantity of fuel will produce, are by no means so numerous as might be expected by those who know not the difficulty of ascertaining the results with precision. People shrink from the task of making accurate trials, either in consequence of the great degree of attention and labour they require, or the

* The latent heat of steam is 1000 (art. 82.) the temperature of low pressure steam is 220°, and the mean temperature of the air being about 52, we have 1000 + 220 — 52 = 1170 nearly; hence dividing the effect in pounds of water heated 1° by 1170 we have the pounds of water that would be converted into steam, and by proportion, the quantity which converts 62·5 pounds of water or a cubic foot into steam.

expense. The adoption of methods arising out of a competent knowledge of the subject reduces both these in a considerable proportion. The following brief collection may however be useful.

Kinds of fuel.		Effect in pounds of water heated 1° by one pound of fuel.	Low pressure steam of 220° from 52°.	
			Pounds of water converted into steam by one pound of fuel.	Pounds of fuel to form a cubic foot of water into steam.
Newcastle or Swansea coal according to Mr. Watt.	From	6950 lbs.	5·93 lbs.	10·5 lbs.
	To	10400 —	8·9 —	7·0 —
	Mean	8675 —	7·4 —	8·75 —
Newcastle coal according to Dr. Black		9230 —	7·9 —	7·9 —
Newcastle coal Wall's End, by my trials		10050 —	8·6 —	7·25 —
Wednesbury coal according to Mr. Watt.	From	5200 —	4·45 —	14·0 —
	To	7800 —	6·68 —	9·34 —
	Mean	6500 —	5·56 —	11·67 —
Pine wood (dry) Count Rumford's experiments		3618 —	3·1 —	20·02 —
Oak wood (dry) ———————————————		5662 —	4·85 —	12·9 —
Peat, compact, from Dartmoor in the ordinary state of dryness, by my trials	}	2400 —	2·05 —	30·5 —
Culm (Glasgow) —————————————————		3330 —	2·85 —	22·0 —
Culm (Welsh) ———————————————————		4175 —	3·56 —	17·5 —

Sleck, or refuse small coal, produces about three-fourths of the effect of good coal of the same species.

We have hitherto considered effect only when fuel gives the whole, or nearly the whole, of its heat, but a certain rate of combustion and perfect management are requisite to obtain this end.

Process of Combustion.

192.—The elementary bodies require very different degrees of heat to cause them to form new combinations. Sir H. Davy has rendered it probable that charcoal and oxygen combine at about 700°, when common air is not present; and hydrogen and oxygen at about 800°. But when the oxygen is afforded by the common air, about 800° for carbon, and 950°* for hydrogen, seems to be nearer the temperatures at which they inflame readily; and when the fuel affords incombustible gases, the intensity will still require to be increased. Hence, we need not be surprised to find, in the common mode of applying heat, that except in as far as it increases the draught through the fire, it is of little or no advantage for a fuel to contain a large proportion of hydrogen. On the other hand, if the intensity of the heat be too great, the earthy parts of the fuel combine with some portion of the carbon and fuse, forming the glassy scoriæ called *clinkers*, by which some combustible matter is lost. We may expect this effect to take place in a considerable degree whenever the heat approaches to about 1500°, and therefore infer that an average heat not exceeding about 1200° is the best for the production of effect.

The circumstances which must be attended to, that the fuel and its products may remain in this temperature till they be consumed, are next to be considered.

193.—First. A quantity of air sufficient to supply the oxygen required for combustion, must have as free access as possible to all the parts of the burning mass, and with as little exposure of the surface of the mass to the cooling effect of other air as the draught of the chimney will allow.

194.—Secondly. The quantity or mass of fuel in combustion must be of such a proportion to the quantity and temperature of the surface to which it communicates heat, that it can only lose as much heat as it generates when it arrives at the best temperature for combustion; allowing for the cooling effect of the surface acted on by the air required in the process.

195.—Thirdly. The flame and smoke must be kept in contact with the vessel as long as it is capable of affording heat.

196.—Fourthly. The fluid to be evaporated should enter so as first to receive the heat where the smoke last acts on the fluid, so that there may be the greatest possible difference of temperature between the smoke and the fluid; and, consequently, that the fluid may deprive the smoke of heat, as it becomes gradually heated to the temperature of the vapour before it arrives over the fire.

Supply of Air and Area of Fire Grating.

197.—The most effective method of supplying fuel regularly with air, that has yet been tried, is that of burning it on a grate placed over a pit to receive the ashes.

* Dr. Thomson says about 1000° from his own experiments. System of Chemistry, Vol. I. p. 224.

And in examining this subject we have first to inquire what quantity of air must pass through the fire for the combustion of each species of fuel. It has been shewn that the different species of fuel require different quantities of oxygen. For the different kinds of coal it varies from 1·87 to 3 lbs. for each pound of coal, and twelve cubic feet of oxygen weigh one pound; also to obtain one pound of oxygen five pounds of air must pass through the fire; consequently, sixty cubic feet of air will be necessary to afford one pound of oxygen. But it is not possible to render the whole of the air effective; part of it will escape unchanged by combustion, and the allowance I have usually made is that only two-thirds is effective; therefore, we require ninety cubic feet of air for each pound of oxygen, and the product when carbon alone is consumed is carbonic acid, and the specific gravity of the air after thus changed by combustion will be 1·05. But a fuel sometimes contains hydrogen, and in that case the oxygen and hydrogen form steam of double the volume of the oxygen; and the bulk of the mixture of air and vapour will be 102 feet for each pound of oxygen combining with hydrogen, and its specific gravity will be 0·9. The last column is computed from the numbers given in the last column of the preceding table, (art. 190.)

Kind of fuel.	Air and smoke for each pound.	Specific gravity of smoke.	Air and smoke for one cubic foot of water converted into low pressure steam.
Caking coal	214	1·03	1780 cubic feet.
Cherry coal	242	1·00	
Splint coal	172	1·02	1780 ————
Cannel coal	315	1·01	
Coke	216	1·05	1950 ————
Ordinary wood	173	0·90	3900 ————

It appears therefore that we may state the quantity of air and smoke in round numbers, for coal and coke, at 2000 cubic feet, for each cubic foot of water converted into steam, and, for wood, at 4000 cubic feet.

198.—The grate must be sufficient to admit the air required for combustion in the state of expansion due to the temperature of the burning fuel; and it is moved through the fire by the joint effect of the draught of the chimney and the ash pit; hence, as deep an ash pit below as possible should be procured, the ash pit narrowing to a uniform breadth, the same width as the grate before it arrives at the fire, the object being to increase the action of the fire without hastening the smoke too rapidly along the flues. By means of the formula (in art. 172.) we easily compute the area of the spaces between the grating under these circumstances. For coals, the quantity for generating the steam of a cubic foot of water is 2000 cubic feet, the temperature not less than 800°, and the height producing the motion h feet. Consequently,

$$\frac{70}{\sqrt{h}} =$$

the area of the spaces; and the bars being usually equal in thickness to the space between them, we have

$$\frac{140}{\sqrt{h}} =$$

the area of the grating for coals in inches; or

$$\frac{1}{\sqrt{h}}$$

the area in feet. But to generate it effectively, double that area should be applied for the steam of a cubic foot of water per hour, or for one horse's power, =

$$\frac{2}{\sqrt{h}}.$$

When the height from the ash pit, to where the smoke enters the chimney, is four feet, then the area is one foot; and one foot of area of grate for each horse power is the common rule of practical engineers.

The proportion of aperture to the solid part of the bar is not always the same, but it ought to be about in the proportion above stated, as air expands to nearly $2\frac{1}{2}$ times in bulk while in the fire.

199.—For burning wood and peat the area of the grate must be

$$\frac{4}{\sqrt{h}}$$

for each cubic foot of water converted into steam; when h is the depth of the ash pit in feet; the increased area being gained by increasing the size of the bars.

Of the Surface of Boiler to receive the Effect of the Fire.

200.—The surface of boiler to produce a given effect must be equivalent to receive the heat which will produce the supply of steam; and as fire, or bottom, surface is the most effectual, that kind of surface should be of sufficient area to receive the whole effect of the fire; while the flue surface, or sides, may receive the effect of the smoke. Hence, we have an easy mode of determining the proportions.

The mean heat of a close fire place may be considered T; and if t be the temperature of the steam, and s the bottom surface, then the heat of conversion of water to steam being 1000, added to its temperature less fifty-two degrees, we have, from an experiment made by Professor Leslie*, $\cdot 828 \, s \, (T - t) = 948 + t$, when one cubic foot of water is to be converted into steam in an hour; or

$$ s = \frac{948 + t.}{\cdot 828 \, (T - t)}, $$

201.—When a mass of fuel is in combustion in a close fire place, we have shewn that it is not desirable for its temperature to exceed 1200°, (art. 192.) Now the surface of the boiler must be at some distance from the fuel, to allow it to develope flame, and therefore the heat having a larger surface to act upon its intensity is less, but at a mean ought not to be less than about 800°, for coal; consequently, we may insert 800 for T. For low pressure steam $t = 225°$, hence,

$$ s = \frac{948 + 225}{\cdot 828 \, (800 - 225)} = \text{2·6 feet, nearly.} $$

For steam of 300°, the force of which is about forty pounds per circular inch above the pressure of the atmosphere, we have

$$ s = \frac{948 + 300}{\cdot 828 \, (800 - 300)} = 3 \cdot 14 \text{ feet.} $$

These examples will be sufficient to shew the increased quantity of surface required for high pressure steam, and we may now proceed to estimate the quantity of side flue.

202.—At an average for coal, 2000 cubic feet of gaseous matter, heated to 800°, is generated, and required for combustion to produce the above effect; and the specific heat of air being $\cdot 00032$, its effect will be equivalent to heating a cubic foot of water $\cdot 00032 \times 2000 \times (800 - t) = \cdot 64 (800 - t)$. Now it will be sufficiently accurate for our purpose to consider the effective excess of temperature to be a little less than the mean between 800 and t; consequently,

$$ \frac{\cdot 828 \, s \, (800 - t)}{2 \cdot 5} = \cdot 64 (800 - t); \text{ or } s' = 1 \cdot 94 \text{ feet.} $$

203.—Comparing $\cdot 64 (800 - t)$ with $948 + t$, we find that the whole energy of the side of flues will amount only to about one-fourth of the effect of the bottom ones; we

---

* Inquiry into the Nature of Heat. Experiment 51 and 52.

may therefore reduce the fire surface found by the rule one-fourth for each cubic foot of water evaporated per hour.   The rule will then become

$$s = \frac{3\,(\,948 + t\,)}{4 \times \cdot 828\,(\,800 - t\,)} = \frac{948 + t}{1 \cdot 1\,(\,800 - t\,)}.$$

204.—But in a steam engine boiler this would barely keep the boiler supplied, whereas it is shewn that there should be a capability of supplying steam with double the rapidity actually required, otherwise the pressure on the piston will be less, and the effect less in the same ratio, (see art. 331—339;) according therefore to this condition the proportion of bottom should be

$$\frac{2\,(\,948 + t\,)}{(\,800 - t\,)} = s.$$

The side flue constantly $3 \times 1 \cdot 94 = 3 \cdot 88$ which may be called four feet.

Common or low pressure steam, temperature 225°	{ Bottom of boiler { Side of boiler flue	4· 1 feet 4·	For converting one cubic foot of water per hour into steam.
2 Atmospheres, temperature 250°	{ Bottom { Side	4·36 4	———————
3 Atmospheres, temperature 275°	{ Bottom { Side	4·6 4	———————
4 Atmospheres, temperature 293°	{ Bottom { Side	4·9 4	———————
5 Atmospheres, temperature 308°	{ Bottom { Side	5·1 4	———————
8 Atmospheres, temperature 343°	{ Bottom { Side	5·65 4	———————

For sea-water, and low pressure or atmospheric steam.

Temperature 230°, it requires	{ Bottom of boiler { Side of do	4·14 4·0	One cubic foot of water per hour into steam.

205.—In comparing these with the usual rules, the sum of the bottom and sides must be taken; and it may be remarked that one cubic foot of steam per hour is so nearly equivalent to the horse power used in steam engine calculations, for the larger kinds of engines, that they may be considered the same in these comparisons. Also a bushel of New-castle coals may be considered equivalent to ten cubic feet of water converted into steam.

Smeaton, with his wonted care, prepared a table shewing the surface of boiler required to be exposed to the effect of the fire and smoke for atmospheric engines, and the quantity of coals to be consumed per hour. His quantity of surface for one bushel per hour is eighty-eight feet, and for thirteen bushels per hour not quite eighty-two feet of surface per bushel.* This is equivalent to 8·2 feet of surface for converting one cubic foot of water into steam per hour. Our deduction, from calculation, is 8·1 feet for low pressure steam.

Mr. Watt says he "finds that, with the most judiciously constructed furnace, it requires eight feet of surface of the boiler to be exposed to the action of the fire and flame to boil off a cubic foot of water in an hour;"† which is only the rule of Smeaton in general terms.

206.—The proportion of the bottom surface, or that within the immediate effect of the fire and flame, seems to have been subjected to no fixed rule; the proportions used in practice vary from three to five feet of bottom surface for each cubic foot of water boiled off per hour. Mr. Millington seems to have first indicated the use of measuring the power of a boiler by its bottom surface; and gives as examples, that a boiler for twenty horse power is usually fifteen feet long and six wide, having ninety feet of surface, or four feet and a half to one horse power: a boiler for a fourteen horse power, sixty feet of surface = 4·3 feet to one horse power.‡ I have observed boilers to be incapable of supplying the proposed quantity of steam when they had less than four feet; and that those were effective which have the proportion assigned by the rule above, provided they also had a proper quantity of flue surface.

207.—In regard to high pressure steam, some interesting trials were made by Mr. Wood,§ with steam carriage engines, which shew the disadvantage of attempting to form steam by intensity of heat instead of quantity of surface.

The first was with a steam carriage boiler eight feet in length, and three feet nine inches in diameter, with a tube twenty inches diameter, passing through its length, which contains the grate for the fuel from whence the smoke passes along to an upright tube at the end, serving as a chimney; the pressure of the steam in the boiler was limited to fifty pounds per square inch above the atmosphere.

The whole surface of the tube forming the fire place and flue would be only forty feet;

* Rees's Cyclopædia, art. Steam Engine.   † Robison's Mechan. Phil. Vol. II. p. 147.
‡ Epitome of Natural Philosophy, p. 266.   § Treatise on Rail Roads. p. 249.

and it was the same in all the trials, but of this not more than two-thirds or twenty-seven feet could be effective as fire surface.

208.—

Time of experiment.	Coals consumed per hour.	Water boiled off per hour.	Pounds of fuel to boil off a cubic foot of water.	Surface of boiler to each cubic foot.
9 hrs. 35 min.	264 lbs.	15·5	17 lbs.	1·74
9 — 27 —	268 —	15·1	17·6 —	1·79
4 — 48 —	323 —	15·8	20·5 —	1·71

The mean intensity of the fire must have been equal to 1200°, to produce this effect; and the fuel consumed is somewhat more than double the quantity which ought to have generated the same quantity of steam.

209.—In another trial the length of boiler was nine feet two inches, its diameter four feet, and the diameter of the tube twenty-two inches, and the force of the steam limited to the excess of fifty pounds per square inch. In this case the whole surface of the tube in contact with the water of the boiler could not exceed fifty-two feet; and two-thirds of this being taken as effective, we have thirty-five feet for the surface.

Time of experiment.	Coals consumed per hour.	Water boiled off per hour.	Pounds of coal to boil off a cubic foot of water.	Surface to a cubic foot of water per hour.
6 hrs. 32 min.	230 lbs.	12·2 feet	18 8 lbs.	2·97 feet
1 — 26¼ —	410 —	23 —	17·8 —	1·52 —

The difference in the results in these trials is chiefly owing to a difference in the density of the steam in the boiler, its state not having been ascertained; and though it might be done in an indirect manner from the number of strokes per minute and the resistance, it would not be accurate enough to furnish us with any useful conclusions.

Of the Space for Steam, and Water in Boilers.

210.—A boiler must obviously contain as much steam as will supply the engine at each stroke without any material decrease in its elastic force; and the space will therefore depend on the manner the steam is to be supplied. If it be admitted to the engine only during part of the time of the piston's descent, there must be so much steam that the use of the quantity required may not lessen the elastic force. If the steam be generated equably, and the space for it only equal to the quantity consumed at each stroke, and all the quantity be wanted during the descent of the piston, the elastic force in the boiler will vary one half, and the loss of effect be very considerable. This subject is therefore worthy of further inquiry, in order that we may see how far the maxims of practice are confirmed by just principles. Without specifying the kind of engine, it is stated, that a boiler should have space for five or six times the volume of steam required for a stroke;[*] others mention eight times; Dr. Young quotes a remark that it should contain ten times the volume,[†] and Prony has stated that it is one of the advantages of a double acting engine, that it requires a smaller boiler than a single acting one.[‡]

211.—Let it be supposed that the action of the fire is uniformly the same, and that during a time 1 it generates a volume 1 of steam, and that this volume is sufficient to supply the engine; but that the whole of it is required in some less time t; and that c is the capacity of the space for steam in the boiler, and p the elastic force at the commencement of letting on the steam. Then $c + t - 1$ is the quantity of steam in the space c at the end of letting on the steam; and the elastic force being inversely as the space, it will be

$$\frac{p\,(c+t-1)}{c}$$

at the time it is shut off; and the variation will be

$$v - \frac{p\,(c+t-1)}{c} = p\left(\frac{1-t}{c}\right).$$

Now in a single acting engine the time t when it acts at full pressure is one half, hence,

$$\frac{p}{2\,c}$$

---

* Millington's Epitome of Natural Phil. p. 251.   † Natural Phil. Vol. II. p. 259.
‡ Architecture Hydraulique, Vol. II. p. 106.

is the loss of elastic force; but if we make $c =$ eight times the quantity required, the loss is only $\frac{1}{16} p$, or the elastic force varies only $\frac{1}{16}$, or about one pound on the square inch.

212.—If the steam be cut off before the stroke be completed, the variation will obviously be greater; for example, in a single engine cut the steam off at half the descent, and the variation of elastic force in the boiler will be

$$p \, \frac{3}{4\,c},$$

one-eleventh, nearly, when the capacity for steam in the boiler is eight times the quantity required for a stroke.

213.—In the double acting engine, the steam acting at full pressure, the time t is nearly the same as the time denoted by 1, and about three times the quantity required for the stroke may be sufficient; but if the steam be cut off at any fractional portion of the stroke, put t equal to that fraction, and it will be found to what the capacity must be increased to render the variation of force inconsiderable. Thus if it be cut off at half the stroke, then

$$p \left(\frac{1 - \frac{1}{2}}{c} \right) = \frac{p}{2\,c},$$

the same as in single engines, and we should not make c less than 8. But it must be remarked that it is in all these cases c times the volume of steam used as it is in the boiler, and not c times the capacity of the cylinder, because during the time the steam acts by expansion there is none entering the cylinder.

214.—For each cubic foot of water converted into steam in an hour by a low pressure boiler, we may assume that one cubic foot of steam is used at a stroke without material error; and if, as agrees with other parts of the arrangement of an engine, the variation be limited to one-thirtieth of the force of the steam, we shall have

$$\frac{1 - t}{c} =$$

one-thirtieth, or $30 \, (1 - t) = c.$

That is, calling the interval 1 from the time the steam valves are opened to the cylinder to a succeeding time of opening them to it; let the fraction of that interval during which the steam valves are open be subtracted, and thirty times the difference will be the space for steam in cubic feet in a low pressure engine.

Thus, let it be a double acting engine where the steam is cut off at two-thirds of the stroke; then, the whole stroke is the distance of the times of opening the steam valves, and two-thirds is the fraction; therefore $1 - \frac{2}{3}$ is $\frac{1}{3}$, and $30 \times \frac{1}{3}$ is ten cubic feet.

215.—In a high pressure boiler the same rule applies; only instead of being the space in feet, thirty times the difference must be divided by the density of the steam compared with the atmospheric steam as unity.

This may be done with sufficient nearness in practice, by dividing by the number of atmospheres equal to the force of the steam in the boiler.

If in a double acting high pressure engine, which admits the steam only during half the stroke, the force of the steam in the boiler be four atmospheres, then for each cubic foot of water the boiler is to boil off per hour there should be

$$\frac{30\left(1 - \frac{1}{2}\right)}{4} = 3\cdot 8$$

cubic feet of space for steam.

216.—Even in a double engine, which is intended to act at full pressure throughout the stroke, there is the time of opening and closing the valves to be deducted, and in some of the usual modes one-fourth of the stroke at least is expended, so that we can scarcely in any case say that less than eight, divided by the atmospheres representing the force of the steam in the boiler, should be allowed as the space in feet for steam for each cubic foot of water boiled off per hour.

217.—*Space for Water in a Boiler*. That there should be water to cover the sides of the boiler a little higher than the flues, is clear; but there is another condition which is less obvious but of considerable importance in effect, and it is particularly interesting in steam boats, where we wish to have neither more of space nor weight than is absolutely necessary.

The quantity of water an engine consumes is not admitted with perfect regularity; it is most equably done when forced in by a pump worked by the engine, and the portion admitted regulated by a float ball. See Plate I. Fig. 2.

The quantity necessary to produce the steam must however be admitted, and its temperature we will suppose to be 100°; now the water in the boiler we will suppose to be 225°, and the proportion of the quantity in the boiler to that admitted ought to be such that the temperature should not be lowered so as to reduce the force of the steam one-thirtieth part; otherwise a manifest disadvantage must take place in the action of the steam. But the depression of the temperature of the water two degrees will diminish its elastic force one-thirtieth, hence, supposing the quantity introduced at each time to be 1, and the quantity in the boiler to be x, we must have

$$\frac{(1 \times 100) + (x \times 225)}{1 + x} = 223;$$

whence we find $x = 62$ nearly; that is, there must be sixty-two times as much water in the boiler as is introduced at one feed, otherwise the force of the steam will be lowered more

than one-thirtieth.   The rule applies to both high and low pressure steam ; for the variation by a change of two degrees of temperature is nearly proportional.   The more frequently the feeding apparatus acts the less water we require, and we also see a stronger
motive for using hot water for the boiler than that of barely saving fuel ; as the colder it
is the more the steam will be reduced.   If a boiler be fed at every stroke it should have
five cubic feet of water, for each cubic foot of steam it is capable of boiling off per hour ;
whether the boiler be high or low pressure.

218.—The self-acting feeding apparatus must be delicately adjusted to reduce its intervals to even twice that time, and therefore such boilers require at least ten cubic feet of
water, for each cubic foot of water boiled off per hour.   But a mode of rendering the self-
acting feed regular is shewn in Plate I. and II.

219.—It is shewn therefore that to limit a low pressure steam boiler of a double acting
engine, with a self-acting feed, to a change of elastic force not exceeding one-thirtieth, we
must have ten feet space for steam and ten for water for each cubic foot of water the boiler
commonly generates in an hour, or for each horse power ; and that if the steam be cut off
before the stroke be completed a greater space must be allowed for steam.

220.—It is usually stated that there should be twenty-five cubic feet of boiler for
each horse power, others say twenty is sufficient, and even so low as eight has been proposed ; while another party state that there is no relation between the cubic contents of the
boiler and the power.   We have now however shewn on unquestionable principles what
ought to determine the least contents of the boiler ; and it appears that to omit the estimation either of the surface to receive heat, or the capacity, is erroneous.   Both should be
considered and determined from the circumstances of the case.

Of the Power of Low Pressure Boilers.

221.—The power of boilers to produce steam is considerably affected by the loss of
heat, and a small boiler more so than a large one.

It is one of those cases which seems to be incapable of being investigated otherwise than
by experience.   In a boiler proportioned to the effect to be produced, the loss of energy
seems to be in the fuel, and it appears to agree very well with practice to consider the loss
proportional to the ratio between the surface and capacity of the quantity of fuel, supposing it to be bounded by similar figures.   In this manner the following table is derived.

Cubic feet of steam per hour equivalent to power of boiler.	Bottom surface for each horse power.	Side surface for each horse power.	Horses' power for low pressure steam.	Quantity of water in boiler with common feed for each horse power.
2·16 cubic feet	8·8 ft.	8·6 ft.	1 horse power	22 cubic feet
1·73 ————	7·1 —	6·9 —	2 ————	17 ————
1·56 ————	6·4 —	6·2 —	3 ————	16 ————
1·46 ————	6·0 —	5·8 —	4 ————	15 ————
1·39 ————	5·7 —	5·5 —	5 ————	14 ————
1·35 ————	5·6 —	5·4 —	6 ————	13·6 ————
1·32 ————	5·4 —	5·3 —	7 ————	13·2 ————
1·29 ————	5·3 —	5·2 —	8 ————	13·0 ————
1·26 ————	5.2 —	5·1 —	9 ————	12·5 ————
1·25 ————	5·1 —	5·0 —	10 ————	12·5 ————
1·22 ————	5·0 —	4·9 —	12 ————	12·2 ————
1·2 ————	4·9 —	4·8 —	14 ————	12· ————
1·18 ————	4·8 —	4·7 —	16 ————	12· ————
1·17 ————	4·8 —	4·7 —	18 ————	12· ————
1·16 ————	4·75—	4·6 —	20 ————	12· ————
1·13 ————	4·6 —	4·5 —	25 ————	11 ————
1·12 ————	4·6 —	4·5 —	30 ————	11 ————
1·10 ————	4·5 —	4·4 —	40 ————	11 ————

When a boiler is made of a larger size than would supply an engine of thirty or forty horses' power with steam, it is much better to make two boilers, and to set them side by side, and besides these there should be a reserve boiler to put in use during repairs. That is, for a forty horse engine I would recommend three twenty horse power boilers; for a sixty horse engine three thirty horse power boilers, and so on; and for smaller engines two boilers, each equivalent to the power of the engine.

Of the Form of Boilers as it depends on Effects.

222.—The quantity of fire and of flue surface having been ascertained, and the capacity, the next object is to consider the form of boiler best adapted for obtaining these proportions in a convenient manner. If we were to consider the strength of the metal alone, they would be nearly spherical, but we well know that a sphere has the least quantity of surface of any solid having the same capacity.

223.—The first boilers used for engines were nearly of a spherical shape. The bottom was next altered to a concave surface, the flue sides were made nearly perpendicular; and the upper part still retained a hemispherical shape. This form was essentially a short cylinder placed on its base, and terminating in a hemispherical head.

224.—*Watt's Boilers.* A rectangular form was adopted by Mr. Watt for the lower portion of the boiler, the upper part he made half a cylinder. The bottom was made concave but the sides flat. For low pressure steam a boiler may be made abundantly strong of this form, and it affords a little more surface without materially increasing the space the boiler occupies. Making the bottom concave towards the fire also may cause the sediment to settle in the angles instead of immediately over the fire. In large boilers a flue was formed through the middle of the boiler, so as to be covered by the water within.

It was justly remarked by Mr. Watt that the sole object of the arrangement of his boilers "was to economise the fuel as much as possible. It is not the shallowness or depth of the boiler that produces this effect; but the making of the boilers of such a shape that the air which passes through the fire shall be robbed of almost all its heat before it can make its escape."* Mr. Watt assured Dr. Thomson that this object is very well attained by the construction he had adopted, and it undoubtedly is so.

225.—When a boiler of a rectangular plan (see Plate I.) is used, the relations of the length, width, and depth to obtain the necessary quantity of surface and of capacity are easily found, when it has no internal flues, and it is doubtful whether any advantage is gained by such flues or not. The following is an approximate rule for the purpose.

RULE. Take the capacity of the boiler for water, and divide it by the quantity of bottom surface (art. 221.) the result will be the depth of water.

Multiply together the bottom and side surface for fire and flue, (art. 221.) and divide the product by twice the capacity for water, less the area of the bottom surface, and the result will be one of the dimensions of the bottom.

Divide the bottom surface by the dimensions found, and it gives the other.

Example. To find the proportions of a boiler for an engine of twelve horses' power, the capacity for water being 12·2 cubic feet for each horse power.

In this case $12 \times 12 \cdot 2 = 146 \cdot 4 =$ the capacity of the boiler for water; and the bottom surface $5 \times 12 = 60$ feet, hence

$$\frac{146 \cdot 4}{60} = 2 \cdot 44$$

feet the depth of water.

Also the bottom surface multiplied by the side surface $= 60 \times 59 = 3540$;

* Dr. Thomson's Annals of Philosophy, Vol. VII. p. 173.

which divided by $(2 \times 146 \cdot 4) - 60 = 232 \cdot 8$ is

$$\frac{3540}{232 \cdot 8} = 15$$

feet nearly for one dimension.

Consequently,

$$-\frac{60}{15} = 4$$

feet for the other dimension; or the boiler should be fifteen feet long and four feet in width.

226.—If the capacity of the top for steam be the same as that for water, and the form a semi-cylinder, the whole depth of the boiler may be found with sufficient accuracy for practice, by making it twice the depth of the water added to one-tenth of that depth; in the example it will be $(2 \times 2 \cdot 44) + \cdot 244 = 5 \cdot 124$ feet.

The proportions given by the rule are different from those commonly used, not much in capacity, but considerably in extent of surface for receiving heat, and in having greater length and less width. Boilers of such proportions are undoubtedly stronger as well as more effective.

227.—*Cylindrical Boilers.* Cylindrical boilers, with the ends rather flat segments of spheres, should always be used for the production of strong or high pressure steam; and even for low pressure steam this form seems best. See Plate II. Many schemes have been suggested for using combinations of cylinders or tubes; but it is extremely questionable whether any plans have been suggested superior to a simple cylinder with convex ends, and applying as many of these as are necessary for the object.

228.—Sometimes the cylinder forming the boiler has the fire wholly within it; and in consequence of this arrangement it is impossible to get surface for the fire to act on, unless the boiler be of such diameter as to render it extremely dangerous. The immense waste of fuel is shewn by the experiments of Mr. Wood, (art. 208,) and yet these boilers have a diameter of four feet, with a pressure of four atmospheres, tending to separate the parts of the boiler with a force exceeding 140 tons, and only a rude safety valve to limit the steam to this force.

229.—RULE *for Cylindrical Boilers.* When a fire is applied externally to a cylinder, which is to contain both water and steam, let the capacity for water and for steam be added together, and also the quantities of fire surface; then divide twice the capacity by the quantity of fire surface, and the result will be the diameter. Also $1 \cdot 27$ times the capacity divided by the square of the diameter will be the length.

Example I. Let the proportions of a high pressure boiler be determined, so that it shall be capable of converting seven cubic feet of water into steam per hour, at a pressure of four atmospheres.

A boiler for this purpose should contain about nine cubic feet of space for each cubic foot of water boiled off per hour; consequently, its whole content will be sixty-three cubic feet. The surface for the fire should be $7 \times (4\cdot9 + 4) = 62\cdot3$ feet, (see art. 204.) Therefore

$$\frac{2 \times 63}{62\cdot3} = 2\cdot03$$

feet, the diameter.

And

$$\frac{1\cdot27 \times 63}{2\cdot03 \times 2\cdot03} = 18\cdot6 \text{ feet.}$$

Example II. Let the boiler be required to boil off twenty-four cubic feet of steam per hour, at three atmospheres, with eleven feet of space in the boiler for each cubic foot boiled off.

Then the content $= 11 \times 24 = 264$ feet. The surface $24 \times (4\cdot6 + 4) = 206\cdot4$ feet; therefore

$$\frac{2 \times 264}{206\cdot4} = 2\cdot6$$

feet, nearly, equal the diameter.

And

$$\frac{1\cdot27 + 264}{6\cdot8} = 50$$

feet nearly; hence, two boilers each twenty-five feet long would be better.

Example III. If a low pressure steam boiler be made cylindrical for a twelve horse power, under the same conditions as a rectangular one, (art. 225.)

Then the content $= 12 \times 2 \times 12\cdot2 = 292\cdot8$ feet. And the surface $= 12 (5 + 4\cdot9)$ $= 118\cdot8$ feet; therefore,

$$\frac{2 \times 292\cdot8}{118\cdot8} = 5$$

feet nearly, equal the diameter; also

$$\frac{1\cdot27 \times 292\cdot8}{5 \times 5} = 14\cdot8 \text{ feet.}$$

The boiler should therefore be $14\cdot8$ feet long, and five feet in diameter, and this I think a better form for boilers than the usual rectangular ones. See Plate II.

230.—The steam pipe S should lead from immediately over the fire, and the water should be admitted at the opposite end at N; and in order that the sediment may be with more certainty deposited where the fire has least force, I would insert a partition O across the boiler, to rise within about four or five inches of the surface of the water. This would prevent cold water checking the steam, and also cause the deposit of sediment to take place where the water entered the boiler; and would confine the cooler parts of its content to where the smoke was of the lowest temperature.

231.—Smaller cylinders, or rather tubular boilers, have frequently been proposed for generating steam; Blakey's has already been mentioned, (art. 25,) and Count Rumford had one put up at the Royal Institution in 1796, for generating steam for warming the rooms. His ideas on the application of his construction to steam engine boilers are worthy of attention.

232.—*Count Rumford's Boiler.* The object of this boiler was to get a larger quantity of surface, and the Count had a model of it made and presented to the French Institute, (October, 1806.) This model, as far as it differs from an ordinary steam boiler, being described, the reader will easily understand how to apply it on the large scale.

The body of the boiler is in the shape of a drum. It is a vertical cylinder of copper, twelve inches in diameter, and twelve inches high, closed at the top and bottom by circular plates.

In the centre of the upper plate there is a cylindrical neck six inches in diameter, and three inches high, shut at the top by a plate of copper, three inches in diameter, and three lines in thickness, fastened down by screws.

The flat circular bottom of the body of the boiler, which as before stated is twelve inches in diameter, being pierced by seven holes, each three inches in diameter, seven cylindrical tubes of thin sheet copper, three inches in diameter, and nine inches long, closed at the lower ends by circular plates, are fixed in these holes, and firmly riveted, and then soldered to the flat bottom of the boiler.

On opening the communication between the boiler and the supply cistern, the water first fills the seven tubes, and then rises to the cylindrical body of the boiler; but it can never rise above six inches in the body of the boiler, for when it has got to that height, the floater is lifted to the height necessary for shutting the cock that admits the water. As the seven tubes that descend from the flat bottom of the body of the boiler into the fire place, are surrounded on all sides by the flame, the liquid contained in the boiler is heated, and made to boil in a short time, and with the consumption of a relatively small quantity of fuel; and when the vertical sides of the body of the boiler, and its upper part, are suitably enveloped, in order to prevent the loss of heat by these surfaces, this apparatus may be employed with much advantage in all cases where it is required to boil water for procuring steam.

And in the case where the boiler is constructed on a great scale, the seven tubes that descend from the bottom of the boiler into the fire may be made of cast iron, whilst the body of the boiler is composed of sheet iron, or sheet copper.

But in all cases where it is required to produce a great quantity of steam, it will always be preferable to employ several of these boilers of a middling size, placed beside each other, and heated each by a separate fire, instead of using one large boiler heated by one fire. For Count Rumford has shewn by experiment, in his Sixth Essay "On the Management of Fire, and Economy of Fuel," that beyond a certain limit, there is no advantage derived from augmenting the capacity of a boiler.

The additional surface obtained by using tubes is unquestionable; and the construction proposed by the Count might be applied with much benefit where much surface is to be gained in a small space. The tubes should, however, have that proportion of capacity necessary for an engine boiler, and not be too small to contain an ascending and descending current.

233.—*Woolf's Boilers.* The idea of cylindrical tubes and a magazine for water and steam, was further expanded by Mr. Woolf into a variety of forms which were successively adopted and abandoned. His first project was to have a horizontal cylinder for containing steam and water, with a series of horizontal tubes below it, crossing it at right angles, and connected to the cylinder by short necks. The lower tubes and half the cylinder to be filled with water, and the flame and smoke to pass alternately over and under the tubes in a waving course. And where very strong steam was required, he had two other smaller cylinders, one on each side, in lines parallel to the large one and above the cross tubes, which are connected alternately to these by short necks; the larger cylinder communicating only with the side cylinders.* The immediate object of this arrangement is to introduce the cold water so as not to interrupt the rising of the steam, which is the fault of both the first arrangement and also of Court Rumford's.

Another mode of application adopted by Mr. Woolf consists in placing the tubes longitudinally, as the larger cylinder, parallel to each other, but in a gently sloping direction; the upper ends of the tubes all open into the large cylinder near to its end. The tubes are about ten inches in diameter, and extend the whole length of the fire place, which is formed below them, and the fire acts directly on the lower surfaces of the tubes, and the flame and smoke on the lower side of the principal cylinder. This plan seems to be the latest he has contrived, and a wonderful stock of ingenuity has been exhausted to very little purpose.

234.—But there is another form given by Woolf to the boiler which is too ingenious

* Philosophical Magazine, Vol. XVII. p. 40.

to be passed without notice. It consists in forming an upper and a lower boiler, and connecting them by short tubes. For a low pressure boiler the arrangement gives much surface, but would be more troublesome to execute than the common boiler, with scarcely any sensible advantage, or at most not more than is gained by making a flue through the boiler.

235.—The reasons for avoiding the complicated forms of the tubular boilers of Rumford and Woolf require very little illustration. We are certain that if a boiler has the proper quantity of surface and capacity it will be effective, and that all that can be done in this respect, by a tubular boiler, is to obtain these proportions, perhaps, in a less space; but if a more simple form will afford them, it certainly claims our preference. As to safety there can be no difference, unless the capacity of the cylinder be reduced to less than would contain the proper store of steam. For it is to be recollected that the stress on the larger cylinder is unalterable by either the disposition or the size of the small tubes; and half the capacity of this cylinder must be capable of holding the store of steam.

Another objection to these boilers is the necessity of using cast iron, but of the defect of this material for boilers, it will be necessary to treat further in giving the rules for the strength of them.

236.—We have now to consider boilers which have the fire within them. They have been long a favourite species with speculative mechanics, and particularly since the high pressure steam engine was brought into use by Trevithick. It seems a most compact and convenient mode of applying heat, and if we could for a moment forget the current of heat blown up the chimney, one might with some people imagine that the whole of the fireplace being within the boiler it must give out its heat to it alone; such an opinion is however absurd.

It is also urged that it is safe, because the part exposed to the heat of the fire, being within the boiler, when it is destroyed the steam will burst inwardly, and this is freely admitted to be true, only it imposes the necessity of having a larger boiler, which of course is more dangerous.

237.—The proportions of these boilers will be found to depend on the following circumstances. That part of the area of the tube appropriated to supply air to the ash pit must be of sufficient size for the purpose; which determines the diameter of the tube. The area of the grating must be considered, (see art. 198,) and then the length of the tube must be at least sufficient to make its superficial contents equal to the surface required for the fire. (See art. 204.) The capacity of the boiler must next be adjusted, so that deducting the space occupied by the tube containing the fire, the quantity remaining will contain the necessary store of water and steam. (See art. 215.)

238.—If the nature of the application admit of a supply boiler being added, to receive and heat the water required to replace that boiled off, then the internal flue should have

T

only the quantity of fire surface, and the smoke should be returned under the supply boiler, as Oliver Evans proposed. When a supply boiler cannot be used, somewhat more than one-fourth of the effect of the fuel will be lost by the smoke escaping at such an elevated temperature.

239.—The construction of boilers for steam boats must be such as will render them secure against danger from the fire, and also with as little of either bulk or weight of materials as possible. When they are low pressure boilers, and I would strongly recommend that no other should be used at sea, the force of the steam does not prevent the use of plane surfaces, to bound the flues and fire. The object then is to arrange the fire-place and flues within the boiler, so as to afford the proper quantities of fire, and flue surface, and of capacity, and admit of being cleaned with facility.

Various methods are adopted, but I have observed that the common tendency of a few years' practice is to simplyfy both the construction, and the means of obtaining effect.

240.—The boiler is sometimes made so as to admit a clear passage of about eighteen inches between the timbers and the boiler, but this excellent practice is by no means so general as it ought to be; for it not only gives a great degree of security against accident by fire, but also renders the examination and repair of the boiler easy and satisfactory.

241.—The grate should not be less than about two feet from the floor, and the sum of the areas of the flues of the fires should be somewhat larger than the area of the chimney, or simply larger than the chimney when there is no more than one fire. It will be an advantage to have as many separate fire-places as is convenient, for several reasons. First, The fire is easier to manage, and a less interruption to the generation of steam is caused by feeding it. Secondly, The flue and fire surface are obtained in less space, because two flues have more surface than one capable of conveying the same quantity of smoke. It is, however, scarcely possible to point out the limits which should determine the choice in different cases, as first expense is too often avoided, under the impression that it is more than equivalent to an unknown loss, which will become as regular as it is certain.

242.—A flue about in the proportion of twelve inches wide, and eighteen or twenty-four inches high, with one of its ends easily accessible, is a good proportion; height rendering the flue more effective than width, in consequence of the hottest part of the smoke pressing against the upper part of the flue, while the bottom gets speedily covered with a coat of sooty matter, which being a bad conductor of heat, the bottom surface has very little effect. Hence, in estimating the quantity of surface the bottom of the flue should not be calculated.

243.—The fire-place is necessarily surrounded by water, but there is no advantage in this; for water is so rapid a conductor of heat that it absorbs it too fast from the fuel which is in combustion, whereas nothing can be more injurious to the perfectness of that process than a rapid abstraction of heat. The sides of the fire ought to be lined with fire bricks as far as the burning fuel extends, and the saving arising from the more perfect combustion of the fuel, and in the duration of the boiler, would more than balance the inconvenience of the construction.

A boiler for a steam boat constructed in this manner is shewn in Plate XVII. Fig. 1, 2, and 3. It differs in some respects from the usual forms, but not in any essential points; the great object is to obtain a sufficient quantity of fire surface, and facility of clearing the flues, is of considerable importance.

244.—*Portable high pressure Boilers.* Boilers for steam carriages, and other purposes where a permanent seat of brickwork cannot be applied, should be arranged in the same manner as those for steam boats, with the exception of the forms being adapted to resist the effect of the steam.

Both the boiler and the flues within it should be cylindrical; the difficulty of the case consists in obtaining even the due proportion of fire surface, without rendering the boiler too large in diameter. Hence, the only thing that seems capable of being done to improve the present construction, is to make the boilers much longer with less diameter; to have the boiler filled with water, and the fire tube larger, with the spaces for steam formed by short vertical cylinders round the steam cylinders.

Of Fire Places.

245.—Various methods have been tried for improving the construction and the mode of supplying the fuel to the fire-places of steam boilers. Smeaton improved them so far that there has been very little more useful effect obtained since, than was done by some of his boilers. The later researches on combustion induced Mr. Watt to add a few further improvements, but experience taught him that what might be done by scrupulous attention and just principles, was not to be expected in ordinary practice.

246.—*Watt's Fire-place.* In improving the furnace Mr. Watt proceeded nearly on the principles of Argand's lamp. The grate and dead plates were laid in a sloping direction downwards from the fire door, at an angle of about twenty-five degrees to the horizon: the fire being lighted in the usual manner, and a small quantity of air admitted through one or two openings in the fire door, so as to blow directly on the blazing part of the fire. The fire at first was kept near the dead plate, and the fresh coals with which it was supplied were laid upon that plate close to the burning fuel, but not upon it. When it needed mending, the burning coals and those upon the dead plate were pushed further down without being mixed, and more coals were laid upon the dead plate, but never thrown on the top of those already on fire, as that would instantly send out a volume of smoke. In this situation they were gradually dried, and the smoke which issued from them consumed by the current of air from the fire door, in passing over the bright burning fuel. The opening or openings, to admit the air, are regulated so as just to admit the quantity which consumes the smoke; more would be prejudicial. He at first constructed these furnaces in a rather different manner; but found the above method the most convenient, and, *when properly attended to*, answers the purpose perfectly with free burning coals, but is more difficult to manage with coal which cakes.

247.—*Roberton's Furnace.* Various methods of construction have been contrived to accomplish the objects proposed by Mr. Watt; that of Messrs. Roberton is perhaps on the whole the best. The opening through which the fuel is introduced into the furnace, is shaped somewhat like a hopper, and is made of cast iron, built into the brickwork, inclining from the mouth downward to the place where the fire rests on the grate. The coals in this mouth piece or hopper answer the purpose of a door, and those that are lowest are by this means brought into a state of ignition before they are forced into the furnace. Below the lower plate of the hopper the furnace is provided with upright front bars, which serve to admit air among the fuel, and to admit an implement to force the fuel back, from time to time, to make room for fresh quantities to fall into the furnace from the hopper. By this arrangement the fuel is brought into a state of ignition before it reaches the further end of the bottom grate, where it is stopped by the rising breast of the brickwork, so that any smoke liberated from the raw coals at the front, must pass over these red hot coals before it can reach the flue.

Below the upper side of the mouth piece or hopper, and at about the distance of three-fourths of an inch from it, is introduced a cast iron plate. This plate is above the fuel, and the space between it and the top of the hopper is open to admit a very thin stream of air, which rushing down the opening, comes first in contact with that part of the fire which is giving off the greatest quantity of smoke, mixes with it before it passes over the hot fuel in the interior, and therefore in passing it inflames and escapes undecomposed. This is the worst part of the apparatus; for air so admitted cools the bottom of the boiler.

The quantity of air admitted to pass over the upper surface of the fire, is regulated by inserting a wedge-formed piece of iron. The front bars are closed by doors which when shut prevent the heat from coming out, and incommoding the workmen.

248.—A considerable improvement was added by Mr. Woolf, to enable them to get rid of clinkers and scoriæ; the contrivance is extremely simple. The combustion of the fuel commences, and is chiefly carried on, on the part of the bottom grate next the hopper, and the fuel is pushed back from time to time along the grate, and at the end, vitrified portions fall into a cavity, the bottom of which is furnished with horizontal slides. These when drawn out by an iron hook applied to the handle of the slide, discharge the clinkers into the ash pit.   (See Plate II.)

249.—The defect of Roberton's method as well as Mr. Watt's, consists in admitting a regular current of cold air when it is not regularly wanted, and where it has an injurious effect in cooling the smoke as it rises against the bottom of the boiler. This is greatly remedied by admitting air by means of small side flues, or at the back of the fire, whence having to pass from the ash pit, through small channels in the hot brickwork, it becomes heated before it issues into the fire-place. But abundance of air will pass the grate if it be properly constructed, and the modification I would recommend is described in Plate II.

Air flues when used should have valves to open or close them, and on the whole very little good is derived from them unless they be attended to with more care than is usually bestowed on the fire of an engine.

250.—*Brunton's Fire-place*. In consequence of the difficulty of supplying a fire equally by hand, so as to sustain the regular demand for steam in a steam engine, it has been attempted to use machinery for that purpose. Several schemes have been tried, but the only one which has succeeded in practice is that invented by Mr. William Brunton.

The method consists in an apparatus for dropping the coals on the grate by small quantities, at short intervals of time (not more than three or four seconds) and in such a manner that the smoke rising from the fresh fuel must pass over that which is in a further stage of combustion, and consequently be consumed; the uniform supply of air for that purpose being admitted.

The machine is also so contrived that a quantity of coal is put on proportioned to the quantity of work, and the air admitted is regulated in a similar manner.

The advantages of such a method are obvious, and the increase of expense of erection not so considerable as might be expected.

A circular horizontal grate which receives the coals is five feet in diameter, and revolves on a vertical axis at the rate of about one revolution per minute. During its revolution the coals fall (from a hopper placed over the boiler) through a vertical narrow rectangular opening, formed through the top of the boiler, of the length and in the direction of the radius of the grate. The quantity discharged at once by the hopper is regulated by the force of the steam in the boiler, and the discharge is made at every fourth or fifth second of time; by this means an uniform fire, regulated by the work it is to perform, is obtained, and with a certainty as absolute as the nature of things will admit. To prevent air being admitted without passing through the channels which are properly regulated, a thin rim on the under edge of the grate, runs in a circular trough filled with sand. These parts, however, will be more clearly understood by a reference to the description of Plate II.

where its application to two boilers, which had been previously erected by Boulton and Watt, is shewn. The saving of fuel by using this apparatus is stated to be about twenty-five per cent; and a grate five feet in diameter burns 260 pounds (three bushels) of New-castle coals per hour, and 336 pounds, or three cwt. of Staffordshire coals; that is, thirteen pounds for each foot of surface of grate for the former; and seventeen pounds each foot for the latter kind. I suppose it requires these quantities also to produce equal effects. The quantity of grate is about two-thirds of that required in the ordinary method.

Apparatus for Boilers.

251.—*Feeding Apparatus.* The use of the feeding apparatus is to supply the boiler with water, in the place of that which is converted into steam. The feed pipe is a vertical pipe passing through the top of the boiler. The lower part of this pipe is turned at the end to prevent steam rising through it, and where it passes through the top of the boiler, it is made steam-tight and fixed very correctly in a vertical position. The top of the pipe terminates in a small cistern head, which is kept supplied with water by a small pump from the hot water cistern; and at the bottom of the small cistern head there is a conical valve, opening upwards, connected by a chain to a lever, which turns on a centre, with a wire attached to the opposite end. This wire passes through an air-tight stuffing box to a flat stone or piece of metal in the boiler, which is so balanced by a weight, on the opposite end of the lever, as to float on the surface of the water. The stone should be so large in proportion to the surface of water as to act sensibly on a very slight depression of the water.

Its action is performed in this manner: when part of the water is evaporated from the boiler, the float descends with the water's surface, and consequently raises the conical valve; now, the small cistern head, being kept constantly full of water, by the pipe from the hot water pump, as soon as the valve is raised, water enters the boiler, and when it is filled to the proper level, it raises the float and shuts the valve, till a repetition of the operation becomes necessary. The surplus water raised by the pump runs off by a water pipe from the cistern head.

252.—The principal circumstance to be attended to in the construction of this apparatus, is, to make the height of the water in the cistern sufficient to balance the strength of the steam. For if this height be too small, the water in the boiler will be forced up the feed pipe by the pressure of the steam, and be driven out at the valve.

For water at 60°, 2·94 feet in height is equivalent to one pound on the circular inch, but the water in the feed pipe will generally be nearly 212°, and then three feet is required. Hence, three feet in height for each pound per circular inch is the proper height.

The stone float should obviously be in that part of the boiler where it will be least disturbed by the formation of steam; and the feed pipe should deliver its supplies as far from the point where the steam is principally generated as possible.

253.—On account of the force of steam required in high pressure engines, an ordinary feed pipe cannot be applied to supply the boiler, without making it of a very inconvenient height; water is therefore supplied to the boiler by a small forcing pump, worked by a lever connected with one of the reciprocating parts of the engine, and this water instead of passing immediately into the boiler, should pass through a pipe or receptacle which traverses back and forward in the steam which escapes from the engine, so as to become considerably heated before it enters the boiler, that it may not check the production of steam. A much better method, however, is to make the smoke pass round and heat a small supply boiler; which should have a communication with the proper boiler. The pump in this case supplies the small boiler.

In supplying a boiler by a pump worked by the engine, the same supply is given at all times, whatever may be the quantity converted into steam and used. Now as the consumption of steam is variable, the quantity injected by the pump must often be in excess. This may be remedied by the use of a float, in land engines. Let A, B, (Fig. 2. Plate I.) be two connected valves, in the box which receives the water from the pump, the one A opening to the boiler, the other B opening to the waste pipe. If the stem of these valves be connected to the lever of a balanced float, as indicated in the figure, the increase of water in the boiler above its proper level will cause the valves to descend, and close the communication to the boiler, while the waste valve opens and admits the superfluous water to run off by the pipe. In this construction the boiler will receive the supply from the pump regularly at all times, except where it is in excess for the quantity used, and then the float F rises, and shuts the passage of entrance to the boiler, and opens the one to the waste, till the quantity no longer exceeds the consumption. This simple arrangement renders the feed regular, which is of much importance.

254.—The same construction applied to the feeding pipe of a low pressure steam engine would be much superior to the common stone float; and I think it would apply as shewn in Plate II. Fig. 2, even to the steam boat; for the oscillation would not prevent either its rise or fall, when an over supply took place or otherwise; and employing the rise instead of the fall of the water to act on the valve, would be a means of safety as well as of preventing irregular influxes of water to check the steam. (See art. 217.)

255.—In a method of admitting water to high pressure boilers invented by Mr. Franklins, the waste water has to raise a loaded valve to escape, and the passage to the boiler is regulated by a balance float placed wholly within the boiler; it is ingenious, but has not the advantage of rendering the supply continuous; it must, as the ordinary feed pipe, stop till the water has descended so as to raise the valve.

Of regulating the Fire of a Steam Boiler.

256.—The force of the steam may be made a means of regulating the fire, either by diminishing the supply of air, or by contracting the chimney by a plate, called the damper. As a means of regulation the former ought to be preferred; it being obvious, that a direct diminution of the quantity of oxygen at its entrance to the fire, must have both a more immediate and a more beneficial effect than contracting the chimney; the effect of the latter being to increase the temperature and force of the smoke in proportion as the aperture is contracted; and, consequently, the smoke escapes at a higher temperature, carrying off a considerable quantity of heat. The regulation by the damper is the kind generally used; the other method is the same in principle, and only differs in being applied to the ash pit instead of the flue.

257.—*Self-regulating Dampers.* Dampers are frequently under the control of those who have the management of the fire; but in the self-regulating damper the fire is made a means of controlling itself, so as to burn with more or less rapidity, as it may be more or less wanted, in the following manner. An iron plate or damper, of sufficient size entirely to close the chimney or flue, slides up and down, vertically in iron grooves, (see Plate III.) with as little friction as possible. To its upper part is attached a chain, which passes over the two pullies *n* and *n*, through a tube in the bottom of the cistern head of the feed pipe, and down the centre of the feed pipe C, to a hollow or bucket-shaped cast iron weight; the feed pipe being made of larger diameter in this part, when a self-regulating damper is applied to a boiler, to admit the weight without blocking up the pipe so as to prevent the descent of the feeding water. The weight is so adjusted by filling it partly with lead, that it may just overcome the weight and friction of the damper plate, chains, and pulleys, when there is no fire under the boiler; consequently, the damper plate will then be drawn up, and the chimney completely open, at which time the weight will rest on the shoulders or projections at the bottom of the feed pipe; the chain being properly adjusted in length for that purpose. Now as soon as a fire is applied so as to generate steam in the boiler, the steam presses upon the surface of the water, and drives it up to a certain height in the feed pipe, and the weight, by becoming immersed in water, has part of its gravitating force balanced, and therefore becomes no longer able to retain the damper plate at its former height; it will consequently descend till equilibrium takes place, and partly closes the chimney, by which the draught of the fire will be checked. Should it move so as to check it too much, less steam will be formed, and the water will rise to a less height in the feed pipe, and part of the force of the weight will be restored so as to raise the damper again: should the fire ever become so fierce as to drive the water up into the cistern head, the weight should be so far raised as nearly to shut the chimney; when a damper shuts perfectly close there is a risk of inflammable air collecting and exploding in the flues.

A hand damper is, however, an appendage which a boiler should always have, for when an engine is not in action it will be useful partially to close it; and no boiler can be considered perfect which has not both a damper, and the means of entirely closing the aperture by which the air enters to supply the fire.

258.—*Self-acting Air Regulator.* The most direct method of governing the action of a fire is to provide the passages which admit air with the means of opening or closing them at pleasure; and it is a still further advantage when this is done by means of the force of the steam, so that as the steam increases beyond its proper strength, it closes the aperture which admits air to the fire. A method of constructing a self-acting regulator of this kind is shewn in Plate I. It is essential in applying it to make all other entrances to the burning fuel to shut as perfectly close as possible.

Safety Valves.

259.—The precautions for safety are of much importance; the boilers of steam engines should never be constructed without them, and they should be done with every care to render them effective in preventing accidents.

Safety valves are called external, or internal, according to the nature of the evil to be prevented.—*An internal safety valve* is to prevent the pressure of the atmosphere crushing in the sides of the boilers or pipes to which it is applied. It is usually an inverted conical valve, retained in its seat by a rod connected to a lever, having a weight at its opposite end, such that the force of the atmospheric pressure will overcome it, when its pressure is three or four pounds on the circular inch greater than the elastic force of the steam in the boiler.

In Plate I. Fig. 1. this valve is shewn as inserted in the man-hole plate; *a* being the valve kept in its seat by the weight on the lever at *b*.

260.—*The external safety valve* is to prevent the risk of explosion, should the steam become stronger than that the boiler is intended to confine; therefore, it is of the greatest importance that it be properly constructed and not liable to derangement.

The application of a loaded valve to limit the force of steam appears to have been first made by Papin to his digesters; and it was applied by Savery to the boilers of his steam engines. It consisted of a conical valve retained in its seat by a weight on a lever; and from its resemblance to a steelyard was called the steelyard safety valve. It is still much used, but it has the obvious defect that the weight may be increased at the will of the workman, or even may be done through the ignorance of a stranger; hence, valves of this form should not be employed unless the lever and valve be wholly inclosed in a box kept locked by the proprietor. Such a box should have a pipe leading into the chimney, to carry off the steam, and a slight wire or chain to lift the valve by, lest it should stick fast by corrosion.

261.—For low pressure steam, the form is rendered more convenient. The conical valve has its load directly upon it, and it ought to be sufficiently large. Its clear area in the narrowest part not being less than is calculated by the annexed rule, and the power of the steam having been determined, a fixed and unalterable weight agreeing with that power should be formed and attached to the rod on the top of the valve, and the whole should be inclosed in a metal box, having a passage larger than the area of the valve to convey the steam away to the chimney or other place.

The greatest power of steam should be a little more than is required to work the engine; suppose it be five pounds on the circular inch, and the diameter of the lowest part of the seat of the safety valve should be 3, then 3 times 3 being 9, the area of the valve in circular inches, and 9 times 5 are 45, which is the required weight in pounds, for the load to be placed upon the valve. This valve will not open till the steam presses it with greater force than five pounds on the inch. The metal box for the valve being locked up, of course no one but the possessor of the key could alter the load on the valve, but a handle passing through the cover is necessary to move it to prevent it rusting fast.

For further security it has been proposed that another safety valve should be placed upon the same boiler; but with rather less load upon it, in order that it may open first, and give notice to the engine man when the steam is likely to become too strong. This should have a stronger handle for moving it, either for letting off the steam when not required, or other purpose; but the handle to raise the locked valve should be either connected by a chain or slight wire, so that it could not be fixed so as to increase the load on the valve. It would be better to rely on the common valve than one locked up till it had become stuck fast with rust.

262.—A conical seated valve does not appear to me to be the best; for the locked valve I would prefer a flat seat, and that the metallic surfaces in contact should be narrow, and of metal not liable to corrosion, nor to fix by unequal expansion.

263.—To prevent the danger of adhering in steam boat boilers, Mr. Nimmo* proposed that the valve should be a hemisphere with its convex surface downwards, to rest in a seat

* Report on Steam Boats; or Partington's Historical Account of the Steam Engine, p. 92.

formed to fit it, and the weight, he proposed, to hang to the lowest part of the valve. See V. Fig. 1, Plate XVII. By this means the motion of the boat would be constantly changing the position of the valve, while its form would render it steam-tight in all positions, without danger of adherence. A chain might be also attached to the upper side of the valve, to lift it without opening the case inclosing it. This method deserves attention. Its defect will most likely be want of stability in its seat.

264 —The most certain and safe method for low pressure boilers is to balance the pressure of the steam by a column of water, of a diameter adequate to allow of the escape of the steam as rapidly as it would be possible for the fire to generate it. A feed pipe is to a certain degree a safety tube of this kind, but neither of the size nor construction which safety requires. The tube or pipe T W should be made recurved at the lower end T, Plate II. Fig. 2, its mouth being not lower than level with the upper edge of the fire flues. At the upper end it should be provided with a pipe U to convey down the hot water without the danger of scalding any one, and the upper part should terminate in a higher pipe V, to convey away the steam. The action of this safety tube is, first to lower the water in the boiler to bring the feeding pipe into action, if it be not so before, and then to allow the escape of the steam. I have had two boilers done in this manner, and the effect of endeavouring to render the steam stronger than it ought to be is completely counteracted; and the boiler restored again to its regular pressure in a few minutes after the tube has discharged its column of water. The discharge of a portion of hot water by the tube, and the admission of colder by the feed pipe, tends to lower the steam, but by the feed pipe alone this does not happen, as the hot water then rises in the feed pipe and prevents the entrance of cold. Another advantage of this construction is, that should the water fall below the mouth of the tube, the steam would escape, and if the noise of its escape did not warn the engineer of the state of the boiler, the want of steam would soon be a motive to look after it.

The height of the tube for different pressures is easily calculated, for the height of a column of water equivalent to one pound on the circular inch, is for ordinary temperatures 3·1 feet; hence, for a pressure of four pounds on the circular inch, 4 × 3·1 = 12·4 feet, the height for the tube, that is, equal to a little more than five pounds per square inch. It is obvious that it is adapted only for low pressure steam.

265.—Other modes have been proposed for constructing valves, some of which are deficient in principle, others are complex in construction, and operation. The solid piston valve proposed by Chevalier Edelcrantz, would either stick fast with high pressure steam or allow a constant escape, and to this difficulty is added the nicety of fitting a solid piston so as to be and remain steam-tight. If an elastic metallic piston be used instead of a solid one, the expense of construction becomes considerable; and a common packed piston is not to be depended on for the purpose, its friction is so irregular.

266.—For high pressure boilers more careful attention to the means of security are necessary than for low pressure ones, as it is easy to shew that the risk is much greater; indeed some most dreadful accidents have had the effect of rendering people more cautious respecting them. Several methods have been used to guard against these accidents by Trevithick, who first brought the high pressure engine into use. He proposed that the safety valve should be inclosed in an iron case and locked, so that no person could get access to it to increase the load beyond what was intended to be employed. He also had a hole drilled in the boiler, which he plugged up with lead, at such a height from the bottom, that the boiler could never boil dry without exposing the lead to be melted, and consequently making an opening for the steam to escape. This contrivance he expected to prevent the boiler being burst by suddenly forcing water into it, when it had been allowed through inattention to boil dry, and become red hot.

267.—A plug of fusible metal riveted into a hole in the bottom of a boiler, so that it may melt and allow the water and steam to escape into the fire, whenever the contents of the boiler attain that degree of heat which produces steam of a dangerous elasticity, is a method of a like nature.

268.—The mercurial steam gauge is generally applied to boilers to shew the state of the steam; it is a curved tube, or inverted siphon, in which the mercury rises by the force of the steam, and indicates the pressure. (See Section VIII.) When this steam gauge is applied to a high pressure boiler, it requires a tube of considerable length; and is an additional security against the bursting of the boiler, because when the steam is too strong the mercury will be displaced into a proper receiver, and the steam escape through the tube when the pressure exceeds that the boiler is designed to sustain. This steam gauge is a most desirable appendage to a high pressure boiler, because it shews at once the state of the steam; but as a means of safety we had better inquire how far either it or metallic plugs are likely to be effective, lest, under an impression of being secure, the reliance may involve us in more of these fatal accidents.

269.—In the first place, it is obvious that the aperture or apertures by which the steam is to escape, should be so large that it may escape as fast as the fire can generate it; if it does not it must accumulate, and eventually explode. Now it is possible to convert a cubic foot of water into steam from somewhat less than 1·5 feet of fire surface, (see art. 200.) and it is making only a small allowance for security to admit that each foot of surface may convert a cubic foot of water into steam.

RULE. Hence, we derive the following rule. Let the density of the steam corresponding to the pressure be found. Then multiply 7·5 times this density, by the square root of the quantity the density is greater than 1, and divide the feet of fire surface by the product; this quotient is the square of the diameter of the narrowest part of the valve in inches. Or, divide the area of the fire surface by the number corresponding to the pressure or temperature, under the head divisor in the following table; and the quotient will be the square of the diameter of the valve in the narrowest part in inches.

Pressure in inches of mercury.	Temperature.	Density of steam.	Divisors.
30	212	1	0
35	225	1·28	5
60	250	2·00	15
90	275	2·85	29
120	293	3·70	45
150	308	4·7	60

The rule, it is to be remarked, is for the smallest aperture that ought to be used, but there is much reason either to use two valves or to double the area determined by the rule.

Example I.   Required the area of a safety valve for a low pressure boiler, fifteen feet long by four feet wide, the fire surface being considered equal to the area of the bottom of the boiler.   In this case $15 \times 4 = 60$, and the divisor is 5 for low pressure steam; hence,

$$\frac{60}{5} = 12,$$

the square of the diameter of the aperture; and the square root is three inches and a half, nearly, for the diameter.   And either two valves of this diameter, or one of five inches diameter, ought to be used.

Example II.   A high pressure boiler with sixty feet of fire surface is used for generating steam of four times the atmospheric pressure, what should be the least diameter of the safety valve aperture.

In this case the divisor is 45, and

$$\frac{60}{45} = 1\ 34 =$$

the square of the diameter, and the square root is 1·16.

270.—Hence, we find that the diameter of a mercurial gauge capable of giving passage to the steam, is not so large as to prevent it being applied in practice, and with success, as a means of rendering boilers safe.   A safety valve in addition should of course be used, as the bends in the pipe would in some degree retard the escape of the steam.

271.—The use of fusible metal plugs I do not think so likely to afford security, for were the plug fusible at the pressure for which the boiler was adapted, it would be so softened by the continued temperature of the working state, as to be incapable of retaining the steam, when made of sufficient magnitude to be useful. Lead would be wholly unfit, its melting point being at 612°, a temperature at which the force of the steam would be about 150 atmospheres. Tin melts at 442°, and at this temperature the force is upwards of 25 atmospheres. Alloys may be formed to melt from 212 to 600°, but we have no evidence that the melting points remain permanent, in alloys which are regularly exposed to a heat so nearly approaching to that which they fuse at when newly formed.

Alloys and Metals.	Melting point.
An alloy of lead 1 part, tin 3 parts, bismuth 5 parts, melts at	212°
Lead 1 —— tin 4 —— bismuth 5 ——————	246
Tin 1 —— bismuth 1 ——————	286
Tin 2 —— bismuth 1 ——————	336
Lead 2 —— tin 3 —— ——————	334
Tin 8 —— bismuth 1 ——————	392
Tin ——————	442
Bismuth ——————	472
Lead ——————	612
Zinc ——————	648

But if a range of about two atmospheres above the working pressure, be necessary to fuse the plug, and with less range than that, it is scarcely probable it will withstand the working pressure, this mode of obtaining safety ought not to be relied upon in practice. As an additional precaution the fusible plug may be adopted, but not as a principal one, certainly not as one in which great dependence may be placed.

272.—It ha salso been proposed to add a pipe to some part of the boiler, of such thin metal that it may burst rather than the boiler, but this plan like that of the metallic plugs, can only be useful in cases where the ordinary safety valves do not act, for if it be made at first so that it would break on a small increase above the working pressure, it would be constantly failing at that pressure, it being well known that a metal strained to near its ultimate force will gradually break. Besides it is exceedingly difficult to determine the strain such a pipe will bear without fracture, within the limits that would render it safe to depend on where life is in hazard.

273.—The risk on high pressure boilers, even at their working pressure, becomes considerable in proportion as that pressure is high, and therefore too much caution cannot be employed about them. At least one good safety valve, and a mercurial gauge of sufficient diameter to allow the escape of the steam, should be applied to each; but it is the practice of careful engineers to apply two safety valves.

On the common safety valve an improvement might be made by constructing it so as to be relieved of part of the load on the valve as it rises.

The Area of Chimneys for Steam Engine Boilers.

274.—Previous to giving particular rules for the area of chimneys, it may be useful to remark that a chimney may afterwards be convenient, if considerably larger than is necessary for the use of the engine it is erected for, while the expense bears a small ratio to the increase of size. Hence, I would recommend that one double the size of that given by the rule should be built, for the rules apply only to one for the actual power of the engine.

The height should not be less than about fifty feet, and should be higher if it be desirable to avoid the nuisance of smoke in the immediate neighbourhood of the chimney. For though by increasing the height of the chimney there is no diminution of smoke, yet it is spread so as to fall over a large surface.*

275.—RULE. The area of a chimney for a low pressure steam engine, when above ten horses' power, should be 112 times the horses' power of the engine, divided by the square root of the height of the chimney.

For less than ten horses' power, it should be 90 multiplied by the number opposite the horse power in the first column of the table, (art. 221.) instead of 112.

Example. Required, the area of a chimney for an engine of forty horses' power, the height of it being seventy feet.

In this case

$$\frac{40 \times 112}{\sqrt{70}} = \frac{4480}{8\cdot4} = 533\cdot2$$

* It is a curious circumstance that when high pressure steam and smoke ascend in the same chimney, the smoke becomes nearly invisible. It seems to have been first observed in Trevithick's engine, when applied to a steam carriage in 1805; and was communicated to Nicholson's Journal (Vol. XII. p. 1.) by Mr. Gilbert, who offers no explanation, but states that the admission of the steam into the chimney improved the draught. Nicholson made an experiment which accounts for the vapour becoming invisible, through the heat of the smoke preventing that degree of condensation which is essential to its being seen. (Journal, Vol. XII. p. 47.) The disappearance of the smoke is not accounted for; but I think it seems to be deposited in consequence of the density being diminished by intermixture with steam, till it becomes incapable of suspending the particles of sooty matter.

square inches. The square root of this is twenty-three inches, which will be the side of a square chimney. Or multiply 533 by 1·27 and extract the square root for the diameter of a circular one.

But in either case I would advise to build a chimney of double the area, or 1066·4 area, that is, make the side of the square thirty-three inches.

In this rule it is supposed that the engine is done in the best manner, and worked with the best coals; that is, one requiring only from nine pounds to eleven pounds of coal per hour, for each horse power, of an engine above ten horses' power. But where fourteen or sixteen pounds of coal per hour is necessary, the flue should be increased in direct proportion to the quantities to be consumed. See the mode of finding the rule in (art. 168.)

276.—When wood is used for fuel, it affords a much larger quantity of smoke, but it is also much lighter, and about one and a half times the area necessary for coals will be sufficient.

277.—The same rules may be applied to high pressure engines; taking the cubic feet of water per hour, or the one-eleventh part of the pounds of coal per hour, instead of the number of horses' power.

278.—The engine chimneys for steam boats and steam carriages are circular, and should not be larger than is absolutely required to give effect to the fuel. This will be about obtained when the square of the diameter is equal to 90 multiplied by the horses' power, and divided by the square root of the height in feet.

But here it must be remarked, that where a chimney is less than about forty or fifty feet in height, the smoke must be allowed to rise at a much higher temperature. It must not therefore be allowed to cool too much by giving its heat to the boiler, otherwise there will be a want of draught. Hence, in low chimneys, the fuel will not produce its full effect.

Different modes of finishing chimney tops, are shewn in Plate I. The least expensive is one of the form of an Egyptian obelisk, and it offers least obstruction to the wind.

Of the Condensation of Steam.

279.—When any substance or body colder than steam itself is put in contact with it, the steam condenses till the temperature of the cold body becomes the same as that of the steam; or till the whole mass of steam be condensed to a degree of elasticity corresponding to the temperature to which the cold body is raised by the heat of the steam. The greater the quantity of the cold body the less its temperature will be raised, and also the colder it is the more the elastic force will be reduced. Hence, to reduce the elastic force of steam as low as possible, the coldness and the quantity of the cooling body should be as great as possible.

280.—Any cold body condenses steam, but that it may be effectively done the body should be capable of presenting a large quantity of surface, and be a good conductor of heat; as when power is to be obtained by condensation the more rapid the condensation is the more power is obtained. It may be easily proved that if steam were so condensed as to lose only equal degrees of elastic force in equal times during the action, half the power would be lost. (See art. 294.) This is the cause of the failure of every method of slow condensation; it cannot be too prompt, unless a sacrifice of power is made in some other way to gain that promptness, and to which the effect gained by condensation is not equivalent.

281.—Water has been found the most effective cold body for condensation; it has great specific heat, perhaps greater than any other body; it is a rapid conductor of heat, and in a jet it applies an immense proportion of cooling surface to the steam.

Now since water is frequently difficult to be procured of a low temperature, and sometimes not in sufficient quantity, it becomes important to inquire what effect is produced by given proportions at given temperatures.

282.—The weight of the water, W, required for condensation, multiplied by the quantity $x - t$ its temperature is raised, gives the heat it absorbs; and, in the steam engine, where the operation is repeated in the same vessels, and at the same temperatures, the excess of the temperature of the steam $T - x$ above that to which the condensing water is raised added to 1000, and the sum multiplied by the weight w of the steam, must be equal to the heat absorbed by the condensing water. That is

$$ W \left(x - t \right) = w \left(1000 + T - x \right)^{*} $$

$$ \text{or,} \frac{w \left(1000 + T - x \right)}{x - t} = W, \text{ and } \frac{w \left(1000 + T \right) + W\, t}{W + w} = x. $$

283.—When the temperature of the condensed water is equal to the temperature of the steam, the quantity of water would be equal to that which simply reduces the steam to water without change of temperature; or

$$ \frac{1000\, w}{T - t} = W. $$

But in this case no effect would be obtained. Any greater quantity of cold water reduces

---

* To make this equation general, let s be the specific heat of the condensing body, and C the heat of conversion, and the specific heat of the body in vapour, then $W\, s \left(x - t \right) = w\, s' \left(C + T - x \right)$.

x

the elastic force, but it must be so far reduced as to render the accession of power more than equivalent to that required to work an air pump, and cover the expense of a supply of water, and the extra cost of the engine.

284.—In low pressure steam T = 220°, and t may be taken at 52° the mean temperature, and if the temperature of the condenser be 100°, then

$$\frac{w\,(\,1000 + T - x\,)}{x - t} = \frac{w\,(\,1000 + 220 - 100\,)}{100 - 52} = 23\tfrac{1}{4}\,w = W.$$

That is 23¼ times the quantity of the water required for steam, will be the quantity of water necessary for condensation. And since a cubic inch of water produces about a cubic foot of steam of the rarity, it is the cylinder of an engine working at this temperature, and one-tenth being added for each foot of the capacity of the stroke, 23¼ × 1·1 is 25⅗ inches for each foot of the contents of the stroke of the cylinder.*

If $x = 130°$, it requires of cold water only fourteen times the weight of the steam to condense it, and for 120° it requires 16·2 times the weight.†

The force of steam at 100° is 2·08 inches of mercury, its force at 130° is 4·81 inches; consequently, the gain of power is 2·73 inches, or about one in thirteen, by condensing at the lower temperature.

If the temperature of the cold water be 70°, and of the condenser 130°, then we find cold water eighteen times the weight of the steam will condense it; and that it requires thirty-seven times the weight to condense at 100°, when the cold water is at 70°.

285.—From these equations the comparative effects of different temperatures may be calculated, and the economy of using or sparing water will be known and acted upon, instead of the usual method of endeavouring to get the greatest power of the steam in places where water is expensive.

When steam is of considerable density it does not condense freely; the reason is obvious, the same surface of injection water acting on steam of greater density, and consequently containing a greater proportion of heat, it abstracts the heat more slowly. To avoid this the condenser should be so large that the steam may expand to the bulk corresponding to a pressure not greater than about one atmosphere and a half. But it is better to make the steam act expansively in the cylinder, by Watt's method, (art. 27.) or expand in a second cylinder by Hornblower's method, (art. 32.)

When a lower temperature than 180° cannot be obtained by condensation, it is not worth the extra expense, and at 180° we have for low pressure steam

$$\frac{w\,(\,1000 + 220 - 180\,)}{180 - 52} = W = 8\,w,$$

---

* Mr. Watt says a wine pint, or 28⅞ inches is " amply sufficient." Robison's Mech. Phil. Vol. II. p. 147.
† The usual temperature is about 120°, or just what the hand can bear.

nearly; or eight times the quantity of water required for steam will be necessary to condense it.

286.—These computations apply to where condensation is made in a separate vessel, the first idea of which we owe to Mr. Watt. When the condensation is made within the cylinder, the metal of the cylinder has to be cooled down to the temperature of condensation as well as the steam, and a large proportion of the steam is lost in heating it again at each stroke. The means of obtaining a maximum of useful effect from condensing in that manner has been shewn, (art. 165.)

287.—To find the quantity of water for injection into an engine condensing in the cylinder, the formula is the same as when a separate condenser is used, the difference being in the quantity of steam required; and the water for condensation is greater than when Watt's condenser is employed by

$$\frac{\cdot 14 \; i \, (T - x)}{x - t}$$

for each stroke, when i is the weight of the mass of iron contained in the cylinder.

288.—The following tabular view of the modes of condensation may perhaps present it in a clearer view to the reader than any other kind of concluding summary.

Steam may be condensed
$\begin{cases} 1. \text{ in the vessel where its power is exerted} \\ 2. \text{ in a separate vessel} \end{cases}$
$\begin{cases} \text{Savery in 1698.} \\ \text{Newcomen in 1705.} \\ \text{Watt in 1769.} \end{cases}$

Steam may be condensed
$\begin{cases} 1. \text{ by projecting a cold fluid against the vessel con-} \\ \quad \text{taining it} \quad - - - - - - - - - \text{Savery.} \\ 2. \text{ by injecting a cold fluid among it} \quad - - - - \text{Newcomen.} \\ 3. \text{ by exposing it to large surfaces of cold fluids or} \begin{cases} \text{Watt.} \\ \text{solids} \quad - - - - - - - - - - - \text{Cartwright.} \end{cases} \\ 4. \text{ by the pressure of cold fluids against the vessels} \\ \quad \text{containing it} \quad - - - - - - - - \text{Perkins.} \\ 5. \text{ by the union of two or more of these methods.} \end{cases}$

SECTION IV.

OF THE MECHANICAL POWER OF STEAM, AND THE NATURE, GENERAL PROPORTIONS, AND CLASSIFICATION OF STEAM ENGINES.

289.—The force of steam when confined, according to its density and temperature, and the circumstances which affect its motion, having been considered, our next object is to investigate the power of steam to produce useful effect, and in this purpose I am desirous of proceeding with the simplicity and fulness this important subject requires.

FIG. 15.

Of the Power of Steam, and the Modes of obtaining it.

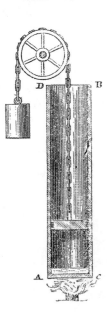

290.—The generation or production of steam, it has been shewn, takes place on the application of heat. Conceive a cylindric vessel, A B, to be placed in a vertical position, with a given depth of water in it; and an air-tight piston on the water balanced by a weight equal to its own weight and friction In this state let heat be applied to the base, A C, then as the water becomes converted into steam, of slightly greater force than the atmospheric pressure, the piston will rise till the whole of the water be in the state of steam. It will be remarked, that the generation of this steam of *atmospheric elastic force* affords no power, the motion being barely produced; it has simply balanced the column of atmospheric air, and excluded it from a given height of the cylinder.

291.—*By Condensation.*—But in this state of things if the steam be suddenly condensed into water again, it is obvious that the piston will be impelled by a force equal to the pressure of the atmosphere on the piston, and through a height equal to that the piston had been raised by the generation of the steam.

292.—It thus appears that the power of steam of the elastic force of the atmosphere, is, when speedily condensed, directly as the space it occupies. That is, multiply the area of the cylinder in inches by the pressure of the atmosphere in pounds on an inch of area, and by the height in feet, and the result, deducting the friction, will be the quantity in pounds the steam would raise one foot in height.

293.—The space occupied by steam of atmospheric elastic force may be increased by raising its temperature above 212°, the increase being equal to the expansion of steam by the given change of temperature; but a quantity of heat nearly equivalent to the increase of volume will be absorbed, and hence, the effect of a given quantity of fuel would not be increased by the expedient.

294.—If the steam be slowly condensed, as it would be by applying external cold, the effect would be much reduced, because the moving force at any period of the stroke would be only the difference between the elastic force of the steam and the atmospheric pressure; and the most rapid condensation leaves a vapour of some elastic force: but as it acts through the same space as the power of the steam, it does not cause a sensible deviation from the ratio of the power, being as the space the steam occupies, when the power is gained by rapid condensation.

295.—*By Generation.* Conceive the same cylinder and apparatus to have heat applied to its base, with only the difference of the piston being loaded with a given pressure per inch of its area. The generation of the steam will raise the loaded piston, but the height through which it will be raised will be less. The steam being acting in opposition both to the pressure of the atmosphere and the load on the piston, the space it will occupy will be in the inverse ratio of the pressures which oppose it in the two cases, supposing the steam of atmospheric elastic force to have been of the same temperature. Thus, if the load on the piston be twice the atmospheric pressure, the piston will be raised only one-third of the height; but on rapid condensation it descends with three times the pressure, and, therefore, whether the steam be generated of atmospheric elastic force, or of a greater force, the power it affords by generation and condensation is the same at the same temperature, and this power is directly as the elastic force of the steam, multiplied by the space it occupies, when the motion of the piston is rectilinear.

296.—But if, as in the last case, a loaded piston be raised, and then a valve be opened which allows the steam to escape, the whole power gained will be equal only to the weight raised descending from the height to which it was raised; and the power which would have resulted from condensation will be lost, and the loss is equal to the pressure of the atmosphere acting through the height to which the piston was raised by the steam. This is the nature of the common *high pressure* steam engine. It is obvious, that the greater the elastic force of the steam, the less is the loss by neglecting to condense it under these circumstances; but, it may be remarked, that unless the valve aperture be equal to the diameter of the cylinder, the steam cannot escape at the necessary rate without part of the load acting to expel it; and so much more of the effective force will of course be lost. The effective power is as the space the steam occupies, multiplied by the excess of elastic force above the atmospheric pressure.

297.—*By Expansion.* Retaining the same loaded piston let it be raised by the conversion of a given quantity of water into steam, to the height which corresponds to the load and temperature; then if the load on the piston be wholly removed at that height, the steam will raise the piston by expanding till it becomes nearly of the same elastic force as the atmosphere, and its condensation will produce the same effect as if the steam had been generated of atmospheric elastic force at first; consequently, the effect in raising the load on the piston is wholly additional, and the joint effect of a high pressure and condensing engine is produced by the same steam. The effective power of steam applied in this manner is equal to the space it occupied, as high pressure steam, multiplied by the excess of its elastic force above the atmospheric pressure, added to the amount arising from multiplying the space it occupies when of atmospheric elastic force by the atmospheric pressure, Hence, by this combination of effect, the power of steam of high elastic force will be nearly doubled.

298.—This is not, however, the mode by which steam can be applied with the greatest advantage; for instead of removing the load on the piston wholly at the height to which it was raised by the generation of the high pressure steam, a part of it may be removed, and then the steam would expand to a height depending on the portion of the load removed; at that height remove a second portion, and so on, successively, till the steam becomes of atmospheric elastic force. In this case, as far as the load was raised in parts by the expansion of the steam, the effect is greater than in the preceding combination; the mode of calculating it will be afterwards shewn, but the principle is that of the expansion steam engines.

299.—The preceding is not the only mode of deriving advantage from expansion; indeed it is only a late discovery, and most probably belongs to Woolf as far as he was capable of understanding it. The methods of Hornblower and Watt only apply to the case now to be considered. Let the piston be raised unloaded, as in the first case, by the conversion of a certain quantity of water into steam of atmospheric elastic force. When the piston is at that height, add a weight equal to half the atmospheric pressure to the line passing over the pulley; then the elastic force of the steam being unbalanced, the piston would rise till that elastic force would be half the atmospheric pressure, or till the piston would be at double its former height. Now conceive the steam to be condensed, and the weight removed from the pulley at the same instant, and the power of the descent, less the power added to produce the ascent, will be one half more than by simply condensing steam of atmospheric elastic force; and even this ratio may be increased by adding the weight in portions to the line over the pulley, and diminishing the elastic forces of the steam. This is the principle of the expansion engines of Hornblower and Watt.

300.—It has been assumed that steam at least of atmospheric elastic force was generated, but this is not a necessary condition, for it frequently occurs that engines work with steam of less elastic force. The same mode of illustration will shew whence this happens. Let half the pressure of the atmosphere on the piston be balanced by a weight over a pulley. Then on the application of heat, steam of half the atmospheric elastic force would be generated, and raise the piston to double the height that it would be raised by steam capable of supporting the atmospheric pressure; consequently, on its being condensed, the descending force will be half the atmospheric pressure acting through double the height; and the steam produces the same effect as before.

We shall have occasion to shew the value of this principle in regulating the power of atmospheric engines.

301.—In all these illustrations of the modes of obtaining power from steam, I have taken the atmospheric pressure as one of the active forces; in some cases steam pressure is employed in practice, but the difference in employing this or that kind of pressure is dependent on other circumstances than its force, such as the rate of cooling and the like, and does not affect the relations of the forces of steam acting with only small alterations of temperature.

Of computing the Power of Steam to produce rectilinear Motion.

302.—If we suppose the force of steam in a cylinder to be equal to the mean pressure of the atmosphere, we may easily compute the power of the steam of a given quantity of water, as far as it possibly can be obtained by condensation, and not acting expansively. Thus the space occupied by steam of 212°, is 1711 times the bulk of the water which produces it, (art. 120.) when it is capable of resisting the mean pressure of the atmosphere, and that mean pressure is 2120 pounds on a square foot; hence, $1711 \times 2120 = 3,627,320$ pounds raised one foot by the steam of a cubic foot of water. Or multiplying by the area of a circle whose diameter is unity, we have 2,860,000 pounds raised one foot for the utmost power of a cylindric foot of water converted into steam. To deduct from this there is the waste, the friction of the piston, and the resistance of the uncondensed vapour. I shall not attempt at present to compute the extent of these deductions, for it would be premature, but shall give an analytical form to the calculation for the purpose of applying it to the expansive and other species of engines.

303.—If f be the force of the steam in inches of mercury, and t be its temperature the weight of a cubic foot in grains is

$$\frac{5700\ f}{459 + t}.$$

Now a cubic foot of water at the lowest temperature it is likely to be when condensed, will be 436,500 grains; hence,

$$\frac{436500\ (459 + t)}{5700\,f} =$$

the bulk of the steam when that of the water is unity; or

$$\frac{76 \cdot 58\,(459 + t)}{f}$$

Now neglecting the taking of an unit for the bulk of the water, we have $70 \cdot 75\,f =$ the force of steam on a square foot; and

$$70 \cdot 75\,f \times \frac{76 \cdot 58\,(459 + t)}{f} = 5418\,(459 + t) =$$

the pounds one cubic foot of water converted into steam, of the temperature t, would raise one foot high, without reduction for loss by friction and uncondensed vapour, or waste.

This conclusion that the power of steam is independent of its elastic force, is the same as resulted from the more popular mode of investigation, (art. 294.)

304.—But if f' be the force corresponding to the temperature of the condensed water, or of the condenser, then

$$\frac{5418\ (459 + t)\ f'}{f} =$$

the resistance.

For the condensed steam is limited to the space which the whole occupied in its elastic state, and therefore offers a resistance proportional to its force acting through that space. But we found the space

$$\frac{76 \cdot 58\,(459 + t)}{f};$$

and the force is $70 \cdot 75\,f'$; consequently, the resistance of the uncondensed steam is

$$\frac{5418\ (\ 459\ +\ t\ )\ f}{f}.$$

305.—For the present let the waste be $1 - w$, and the friction of the piston be denoted by F, then the power of the steam of a cubic foot of water of the temperature t, is

$$5418\,(\ 459\ +\ t\ )\,w - \frac{5418\,(459+t)\,(f'+\mathrm{F})w}{f} =$$

$$5418\,(\ 459\ +\ t\ )\ w\ \times\ \left(\ 1 - \frac{f'+\mathrm{F}}{f}\ \right).$$

306.—We may next ascertain the effect of *expansion;* which is easily computed, for, when the temperature does not sensibly alter during the action, the force of the steam is inversely as the space it occupies; therefore if b be its bulk, and p its force, and x any variable increase of bulk, and $\dot{x}$ its fluxion; then

$$b\ +\ x\ :\ b\ ::\ p\ :\ \frac{p\ b}{b+x}\ ;\ \text{and}\ \frac{p\ b\ \dot{x}}{b+x} =$$

the fluxion of the power developed in expanding through the space x; and the fluent of this quantity is $p\,b \times$ hy. log. $(\ b\ +\ x\ ) + \mathrm{C}$; but when $x = o$, $p\,b \times$ hy. log. $(\ b\ +\ x\ )$ $+ \mathrm{C} = o$; hence, the power is

$$p\,b\ \times\ \text{hy. log.}\ \frac{b\ +\ x}{b}.$$

But, to return to our previous notation, make $x = \overline{n - 1}\ b$; and $b =$

$$\frac{76\cdot58\ (\ 459\ +\ t\ )}{f}\ ;$$

and $p = 70\cdot75\,f$; consequently,

$$p\,b \times \text{hy. log.}\ \frac{b+x}{b} = 5418\ (\ 459\ +\ t\ )\ \times\ \text{hy. log.}\ n = \epsilon$$

= the additional power gained by the steam expanding.

307.—When the expansive principle is employed, that is, if the steam be to expand during its action on the piston, an increased length of cylinder becomes necessary; and the reduction of effect which must follow from this cause has been totally overlooked. If n be the bulk of the steam in its expanded state, when its bulk corresponding to the force f and temperature t is unity, then

Y

$$5418 \ (459 + t) \ w \times \left(1 - \frac{n \ (f' + F)}{f} \right) + \epsilon =$$

the mechanical power of a cubic foot of water, when ϵ is the additional power gained by employing the expansion of the steam.

On the value of ϵ being inserted in the equation it becomes

$$5418 \ (459 + t) \ w \times \left(1 + \text{hy. log. } n - \frac{n \ (f' + F)}{f} \right) =$$

the mechanical power of a cubic foot of water, when the expansive force of the steam is employed.

308.—This equation has a maximum, which will be when

$$\text{hy. log. } n - \frac{n \ (f' + F)}{f} =$$

a maximum. That is, when

$$n = \frac{f}{f' + F}.$$

Consequently, we shall have the greatest possible quantity of mechanical power when

$$\frac{f}{f' + F}$$

is inserted for n; or

$$5418 \ (459 + t) \ w \times \text{hy. log. } \frac{f}{f' + F} =$$

the mechanical power.

And where a table of hyperbolic logarithms cannot be conveniently referred to, the result may be obtained by multiplying the logarithm of

$$\frac{f}{f' + F}$$

found from the common tables of logarithms, by 2·302585, which will give the corresponding hyperbolic logarithm.

309.—In the best constructed engines, the waste of steam is not less than one-tenth; and, to get the extreme power of the steam of a given quantity of water at this rate of waste, we have $1 - \frac{1}{10} = w = \cdot 9$; and the equation becomes

$$4876 \ (459 + t) \times \text{hy. log. } \frac{f}{f' + F} =$$

the greatest possible power the steam of a cubic foot of water can afford, when acting expansively.

310.—In like manner taking the same loss by waste, we have from (art. 304.)

$$4876 \ (459 + t) \times \left(1 - \frac{f' + F}{f} \right) =$$

the greatest possible power of the steam of a cubic foot of water, when the expansive power of the steam is not used.   Consequently,

$$4876 \ (459 + t) \left\{ \left(\text{hy. log. } \frac{f}{f' + F} \right) - \left(1 - \frac{f' + F}{f} \right) \right\}$$

is equal to the gain by employing the expansive power.

311.—Though these equations shew us the limits of steam power, and are fittest for illustrating the advantages or disadvantages of difference of temperature, and elastic force, clearly exhibiting the economy of using steam of considerable elastic force, yet they still require to be applied to engines of different species.*   This will be done in Sect. V. and VI. but before I quit the illustration of general principles, it will be desirable to investigate the rotary action of steam.

Of computing the Power of Steam to produce Rotary Motion.

312.—In a great variety of the cases where steam is employed, a continuous circular motion is to be produced, and it is very generally imagined that a great advantage would be gained if the rotary motion were produced by the direct action of steam, instead of being obtained by the intervention of moving parts, for converting the rectilineal motion produced by steam into a rotary one.

But the fact of every person who has attempted to produce an engine acting by the rotary power of steam, having in a greater or less degree failed in rendering it as effective as a reciprocating engine, makes the theoretical principle of rotary action, an interesting subject of investigation.

* A series of tables calculated by these formulæ were published in my Treatise on Rail Roads, p. 161—166.

313.—Conceive a piston, D E, to be fitted to a regularly curved vessel A B, so that it may move round C, the centre of curvature of the vessel, and consequently the centre of motion. Now whether the piston be moved by the force of high pressure steam, or otherwise, the pressure on an inch of area of the piston, will be equal on all its parts; that is,

FIG. 16.

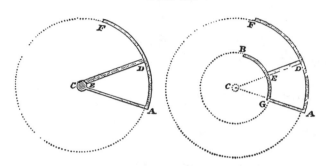

the pressure on an inch, at the most distant part D from the centre of motion, is the same as the pressure on an inch at the part E, nearest to that centre. But since the piston is constrained to move in a circle, the effects of these equal pressures are as their distances from the centre of motion, and limited by the effect of the pressure at the most distant part D. Hence, if the effective pressure of the steam be ten pounds on the inch, we have

$$D C : E C : : 10 : \frac{10 \times E C}{D C} =$$

the effect at E, that at D being 10. If the centre of curvature C, were nearer to the side of the vessel, the effect at E would be less; therefore, the effect of the pressure to produce motion, is less than in a straight vessel, having the same base; and if the bases be the same, the space the pressure acts through will be as the quantity of steam. Consequently, the quantities of steam being equal, the power of rotary action will be less than that of rectilineal action.

314.—If a rectangular piston, D C, revolve round a centre C, then nearly half the power of the steam will be lost.

This rough inquiry will be sufficient to shew that much is lost by attempting to employ the rotary action of steam, besides the various other objections arising out of the excess of friction, and the difficulties of executing the parts so as to act properly; usually called practical difficulties.

315.—To conduct the inquiry so as to reduce the effect to more accurate measures; put

$d = $ D E the diameter of the piston ;

$r = $ E C, the radius of the interior circle;

$x = $ any variable portion of the diameter of the piston counted from E ;

$y = $ the breadth of the piston, and

$f = $ the force of the steam on an inch of area.

Then $r + d = $ D C; and as

$$r + d : r + x : : f : \frac{f(r + x)}{r + d} =$$

the force at any point at the distance x from E; and

$$\frac{f y (r + x) \dot{x}}{r + d} =$$

the fluxion of the pressure at that point; and the space described being $2 p (r + x,)$ we have

$$\frac{2 p f y (r + x)^2 \dot{x}}{r + d} =$$

the fluxion of the power.

When y is constant the fluent is

$$\frac{2 p f y (r + x)^3}{3 (r + d)} + C;$$

and making this equation nothing, when, $x = o$, that is when the power is nothing, we have

$$\frac{2 p f y}{3 (r + d)} \times (3 r^2 x + 3 r x^2 + x^3);$$

and when $x = d$,

$$\frac{2 p f y d}{r + d} \times (r^2 + r d + \tfrac{1}{3} d^2) =$$

the power of the steam acting in a rotary direction, the piston being a rectangle $d y$.

316.—If the piston D C revolve on an axis in the centre C, then $r = o$, and

$$\frac{2 p f y d^2}{3} =$$

the rotary power

But the space occupied by the steam is $p\,d\,(\,2\,r + d\,)\,y$, and its rectilineal power is $p\,f\,d\,(\,2\,r + d\,)\,y$. Hence the rectilineal is to the rotary effect of the steam as

$$(\,2\,r + d\,) : \frac{2\,(\,r^2 + r\,d + \tfrac{1}{3}\,d^2\,)}{r + d};$$

or as

$$2\,r^2 + 3\,r\,d + d^2 : 2\left(\,r^2 + r\,d + \tfrac{1}{3}\,d^2\,\right).$$

When $r = o$, or the piston revolves on a centre, then the ratio becomes $3 : 2$, or one-third of the power is lost; the same conclusion resulting however the steam acts.

317.—We have supposed the piston to be of parallel width, but in some schemes it has been made circular; and in such a case the valve of $\tfrac{1}{2}\,y$ is $\sqrt{d\,x - x^2}$. Consequently,

$$\frac{4\,p\,f\,(\,r + x\,)^2\,\sqrt{d\,x - x^2}\,x}{r + d} =$$

the fluxion of the rotary power. Its fluent is

$$2 \times \cdot7854\,p\,f\,d^2\left(\frac{r^2 + r\,d + \frac{5}{16}\,d^2}{r + d}\right) =$$

the power when $x = d$.

This is a little less than the effect of a rectangular piston. When the piston revolves round an axis in its edge the rectilineal power of a given quantity of steam is to its rotary power as $3\cdot2 : 2$. In the rectangular one it was as $3 : 2$. Hence, we see there is no possibility of applying steam with the same advantage in a rotary, as in a rectilineal engine; and, even to approximate to it, the radius of the circle described must be great in comparison with the diameter of the piston, and consequently difficult to execute. To employ any other than a circular form for the piston, would cause more friction, and expose a larger portion of surface to the cooling effect of the atmosphere. These are radical objections to the rotary action of steam that cannot be removed by art.

318.—It is so obvious, that it is not necessary to shew, that the impulse of steam cannot be employed without great loss of fuel, we may, however, take a general view of the modes in which the action of steam may be applied.

Modes of applying the Power of Steam.

319.—The arrangement being presented in a tabular form will be more clear than in continued description; the modes of obtaining the different species of power, and the measures of their effects, have already been explained; for condensation in (art. 291.) for generation in (art. 295.) and for expansion in (art. 297.)

320.—The action of steam as a moving force is derived from
{
1. the generation of steam. (Worcester.)
2. the expansion of steam. (Hornblower.)
3. the condensation of steam. (Savery.)
}

Of the species of action there may be used
{
separately { generation.
condensation.
or jointly { 1 and 2.
1 and 3. (Savery.)
2 and 3. (Hornblower.)
1, 2, and 3. (Woolf.)
}

The action may be { by pressure
by impulse.

The action may be exerted on { a solid, (Newcomen,)
a fluid, (Worcester,) } it may be { continuous.
successive.

The motion of the surface acted upon may be in a { straight line.
curved line.

321.—The pressure of steam is the kind of action which is employed in practice; and the reasons for giving the preference to rectilineal motion have been shewn, (art. 317.) In order that it may be economically employed, it is found that a solid is best adapted to receive its action, fluids being liable to decompose by contact with hot steam, or to condense and waste the steam. And, in consequence of a cylinder being the figure adapted to the object, and possessing the greatest capacity with the least surface, and therefore having the least loss both by cooling and by friction, it is almost universally employed. The action is necessarily successive to render it rectilineal; but all the species of action are used either jointly or separately. From these species of action therefore the engines may be classed.

322.—It will be remarked that steam must be either condensed, or generated under pressure, to afford power by expansion; hence, engines may be divided into two classes, depending on condensation being used, or not; this arrangement being most convenient.

Classification of Steam Engines.

323.—I.   Non condensing engines acting by the { 1. generative power of steam.
                                                  { 2. generation and expansive power of steam.

324.—II. Condensing engines acting by the { 1. condensation of steam.
                                             { 2. condensation and expansion of steam.
                                             { 3. generation and condensation of steam.
                                             { 4. generation, expansion, and condensation of steam.

325.—All the engines of the first class, and the third and fourth kinds of the second class, require high pressure steam.   Engines of the first class are remarkable for simplicity of construction, but they never give the whole of the power of the steam.   Engines of the second class require a considerable quantity of cold water for condensation, and therefore in some cases cannot be applied.   The greatest effect is obtained by the second and fourth kinds of the second class; or rather it is only in these two species, that the whole power of the steam is obtained.

326.—In both classes there are certain proportions between the length of the stroke, and the diameter of the cylinder, and between the length of the stroke and the velocity; which give a maximum of useful effect to a given quantity of steam.   These being considered, and also the proportions of the additional parts required in condensing engines, the general rules for the power of engines may be given.

Of the Ratio between the Length of the Stroke, and the Diameter of the Cylinder.

327.—The relation between the diameter of a steam cylinder and the length of the stroke, and consequently the proportions of the cylinder, have now to be considered.   If all the apertures, and all other parts be duly proportioned, and the velocity regulated so as to be esteemed uniform, then there is no circumstance relating to the motion, which has any influence on the proportions of the steam cylinder, excepting the small difference arising from the friction not increasing exactly in the same proportion as the square of the diameter; and this difference is so small in ordinary proportions that we may safely neglect it.

328.—The only other circumstance which renders it necessary to attend to the proportions of a cylinder is, the quantity of cooling surface to which the steam is exposed during its action. This surface ought to be the least possible; for its effect in condensing, and therefore destroying the power of the steam is considerable. (See art. 156.)

The quantity of surface consists of one end of the cylinder, one side of the piston, and the concave surface of the cylinder; but the latter is only gradually brought in contact with the steam during the stroke, and its effect, therefore, only equivalent to half the effect on an equal surface bounding the steam during the whole of the stroke. Now the power of an engine is greatest when the effect of a given quantity of steam is the most possible; hence, the question is, to find the least surface capable of confining a given quantity of steam during its action.

329.—When the length of the stroke is twice the diameter of the cylinder, a given quantity of steam is bounded by the least possible quantity of surface during its action in the cylinder;* hence, I conclude it is the best proportion for the cylinder of a steam engine,

---

* Let the diameter of the cylinder be x, its length l, its capacity C, and $p = 3\cdot1416$. Then,

$$C = \frac{p\, l\, x^2}{4};$$

and therefore

$$l = \frac{4\,C}{p\,x^2}$$

Now the sum of the areas of the bases is

$$\frac{p\,x^2}{2};$$

and the area of half the concave surface is

$$\frac{p\,l\,x}{2} = \frac{2\,C}{x}\;;$$

hence, the whole surface of the steam exposed to cooling surfaces during its action is

$$\frac{p\,x^2}{2} + \frac{2\,C}{x}\;;$$

and this surface is to be a minimum, which is determined by taking its fluxion and making it equal to zero. That is,

$$p\,x\,\dot{x} - \frac{2\,C\,\dot{x}}{x^2} = 0;$$

whence,

z

except when the space for the engine limits the length of the stroke; and the same conclusion applies to both atmospheric and steam pressure engines.

330.—If we refer to the practice of engine makers, we find no indication of a settled rule for the proportions of the cylinder, when the length of the stroke is unlimited by convenience. The proportions followed at different times by Boulton and Watt, in cases where the stroke was not limited, vary from 1 ¾ to nearly 3 to 1, the most common about 2·7 to 1, the changes having no regularity. In Smeaton's table of the proportions of atmospheric engines*, the length of the stroke is made to vary nearly as the square root of the diameter, and commences at the lower part of the scale with the proportion of 4 to 1; why the square root of the diameter was fixed upon does not appear. Equally irregular are Maudslay's proportions but approaching to 2 to 1; Fenton, Murray, and Wood's about as 2 ½ is to 1. The object seems to have been to render the velocity nearly the same in all engines; the circumstances which regulate the velocity may therefore next be considered.

Of the Maximum of useful Effect in Steam Engines.

331.—In steam engines there is a certain velocity for the piston which gives a maximum quantity of useful effect.

In an engine already constructed, the velocity which gives the most useful effect that the engine is capable of producing, is limited by the proportions which have been given to the parts of the engine.

But in an engine to be designed, all the parts should be arranged to agree with the velocity which gives the maximum effect of a given quantity of steam : the difference between these cases is considerable; but in illustrating each by example, I shall have an opportunity of shewing that a general rule could not be derived from experiments on a particular engine.

$$x^3 = \frac{2\,C}{p};$$

and substituting for C its equal, we have $2\,x = l$; or a cylinder is of the best proportion when its length is twice its diameter.

Rees's Cyclopædia, Art, Steam Engine.

Of the Maximum for Engines equalized by a Fly.

332.—Our most simple case for consideration is that where the pressure on the piston is the same throughout its stroke; and we must suppose the fly, conjointly with the mass of matter in the engine, to be so proportioned as to render its velocity as nearly as possible uniform.

Then the greatest uniform velocity the engine could possibly acquire, would be equal to half that which a falling body would acquire in descending the length of the stroke; and with this velocity the work done would be nothing; as the whole force of the steam would be expended in keeping the engine moving at that velocity.

It must be evident that a regulated or uniform velocity cannot be greater than half the velocity a falling body would acquire in descending the length of the stroke, because with any other velocity the mass moved would not be capable of receiving and imparting equal quantities of motion in equal times, a circumstance essential to the uniform motion of an engine moved by an uniform force.

333.—As at the greatest possible velocity an engine has no useful power, and, on the other hand, if the resistance be equal to the pressure of the steam, it will have no velocity, there must be an intermediate velocity which is the best possible for the engine to work with; and this velocity is one-half the greatest uniform velocity.* Now the velocity a falling body would acquire in descending through the length of the stroke is equal to eight times the square root of the length of the stroke, in feet per second; therefore the velocity which corresponds to the maximum of useful effect, being one-fourth of this velocity, is twice the square root of the length of the stroke in feet per second, and 120 times the square root of that length in feet per minute.

334.—Hence, for engines regulated by a fly, if the pressure on the piston were the same throughout the stroke, the best velocity for the piston in feet per minute would be 120 times the square root of the length of the stroke in feet.

---

* Let V be the greatest uniform velocity; m the force producing it; and $w =$ the mass of matter by which it is rendered uniform ; v being any other velocity. In this case $m\,v - w\,v =$ the effective action ; and since $m\,v = V\,w$, it is

$$\frac{m\,V\,v - m\,v^2}{V} =$$

the effective part which is to be the greatest possible.

The fluxion of the variable part is $V\,\dot{v} - 2\,v\,\dot{v} = o$ when the expression is a maximum ; whence, $V = 2\,v$.

That is, if the length of the stroke be four feet, the square root of four being two, the velocity for a four feet stroke is $2 \times 120 = 240$ feet per minute; but the action of the valves not allowing this perfection, almost all engines belong to the next case.

335.—If the steam act expansively the velocity must be less, because the pressure on the piston varies, and the uniform motion the steam would generate in the length of the stroke would be less.

336.—In a steam engine where the steam acts expansively, the supply of steam being cut off at the

$$\frac{1}{n}$$

part of the stroke, the best velocity for the steam piston will be found by multiplying the

$$\frac{1}{n}$$

part of the length by ·7 added to 2·3 times the logarithm of n; then 120 times the square root of the product is the velocity in feet per minute.*

Example. Let the steam be cut off at one-fourth of the stroke, then $n = 4$; and let the length of the stroke be eight feet. The logarithm of n is 0·60206; therefore, $\overline{0·60206 \times 2·3} + ·7$ is 2·0845, which multiplied by one-fourth of the length, or 2, is 4·169.

* It has been shewn (art. 306,) that the expanding power of steam is

$$p\, b \times \text{hy. log. } \frac{b + x}{b},$$

which added to the uniform portion, and the resistance r from friction and uncondensed vapour being subtracted, it is

$$p\, b \left(1 + \text{hy. log. } \frac{b + x}{b} - r \right) =$$

the power; but $b + x = l$ the length of the stroke, and

$$b = \frac{l}{n};$$

hence,

$$\frac{p\, l}{n} \left(1 + \text{hy. log. } n - r \right) =$$

the power. In ordinary circumstances $r = ·3$, consequently,

$$120 \sqrt{\frac{l}{n} \left(·7 + \text{hy. log. } n \right)} =$$

the velocity in feet per minute.

The square root of 4·169 is 2·04; consequently, 2·04 × 120 = 245 feet per minute, the velocity for an eight feet stroke when the steam is cut off at one-fourth of the stroke.

337.—In the usual construction of engines not intended to act expansively, put $n = \frac{1}{4}$; and then 103 times the square root of the length of the stroke.

For steam engines working expansively at the ordinary pressure of about eight pounds on a circular inch of the safety valve of the boiler, the best proportion for cutting off the steam is about half the stroke, and then the rule becomes 100 times the square root of the length of the stroke in feet, for the best velocity in feet per minute for the steam engine.

Other general rules may be easily derived from the investigation in the notes.

338.—In single acting engines, regulated by a fly, the same relation would obtain between the length of the stroke, and the velocity of the piston; but such engines cannot be used with advantage for producing a continuous motion.

Of the Maximum of useful Effect in Engines for raising Water.

339.—In single engines for raising water we have two strokes to consider, of different species; the piston being caused to ascend by a counter weight, which should be capable of raising the piston in a short time without adding materially to its load in the descent: and the descending stroke should not be slower than gives the maximum of useful effect, because in both cases, a considerable loss of the power of the steam is taking place during its action.

340.—In the ascent of the piston it must be evident that it should never acquire a greater velocity than one which the steam can follow, so as to press it with a force nearly equal to the pressure of the atmosphere; and when the apertures for the steam are arranged for the descending stroke, the ascending one will be regulated by the passage of the steam through the same apertures, and if this be the case, and in the present construction of these engines it always is so, our inquiry may be confined to the descending stroke.

341.—The descent of the piston, if the effect of the steam alone were considered, it is obvious, should be determined by the condition which gives the greatest effect by a given quantity of steam; but it is dependent on the resistance of the water increasing as the square of the velocity, and the decreasing effect of the steam in the simple ratio of the increase of the velocity.

342.—Now we may be allowed to consider the motion an uniform one, as it is nearly so during the greater part of the stroke, and then when the steam acts at full pressure during the whole of the descent, the velocity in feet per minute should be ninety-eight times the square root of the length of the stroke.*

343.—When the steam acts expansively, the velocity may be found from that of an expansive engine regulated by a fly, (art. 336 to 338.) as 0·8 times that velocity will be the proper velocity for an engine for raising water.

Thus we found the velocity for an eight feet stroke, in an expansive engine where the steam was cut off at one-fourth of the stroke, to be 245 feet per minute, (art. 336.) and 0·8 of 245 is 196 feet per minute.

In these investigations I have not attempted to enter into those minute particulars which embarrass the calculation, without producing any material effect on the result.

Of the Proportions of Air Pumps and Condensers for Steam Engines.

344.—The water used for producing steam, and for condensing it, contains a considerable quantity of air, and sometimes carbonic acid, and other gases. These gases se-

Suppose the arms of the beam to be of equal length; the steam apertures being the same, the ascending and descending strokes should be made in equal times. The greatest possible velocity V will be generated when the resistance to the water in the pumps is equal to the counter weight; and as the forces in both directions are to be equal, if m be the force producing the motion; $\frac{1}{2} m$ = the resistance at the velocity V; and the resistance to motion in pipes being as the square of the velocity,

$$\frac{m\,v^2}{2\,V^2} =$$

the resistance at v, to which the counter weight must be equal; consequently,

$$m\,v - \frac{m\,v^3}{V^2} =$$

the effective power which is to be the greatest possible.

The maximum takes place when $V^2\,\dot{v} - 3\,v^2\,\dot{v} = o$, that is when $3\,v^2 = V^2$, or $v = \cdot577$ V. To determine V, it may be remarked, that the motion commences with an excess of power, which diminishes by the increase of resistance, till the motion becomes uniform; that the area of the passages of the valves are only half the area of the pump, and that the mass of matter moved is twice the excess of moving force; hence, $V^2 = 8\,s$, and since $3\,v^2 = V^2$, it is

$$v = \sqrt{\frac{8\,s}{3}} = 1\cdot633\,\sqrt{s}.$$

When the velocity is in feet per minute, it is

$$60 \times 1\cdot633\,\sqrt{s} = 98\,\sqrt{s}.$$

parate when water is boiled, and rise with the steam, hence, were there not some method provided to take the air away when the steam is condensed, the cylinder of a steam engine would become filled with hot air so as to impede, and in the end resist, the pressure of the steam.

345.—To estimate, therefore, the proper size for an air pump, the quantity of air or other gas contained in water should be known.

Experiments on this subject have been made by Mr. Dalton,* Dr. Henry, M. Saussure, and Dr. Ure. M. Saussure,† ascertained that boiling alone was not capable of freeing liquids completely from air, but that it may be done by the joint action of heat, and the air pump. In a steam engine both these causes operate in extracting air from the water introduced into the engine. According to his experiments 100 volumes of water absorb about five volumes of atmospheric air.

In an experiment made by Dr. Henry‡ on spring water, he found that it afforded by boiling 4·74 per cent of gaseous matter; of which 3·38 per cent was atmospheric air, and 1·38 per cent carbonic acid; but as it is probable that this water was fully saturated, it follows from Saussure's remarks, that a greater proportion would have been obtained if it had been subjected to the combined action of boiling and the air pump; and the whole proportion of gaseous matter in spring water could not be estimated at less than seven per cent.

Dr. Ure's‖ experiments were also made by boiling different kinds of water, and measuring the result by a pneumatic apparatus, and at the temperature of about 55°; the proportions of gaseous matter in 100 volumes of water were found to be as stated below.

Canal water (in winter) - - - - - - - - -	2·67 per cent
Filtered river water supplied to Glasgow by the pipes of the Cranstonhill Water Company -	2·52
Filtered river water from the pipes of the Glasgow Water Company - - - - - -	2·50
Water from the river Clyde, when swollen by winter rains - - - - - - - - - -	2·80

It cannot be supposed that in these experiments the whole of the gaseous contents of the water were obtained, but assuming that two-thirds of the total gaseous contents are obtained by boiling, the quantity will vary from 3·75 to 4·2 per cent; and therefore, for river

---

* Philos. Mag. Vol. XXIV.
‡ Thomson's Chemistry, Vol. III. p. 204.
† Annals of Philosophy for 1815, Vol. VI. p. 329.
‖ Quarterly Journal of Science, Vol. XXI. p. 71.

and canal water, we may assume that water contains five per cent of air, or $\frac{1}{20}$ of its volume.

The action of pumping it appears, from Dr. Ure's researches, expels a portion of air from water.

346.—The preceding articles afford us data sufficiently accurate for the general purposes of inquiry, and in such inquiry we may suppose that, at the mean temperature and pressure,

> River or canal water contains $\frac{1}{10}$ of its volume of gaseous matter,
> Spring or well water　-　-　$\frac{1}{14}$

347.—The quantity of water which enters into a steam engine will all of it give out nearly the whole quantity of air it contains; therefore, calculating the volume of water used for steam at each stroke of the engine, and adding to it that used for injection in the same time, we have one-twentieth part for the volume of air at 60°; but in the condenser it will be of the temperature of the hot well, or about 120°. and the quantity the air expands by this increase of temperature being calculated, (see art. 119.) the bulk is found to be 5·6 per cent of that of the water, or $\frac{1}{18}$ of the bulk of the water nearly.

348.—Let the injection water added to the water of the condensed steam be　　of the volume of the cylinder for each stroke, then $\frac{1}{18}$ of $\frac{1}{20}$, or $\frac{1}{360}$ of the cylinder's volume of air, would accumulate at each stroke if there were no air pump. Now a cubic foot of air mixes with a cubic foot of steam, when both are of the same force and temperature; (art. 122.) consequently, this air must accumulate and fill half the capacity of the condenser after a few strokes, and the capacity of the air pump must be such as will remove $\frac{1}{360}$ of air, and $\frac{1}{360}$ of vapour, equal $\frac{1}{180}$, in order to get rid of the air which enters at each stroke, when it is of such a degree of density that its force is equal to the force of steam corresponding to the temperature of the hot well.

349.—If it be assumed that the elastic force of the uncondensed vapour is equal to two inches of mercury, and the air pump be equal to the condenser, then the bulk of the air and vapour being shewn above to be $\frac{1}{180}$ at the pressure of thirty inches, and the bulk being inversely as the pressure, we shall have $2 : 30 : : \frac{1}{180} : \frac{1}{12}$.

Now the air pump has to clear the engine of $\frac{1}{12}$ of the volume of the cylinder of air and vapour, and of $\frac{1}{35}$ of its volume of water; the sum of these is, in the nearest fraction, $\frac{1}{18}$ of the capacity of the cylinder for each stroke.*

* Let

$$\frac{a}{n}$$

be the volume of air contained in the injection water and the steam, t' = the temperature of the condenser, and f' the force of steam corresponding to it, f being the force in the cylinder. Then

350.—In a double engine the air pump makes only one stroke for each cylinder full of steam, but since the condenser receives a new quantity to replace that taken by the pump there is no expansion: hence, $\frac{1}{18}$ part of the capacity of the cylinder of a double engine is the least proportion for the air pump, so that the engine may work effectively in the same manner as for a single engine. In both cases the condenser and air pump are supposed to be of equal size to render this proportion applicable; and that river water is used.

351.—For well water the same mode of calculation gives about one-twelfth for the relation between the capacity of the air pump and the cylinder. The usual proportion in Boulton and Watt's practice is one-eighth; and as I have made no allowance for leakage nor imperfect action of valves, this proportion appears to be nearly correct for the case considered.*

352.—There is one thing very evident in this operation. It is that an air pump half the size would be as effective as the present construction, if we could condense in the pump itself; and I see no difficulty in doing so, and propose to shew its application to a simple atmospheric engine. (See art. 400.) The advantage, however, will be better understood if we shew the power an air pump requires to work it.

Power required for working an Air Pump of a Steam Engine.

353.—Let $v =$ the velocity in feet per second; $p' =$ the force of the steam or vapour

$$\frac{f}{f'}\frac{a}{n}\left(\frac{459 + t'}{511}\right) =$$

the volume of air, and $a =$ that of water for each stroke.

The condenser must contain both these quantities, and also what the pump leaves; and with an allowance of half for leakage and imperfect action of the valves, its least capacity must obviously be

$$\frac{3fa}{f'n}\left(\frac{459 + t'}{511}\right) + 2a = 3a\left(\frac{f}{f'n}\left(\frac{459 + t'}{511}\right) + \cdot 67\right).$$

When the pump ascends the air will saturate with vapour, and become of twice its former volume, hence, if the air pump and condenser jointly contain it in its state, they will be of equal size, and the quantity required will be removed at each stroke of the pump. Putting $= 100, f' = 2$ in. and $f = 30$, we have

$$3a\left(\frac{16\cdot5}{n} + \cdot 67\right) =$$

capacity of air pump $=$ condenser.

If $n = 20$, as for river water, then $4\cdot48\ a =$ capacity of pump.

If $n = 14$, as for well water, then $5\cdot55\ a =$ capacity of pump.

* In some instances air pumps for double engines have been made about two-thirds   the diameter of the steam cylinder, and half the stroke; such pumps are undoubtedly too large.

A A

in pounds per circular inch; $r =$ the friction of the piston and piston rod and resistance of the valves; $a =$ the diameter of the pump in inches; and

$$\frac{a^2}{n} =$$

the area of the valves. The head capable of producing the velocity $n\,v$ through the valves, (see art. 136,) is $n\,v = 6\cdot5\sqrt{h}$; and

$$\frac{n^2\,v^2}{42} = h.$$

In a mixture of air and steam, at the mean force in such a pump, twenty-one feet in height is equivalent to a pressure of one pound per circular inch; hence,

$$\frac{n^2\,v^2}{21 \times 42} =$$

pressure in pounds, $=$

$$\frac{n^2\,v^2}{882}.$$

Put $l =$ the length of the stroke:

The resistance to the *aescent* of the piston will be

$$r\,a^2\,l + \frac{n^2\,v^2\,a^2\,l}{882} = a^2\,l\left(r + \frac{v^2\,n^2}{882}\right).$$

The resistance to the *ascent* of the piston will be found by considering that the air and vapour is compressed till its elastic force becomes of such an excess above the atmospheric pressure, that it escapes through the valve at the velocity corresponding to the motion of the piston. The friction of the piston and weight of the water is to be added; and the force of the vapour in its expanded state may be considered equal to the sum of the forces necessary to cause it to pass the valves.

By (art. 306,) the resistance of the air and vapour is

$$p\,b \times \left(1 + \text{hy. log.}\ \frac{b+x}{b}\right);$$

and making $b + x = l$, we have $b =$

$$\frac{p'\,l}{p},$$

therefore,

$$l\,p'\left(1 + \text{hy. log. } \frac{p}{p'}\right) =$$

the power; and when the pressure of the atmosphere $p = 11\cdot55$ lbs. and the force of the vapour $p' = \cdot77$ lbs. or two inches of mercury, then

$$l\,p\left(1 + \text{hy. log. } \frac{p}{p'}\right) = 2\cdot85\,l,$$

very nearly. The quantity of water will be one-sixth of the capacity, or $\cdot055\,l^2\,a^2$ pounds raised one foot. Hence, the whole power required for the ascending stroke will be

$$a^2\,l\left(2\cdot85 + \cdot055\,l + r\right).$$

354.—The whole power to work the pump is therefore

$$\frac{a^2\,v}{2}\left(2\cdot85 + \cdot055\,l + \frac{n^2\,v^2}{882} + 2\,r\right) =$$

the pounds raised one foot per second.

Example. Let the velocity be $1\cdot8$ feet per second; the diameter of the pump twenty-four inches; the length of its stroke four feet; the friction two pounds per circular inch; and the area of the valves half the area of the pump. These numbers inserted we have

$$\frac{24 \times 24 \times 1\cdot8}{2}\left(2\cdot85 + \overline{\cdot055 \times 4} + \frac{2 \times 2 \times 1\cdot8 \times 1\cdot8}{882} + 4\right) =$$

$$518\cdot4\left(2\cdot85 + \cdot22 + \cdot0146 + 4\right) = 3670\,\text{lbs.}$$

raised one foot per second.

As 550 pounds raised one foot per second, is the steam engine horse power, hence,

$$\frac{3670}{550} = 6\tfrac{1}{4}$$

horses' power nearly. The pump would answer for a double engine of about 134 horses' power; therefore in this case about one-twentieth part of the power of the engine is required for the air pump; or one-tenth in the case of a single engine of the same sized cylinder; or a loss equivalent to thirteen horses' power in an air pump of the size in the example. To reduce this loss one-half by the mode proposed (in art. 352,) is certainly worthy of attention.

355.—It is important to remark the circumstances which contribute to this loss of power.   The loss is proportional to the capacity of the pump, therefore the smaller it is the better, provided it be sufficient to take the air.   The friction is four-sevenths of the power, the actual resistance of the vapour nearly three-sevenths, and that of the water about one thirty-second.   The resistance is greater the smaller the passages and valves are, but such increase does not affect the whole power in a material degree.   The increase of the size of the air pump, beyond the proportions I have given, can give advantage only in an ill constructed and leaky engine; but its decrease, after a very short range, reduces the power considerably.

SECTION V.

OF THE CONSTRUCTION OF NONCONDENSING ENGINES.

356.—Noncondensing engines, usually called *high pressure engines*, are moved by steam generated under a considerable degree of pressure, and it is the excess of this pressure above the pressure of the atmosphere, which constitutes their power to produce motion. From thirty to forty pounds on a circular inch is the excess above atmospheric pressure, commonly employed in this country.

357.—The working parts of the engine consist of a cylinder, having passages provided with cocks or valves for steam to enter into it, either at the top or at the bottom; and also the means of letting out the steam to the atmosphere, either at the top or bottom. The cylinder has an air-tight piston, to be moved from one end to the other by the pressure of the steam, with a rod fixed to it, called the piston rod, which slides through an air-tight box at the top of the cylinder, to give motion to a crank or some other piece of machinery.

358.—Now, with steam in the boiler having a force of thirty pounds to the circular inch, if the piston be at the bottom of the cylinder, and the passage from the boiler to the bottom, and that to the atmosphere at the top, be both open, and the rest shut, the steam will exert a pressure of nearly thirty pounds on each inch of the area of the piston, and cause it to ascend. A little before it arrives at the top the cocks must be shut, and the moment it has got to the top the other two cocks should be opened. The steam from the boiler will then press the piston downwards and the steam before let in will flow out into the open air. Again the passages must be closed a little before the completion of the stroke, and in this manner the operation may be continued.

359.—The close of the cocks before the termination of the stroke prevents either concussion against the end of the cylinder, or strain on the crank shaft, and when properly managed the elasticity of the steam destroys the momentum of the piston, and recoils it back without loss of force.

This will afford the reader a general notion of the action of steam in noncondensing engines, and prepare him for entering more closely into their minutiæ. I have divided them into two kinds; and of the varieties depending on different forms of construction there is an immense number.

360.—Noncondensing engines.

| Acting by | 1. the generative force of steam. | Leupold, (art. 12,) 1720. Watt, (art. 26,) 1769. Trevithick, (art. 56,) 1802. Evans, (art. 58.) |
| | 2. the generative and expansive force of steam. | Oliver Evans, (art. 58.) Taylor and Martineau. |

361.—*First Species.* When the power is derived solely from *generating the steam under pressure,* (art. 295,) the construction of the parts constituting the engine is very simple. The common method is represented in Fig 1, Plate IV. With the object of losing as little heat as possible by the cooling of the cylinder, it is generally placed partly within the boiler, and the steam is admitted and let out by a four-passage cock A, placed just without the boiler, with a throttle valve V to regulate the entrance of the steam. The steam escapes to the atmosphere by a pipe E, which is generally surrounded by water W, for the supply of the boiler, which has the effect of partially condensing the escaping steam, and facilitating its escape from the cylinder, as well as of increasing the temperature of the water before it be admitted to the boiler.

362.—This construction is defective, in as far as there must be an absolute loss of all the steam in those parts of the admission pipes which are between the cock and the cylinder, and the great density of high pressure steam renders the loss of power considerable. To avoid it, there should be two double-way cocks, one at the bottom and one at the top of the cylinder : or the passages may be opened and closed by a slide, as shewn in Fig. 2. where it will be obvious, that the spaces between the stops and the cylinder are as small as possible.

363.—If we now trace the action of the steam, and the opening of the passages, we shall find to what points to attend in perfecting the operation of the engine. In Plate IV. Fig. 1. represents an engine, of which C is the cylinder, and P the piston at the top ready for descending. The motion of the cock A might end with the end of the stroke, but the steam would be cut off, and indeed all the passages stopped when it is half turned. The closing, when quickly done, commences sufficiently before the end of the stroke, to effect the recoil of the piston; (art. 359.) and at the instant of its change of motion the steam is fully on it. The compression which the steam left in the cylinder receives when the cock is closed, is not only a means of changing the motion without loss of force, but also occupies the space at the end of the stroke, so as to require only a small quantity to refill it with steam. We might arrange the motion so that the cocks would be half turned, and all the passages closed just at the end of the stroke; this, however, would not be so good a method, as when the cock turns with proper quickness, there would be no sensible accumulation of steam to recoil the piston, and the force of that in the boiler would not be fully on, till a part of the stroke was made, and the waste at the terminations of the strokes would be greater. Hence, to complete the motion of the cock with the termination of the stroke, is the better method.

364.—In the construction, Fig. I. at every double stroke there is a loss of the force of all the steam contained in the passages between the cock and the cylinder. This defect may be avoided by the use of the slide, Fig. 2, and 3. The motion of the slide should terminate with the stroke in the same manner as with a cock, and in this construction the recoil of the compressed steam is greater, because it has less space of passage to retreat into. Valves may be placed to give similar advantages, but slides or cocks are in my opinion better adapted to high pressure engines.

365.—The modes of giving motion to the cocks, slides, or valves are various; they depend chiefly on the nature of the action the engine is intended for. The same methods are applicable to engines of all species, and therefore are described together, (see Sect. VII.) The power is usually regulated by a throttle valve; but more perfectly by means of Field's valve, (see Sect. VIII.)

366.—*The Proportion of Parts.* The length of the stroke of the steam piston should not if possible be less than twice its diameter, (art. 327.) The velocity in feet per minute should be 103 times the square root of the length of the stroke in feet, (art. 337.) And, as 4800 is to the velocity thus found, so is the area of the cylinder, to the area of the steam passages, (art. 154.) The strength, proportions, and construction of the parts are given in Sect. VII. and the methods of equalization and regulation in Sect. VIII.

367.—*The Power of a Noncondensing Engine* may be calculated with considerable accuracy, from knowing the excess of the force of the steam in the boiler, above the atmospheric pressure, as shewn by the steam gauge, the diameter of the cylinder, and the velocity of the piston. The effective pressure on the piston is less than the force in the boiler when that force is represented by unity,

First, by the force producing motion of the steam into the cy-

    linder, (art. 154.) - - - - - - - - - -   ·0069

Second, by the cooling in the cylinder and pipes, (art. 158.)   ·016

Third, by the friction of the piston and waste - - - - -   ·2000

Fourth, by the force required to expel the steam into the at-

    mosphere, (art. 154.) - - - - - - - - - -   ·0069

Fifth, by the force expended in opening valves and friction of

    the parts of the engine - - - - - - - - - -   ·0622

Sixth, by the steam being cut off before the termination of the

    stroke, (art. 363.) - - - - - - - - - - -   ·1000

                                               —

                                           ·3920

We may consider this 0·4, and then the effective pressure is 0·6 of the force of the steam in the boiler, diminished by the pressure of the atmosphere, whence we have the following rule for the power of an engine of this species.

368.—RULE. *For noncondensing engines working at full pressure.* Multiply six-tenths of the excess of the force of the steam in the boiler, less four-tenths of the pressure of the atmosphere in pounds on a circular inch, by the square of the diameter of the cylinder in inches, and by the velocity of the piston in feet per minute. The product is the power of the engine in pounds raised one foot high per minute.*

To find its equivalent in horses' power divide by 33,000.

Example. Let the diameter of the cylinder be eleven inches, and the length of the stroke 2·5 feet, the number of strokes per minute thirty-three, and the force of the steam in the boiler twenty-four pounds per circular inch above the atmospheric pressure. In this case the velocity is $2 \times 2 \cdot 5 \times 33 = 165$ feet, and $\overline{(24 \times 0 \cdot 6 - \overline{11 \cdot 5 \times \cdot 4})} \times 121 \times 165 = 195,657$ lbs. raised one foot per minute.

Also

$$\frac{195657}{33000} = 6$$

horses' power very nearly.

---

* Put d = the diameter of the cylinder in inches, v = the velocity of the piston in feet per minute, and f = the force of the steam in the boiler in inches of mercury; then

$$\frac{0 \cdot 6 f - 30}{2 \cdot 6} \times d^2 =$$

the power in pounds raised one foot per minute.

369.—If the area of the cylinder in feet be multiplied by the velocity of the engine per minute in feet, it will be the volume of steam consumed when of the density of that in the boiler; and dividing by the volume of steam which a cubic foot of water forms at the temperature or force in the boiler, (art. 121, or tables at the end,) the result will be the cubic feet of water consumed per minute, when the quantity of water, and consequently the quantity of fuel, (art. 190,) will be known; but the supply of water should be a little in excess.

370.—The purposes to which noncondensing engines of this kind have been applied, are to impelling steam carriages, moving materials within deep mines, draining mines in places difficult of access, driving machinery in places where water cannot be obtained at a moderate expense, and in various instances where low pressure steam was equally available; but for most of these purposes it is inferior to the next species: the sole advantage it possesses being that of uniformity of moving force in every part of the stroke; which in some instances is desirable, in others hurtful.

Noncondensing Engines to work by Expansion.

371.—*Second Species.* The only difference required in the construction of a noncondensing engine to enable us to use the expansive power of the steam, is in the arrangement for opening and closing the steam passages. The steam must be admitted from the boiler only during a part of the stroke, and then shut off, but the passage for the escape of the steam should be open during the whole of the stroke. When the passage from the boiler is shut the steam acts by expansion, and the power it affords by expansion is wholly in addition to that which is obtained by the preceding species, whence the economy of this method.

372.—The most important question, is to determine that point in the length of the stroke at which the steam should be cut off, so that it may afford the greatest quantity of useful effect from a given quantity of steam; for then a given quantity of fuel produces the greatest useful effect. Now we have shewn the resistance from friction, &c. to be nearly, if not exactly, 0.4 of the whole force of the steam in the boiler, (art. 367,) and it is obvious, that when the steam has expanded till its excess of force be equal to this resistance, it will produce no further useful effect; and also that as far as the expansion exceeds this limit, there must be a decided loss of power. Hence, if we consider the capacity of the cylinder to be 1, the force being inversely as the space the steam occupies, it must be, as the whole force of the steam in the boiler is to 1, so is the whole force on the piston, when it is just equal to the friction, to the portion of the stroke when the steam should be cut off.*   That is, if the whole force in the boiler be 120 inches of mercury, the atmospheric pressure being 30, the resistance is 120 × ·4 = 48 = the inches equivalent to the friction; and 30 + 48 = 78 = the whole force on the piston, consequently,

$$120 : 78 :: 1 : 0.65 = \frac{1}{1.54} =$$

the portion of the stroke to be made before the steam be cut off, when its force in the boiler is 120 inches, or ninety inches in excess above the atmosphere.

373.—The excess of force in the boiler must be about four-tenths of the pressure of the atmosphere, or twelve inches of mercury, to cause motion at the proper velocity; but the absolute friction being only about half this force, the engine may begin motion with about a force of six inches.

374.—The most common mode of cutting off the communication between the cylinder and boiler, at the proper period of the stroke, is, to give a slide two motions; the first shuts off the steam, and the second motion lets it on at the opposite end, and opens the other to the atmosphere.   Such a construction is shewn by Plate IV. Fig. 4, the position of the slide being shewn when the steam is cut off by its first motion.   This construction represents the principle followed by Messrs. Taylor and Martineau, but they place the axis of the cylinder horizontally, and construct the pistons of the cylinder, and of the slides, in a rather different manner from those drawn.

A horizontal piston rod never works well, and the expense of such a frame as enables us to use it in a vertical position can rarely be more than equivalent to this defect; nevertheless, in mountainous districts, where mines are difficult of access, a horizontal cylinder has the advantage, being very easily fixed.†

375.—The power of an engine of this kind should be regulated by altering the time of cutting off the steam; its power may vary from full pressure through the stroke to that obtained by cutting off at the point above determined : the average state that will give the greatest advantage being to cut off at a mean between the point which gives the maximum of effect, and that which gives the greatest power required for the work; for a loss of

* Put f = the force in the boiler in inches of mercury, and

$$\frac{1}{n} =$$

the portion of the stroke made before the steam is cut off, then

$$\frac{f}{\cdot 4\,f + 30} = n.$$

† Belidor shews a method of constructing a piston for a horizontal cylinder by the addition of friction rollers.   Archi. Hydrau. Tom II. p. 240.

power arises from cutting off sooner than is indicated by the rule when the right amount of friction is calculated upon. For modes of giving motion to the slide, (see art. 478.)

Oliver Evans made a rude attempt to investigate the advantage of cutting off the steam in high pressure engines, claiming the principle as his own; but the engine he describes is not arranged for that purpose, he uses valves for the steam passages;* the objection to valves above a certain size is the difficulty of opening them.

376.—The proportions of the parts for expansion engines may be ascertained by the same rules as for full pressure, (art. 366,) excepting that the velocity should be found by the rule, (art. 336.)

377.—To determine the *power of a noncondensing engine working expansively*, it will be most useful first to ascertain the mean effective pressure on the piston, and from thence the power.

To find the mean pressure. Let the steam have to be cut off at the

$$\frac{1}{n}$$

part of the stroke. Add 1 to 2·3 times the logarithm of n; divide the sum by n and sub-tract 0·4 from the result, the remainder multiplied by the whole force of the steam in the boiler in pounds per circular inch, and 11·55 subtracted from the product for the pressure of the atmosphere, the result is the mean effective force of the steam on the piston in pounds per circular inch.†

---

* Steam Engineers' Guide, p. 30, and 67. Philadelphia, no date.

† Making $b + x = l$, and

$$b = \frac{l}{n}$$

we have by (art. 306,)

$$p\, b\, \left(1 + \text{hy. log.} \frac{b + x}{b} \right) = \frac{p\, l}{n}\, (1 + \text{hy. log. } n)\,;$$

and therefore the power of a cylinder d inches in diameter, working at a velocity of v feet per minute is

$$\frac{p\, v\, d^2}{n}\, (1 + \text{hy. log. } n)\, -$$

friction and resistance of the atmosphere. The latter is $\cdot4\, p\, d^2\, v + 11\cdot55\, d^2\, v$; hence

$$d^2\, v\, \left(p\, \left(\frac{1 + \text{hy. log. } n}{n} - \cdot4 \right) - 11\cdot55 \right) =$$

To find the power, multiply the mean effective pressure by the square of the diameter of the piston in inches, and by the velocity in feet per minute; the product is the pounds raised one foot per minute.

To find its equivalent in horses' power divide by 33,000.

378.—Example. If an engine work expansively, the steam being cut off at the

$$\frac{1}{1 \cdot 5}$$

part of the stroke, the cylinder being twelve inches in diameter, the velocity 160 feet per minute, and the whole force of the steam in the boiler 120 inches of mercury, or forty-six pounds per circular inch.

First.   2·3 × log. 1·5 = 0·405
        Add 1   1·
        _____
        Divide by 1·5)1·405
        _____
           0·936
    Subtract   0·4   for the friction and loss.
        _____
           0·536
    Multiply by   46   = the pressure of the steam.
        _____
          24·656
    Subtract   11·55   for the resistance of the atmosphere.
        _____
        13·106 lbs. the mean effective pressure on the piston.

the pounds raised one foot per minute.

The hyperbolic logarithm of n is equal 2.30285 times the common logarithm of n; whence the rule.

When n is fixed by the rule, (art. 372,) the formula reduces to the more simple form of

$$\frac{d^2\ r\ p}{n}\left(\text{hy. log. } n\right) =$$

the power in pounds raised one foot high per minute . and

$$n = \frac{p}{\cdot 4\ p\ +\ 30}.$$

$$12 \times 12 = 144 = \text{the square of the diameter.}$$
$$\text{Multiply by} \qquad 160 = \text{the velocity,}$$
$$\overline{23040}$$
$$\text{and again by} \quad 13.1 = \text{the mean pressure.}$$
$$\overline{301824} \text{ lbs. raised one foot per minute.}$$

And

$$\frac{301824}{33000} =$$

9·146 horses' power.

379.—When the steam is cut off at the

$$\frac{1}{n}$$

part of the stroke, the quantity of steam consumed in cubic feet per minute, will be found by multiplying the area of the cylinder in feet by the velocity in feet per minute, increased by one-tenth, and dividing the product by n. If this result be divided by the volume which a cubic foot of water occupies when in the state of steam of the force in the boiler, (see art. 121, or tables,) the quotient will be the quantity of water required per minute, and the equivalent quantity of fuel will be found by (art. 190.) The actual supply of water, and the power of the boiler should be n times that quantity; the engine will then either work at full pressure or expansively as occasion may require.

380.—In an expansive engine the moving force varies from full pressure to nothing in the course of the stroke; for some objects this variation is desirable, because the motion of the piston is not accelerated so much towards the end of the stroke. It may be used for any of the purposes to which steam power has been found applicable, and where water is not easily procurable it becomes the most economical species of engine.

381.—*Double cylinder expansive engine.* An engine of the noncondensing kind may be worked expansively by means of a double cylinder, according to Hornblower's method, (art. 32.) In Plate IV. Fig. 5, C is the cylinder for the strong steam, and B that in which it acts by expansion. The steam enters at S from the boiler, and passing through the passage t at the top of the small cylinder, forces down the piston; the steam previously in the cylinder C passes through b, and ascending by the pipe e, enters the large cylinder B at a, and by its expanding force causes the piston to descend, the expanded steam below the piston escaping to the atmosphere by the passage c, and through d. The pistons in the passages being moved by the rods g, h, to the other sides of the apertures to the cylinders, the pressures are reversed, and the expanded steam escapes to the atmosphere by the passage a, through the aperture f. This construction is not very complex to obtain the motion of both pistons in the same direction; but it obviously could be done by one slide if the pistons had contrary motions, and I see no sound objection to their motions being contrary; and then the axis of motion should be between them. The effect of this mode of applying steam is the next point of consideration.

382.—Let $0.385 f$ be the force on a circular inch on the small piston, and a its area, and l the length of its stroke. Also, let $m\,a$ be the area of the large piston, and $n\,l$ the length of its stroke. Then at any portion x of the descent of the piston, in the small cylinder, $n\,x =$ the descent in the large one. The original space of the steam being $l\,a$, and its pressure being inversely as its bulk;

$$(l - x) \, a \dotplus m\,n\,a\,x : l\,a :: f : \frac{f\,l}{l - x + m\,n\,x} =$$

the elastic force of the steam between the pistons. And if $.385 f'$ be the resistance from friction, loss of force, and the resistance of the atmosphere, we have

$$.3\!\cdot\!5\,f\,a \left(1 - \frac{l}{l + (m\,n - 1)\,x} + \frac{m\,l}{l + (m\,n - 1)\,x} \right) - .385\,m\,a\,f' =$$

the forces of both the cylinders; and the fluxion of the power is

$$.385\,f\,a\,(\,\dot{x} + \overline{n\,m - 1}\; l \left(\frac{\dot{x}}{l + (m\,n - 1)\,x} \right) - .385\,n\,m\,a\,f'\dot{x}.$$

Its fluent is

$$385\,f\,a \Big\{ x + l \text{ hy. log. } (l + \overline{m\,n - 1}\;x) \Big\} - .385\,n\,m\,a\,f'\,x$$

which, corrected, becomes when $x = l$

$$.385\,f\,a\,l \Big\{ 1 + \text{hy. log. } m\,n \Big\} - .385\,n\,m\,a\,f'\,l.$$

Or

$$.385\,f\,a\,l \left\{ 1 + \text{hy. log. } m\,n - \frac{m\,n\,f'}{f} \right) =$$

power.

383.—The ratio of the capacity of the large cylinder to the small one, is dependent on the amount of friction and loss of force. In the small cylinder the loss must be the same as in the cylinder of an engine working at full pressure; this appears from our mode of inquiry to be 0·4 of the force of the steam in the boiler, (art. 367.) And in the second cylinder the friction of the piston, the cooling of the cylinder, and the excess of force required to expel the steam into the atmosphere added together make $016 + \cdot2 + \cdot007 = \cdot223$ of the remaining force, or $(\cdot223 \times \cdot6) + \cdot4 = \cdot5338 =$ the whole loss in the two cylinders. Hence, as

$$\cdot5338\,f + 30 : f :: 1 : m\,n = \frac{f}{\cdot5338\,f + 30} =$$

the capacity of the large cylinder, when that of the small one is unity.

If $f = 120$ inches, then

$$\frac{f}{\cdot5338\,f + 30} =$$

1·28. = the capacity of the large cylinder, that of the small one being one. And in all cases the value of $m\,n$ must be less than that of n in the note to (art. 377.) Also, since $\cdot5338\,f + 30 = f'$ we have from (art. 382.)

$$\cdot385\,f\,d^2\,b \left\{ 1 + \text{hy. log. } m\,n - \frac{m\,n\,f}{f} \right) =$$

$$\cdot385\,f\,d^2\,v \times \text{hy. log. } \frac{f}{\cdot5338\,f + 30} . =$$

the power of the engine, of which the velocity of the small piston is v feet per minute, and its diameter d inches, the whole force of the steam being f inches of mercury in the boiler.

Consequently, the power is less in an engine with a double cylinder, than in one with a single cylinder, in the ratio of the hyperbolic log. of

$$\frac{f}{\cdot5338\ f\ +\ 30} \text{ to that of } \frac{f}{\cdot4\,f + 30} .$$

A decrease of power and a more complex arrangement renders the double cylinder engine inferior in every respect, except that of the moving force being more equal than in a single cylinder.

384.—*Of the best force for the steam of noncondensing engines.* The circumstances determining the choice of the force of the steam are almost entirely of a practical nature. As far as regards the production of the steam itself a greater quantity of fuel will be required to generate strong steam, and there will be more loss of heat in the operation ; consequently, as far as the generation of the steam is concerned, the lower the force the better. But in the noncondensing engine the steam has to work against the pressure of the atmosphere, and loses so much of its effect; hence, the more its force exceeds the atmospheric pressure, the greater will be the effect of a given quantity of steam in proportion to this loss.   On the contrary when the strength of the steam is considerable there is much waste by leakage,* which with the extra expense of fuel tends to counterbalance the advantage of increasing the force; and considering these circumstances, with the danger of strong steam, it appears to me that steam of four or five atmospheres is about the best force for these engines.

* The rate of increase of loss by leakage may be estimated, for it depends jointly on the goodness of the workmanship, and the force tending to separate the parts.   Now a good workman may fit the parts so that they would not exceed, under a strain of one atmosphere, a continued aperture of the 5000th part of an inch in breadth; and then if f be the force in inches of mercury, and d the diameter in inches, the magnitude of the joint will be

$$\frac{3\cdot1416\ d\,f}{150000}$$

square inches.

The velocity of escape will be $6\cdot5\ \sqrt{86\cdot5\ (459 + t)} = 60\ \sqrt{459 + t}$; (art. 136,) consequently, the quantity lost per second, is

$$\frac{3\cdot1416\ d\,f\ \sqrt{459 + t}}{2500}$$

If v be the velocity of the piston in feet per second, the steam required in the same time will be $\cdot7854\ d^2\,v$; hence, the quantity required being unity, the loss will be

$$\frac{3\cdot1416\ d\,f\ \sqrt{459 + t}}{\cdot7854\ d^2\,v\ \times\ 2500} = \frac{f\ \sqrt{459 + t}}{625\ d\,v}.$$

When $v = 4$, $d = 10$, $f = 133$ and $t = 300$◦

$$\frac{f\ \sqrt{459 + t}}{625\ d\,v} =$$

one-seventh, nearly.

SECTION VI.

OF THE CONSTRUCTION OF CONDENSING ENGINES.

385.—The distinguishing feature of this class of engines is, that of condensing the steam to the state of water. The moving force is nearly equivalent to the force of the steam, as in the boiler, moving through the difference between the space in the state of steam and that in the state of water. Different systems of construction render the effective or useful power more or less, but I will endeavour to give the general principles in a few words, and then proceed to more minute detail.

386.—The essential parts of a *single condensing engine* consist of a *cylinder* having a passage to admit steam at the top, and one from the bottom to convey the steam to another cylinder, called a *condenser*. The condensing cylinder has a passage from the lower part of it to an air pump; and both the air pump and the condenser are immersed in a cistern of cold water, a jet of which plays into the condenser. The cylinder has an air-tight piston fixed to a rod, which moves in an air-tight box in the top of the cylinder; and conceive there to be a valve in the piston, which, whenever the piston arrives at the bottom, opens and allows the steam to pass from the upper to the lower side of the piston. Then, let the jet be stopped, and the cylinder and condenser be filled with steam from the boiler, and the piston be raised to the top of the cylinder by a counter weight at the other end of the beam, to which the piston rod is fixed. The cock being open to admit the steam from the boiler to the cylinder, if the jet of cold water be allowed to play into the condenser, nearly the whole of the steam in the condenser, and in that part of the cylinder below the piston, will be reduced to water, and the pressure on the top of the piston being equal to the elastic force of the steam in the boiler, while the elastic force of the vapour remaining below it is very small, the difference of the forces will press down the piston, and, consequently, raise an equivalent weight at the other end of the beam.

c c

When the piston arrives near to the bottom of the cylinder, the passage to the condenser is shut, and the valve in the piston opens, then the steam above the piston passes through to below it, as the piston rises by the action of the counter weight; and, being at the top, the valve in the piston closes, and the valve to the condenser opens, and another stroke is made, and so on successively.

But since a large quantity of water is used at each stroke, and the water contains a considerable quantity of air, the condenser would soon become filled with air and water, and the engine would cease to work; to avoid this the *air pump* is added, which, being worked by the beam, makes a stroke at each stroke of the steam piston, and clears the condenser of air and water.

387.—In an atmospheric engine with a condenser, the principal difference consists in the steam being let both into and out of the cylinder by passages at the bottom, and the descent of the piston is caused by the pressure of the atmosphere on its upper surface which is open to the air.

388.—But in the atmospheric engine, as constructed before Mr. Watt's improvement of condensing in a separate vessel, the jet of cold water was thrown into the cylinder itself at each stroke, and, hence, the cylinder required to be heated and cooled at each stroke at a great expense both of fuel and cold water.

The addition of a separate condenser was the most valuable of Mr. Watt's improvements, his next in importance was the double acting engine. The saving from the concentration of power which results from these improvements can be judged of only by those who are intimately acquainted with the employment of mechanical power. To them the merit of his invention must be known and duly appreciated, and by their estimation it must ultimately be valued in public opinion.

389.—The double acting engine, in general construction, resembles the single one described in the preceding article, (art. 386.) It differs in having a passage from the boiler both to the top and the bottom of the cylinder, and a similar passage from both to the condenser. Hence, it does not require a counter weight to raise the piston, nor that the steam should pass from the upper to the lower side. The force of the steam impels the piston in both directions, and compared with a single engine of the same size, a double quantity of steam is used, and double power is exerted in the same space and time.

390.—In any of these species steam may act expansively, whether the atmospheric or steam pressure be used, but the moving force may be rendered more uniform by using two cylinders of different sizes; in the smaller of these cylinders the steam acts with all its force throughout the stroke, and in the other it gradually expands as the stroke proceeds, and therefore the moving force is variable: but as the forces on both pistons jointly constitute the moving force, it is never less than the force of the steam on the smaller piston. This arrangement was devised by Hornblower, (art. 32.) The engines may be of either single or double power, but whether the engine has double or single power, it is, in the usual construction, a complex piece of machinery.

391.—We may now proceed to arrange the different species, and shew the proportions adapted to particular cases.

Second Class.　Condensing Steam Engines.

I. By condensation	1. Atmospheric pressure		Newcomen, 1705.
	2. Steam pressure	Single	Watt, 1769.
		Double	Watt, 1782.

II. By condensation and expansion	1. Atmospheric pressure		
	2. Steam pressure	Single	Watt, 1782.
		Double	Watt, 1782.
		Combined cylinder,	Hornblower, 1781.

III. By generation and condensation.

IV. By generation, expansion, and condensation	Single	Cornish engineers on Watt's construction. Woolf, 1804.
	Double	Cornish engineers on Watt's construction. Woolf, 1804.

Of the Construction of Engines working by Condensation.

392.—Of the engines working by condensation alone, two kinds may be distinguished; in the one kind the moving force is the pressure of the atmosphere, in the other it is the pressure of steam; the former may be further divided into those which condense in the cylinder, and those having separate condensers; and the latter into single and double acting engines.

Atmospheric Engines.

393.—*The common atmospheric engine.*　In the atmospheric engine, as it is usually constructed, condensation is effected in the cylinder. The parts required for this object are, a cylinder C, Plate VI. Fig. 1, close at the bottom and open at the top; a piston P; a passage S for the steam from the boiler to the bottom of the cylinder, provided with a valve V, or cock; a passage for cold water to condense the steam, to inject into the cylinder at I, with a cock D; and a passage E for the water used for injection to run out at, provided with a self-acting valve F to prevent it flowing back; a valve G for the air contained in the water to escape at is also necessary. Mechanism for opening and closing the valves is connected to the engine beam, and a small supply of water by a pipe, is constantly furnished to the top of the piston to keep the packing saturated so as to be steam-tight.

394.—The operation is simple, the beam is so balanced that the steam being admitted by the steam valve V below the piston, it rises to the top of the stroke; the steam valve is then shut, and the injection cock D is opened, and a jet of cold water rises through I, which condenses the steam to a lower degree of elasticity, and the water runs out by the passage E at the valve F; the pressure of the atmosphere on the piston P being unopposed it forces it down, and the air extricated by the water is expelled towards the termination of the stroke at the valve G.

The steam and injection cocks are moved by tappets on a bar moving vertically, and connected to the beam. The steam valve should close and the injection cock open just when the up stroke is completed, and the period of closing the injection cock should be adjusted to the power the engine is to exert; the steam valve ought to open with the rise of the piston.

395.—*The proportions of the parts.* The length of the cylinder should be twice the diameter, (art. 329.) The velocity in feet per minute should be ninety-eight times the square root of the length of the stroke in feet, (art. 342,) the engine being supposed to be applied to raise water. The area of the steam passages will be found by this proportion: as 4800 is to the velocity in feet per minute, so is the area of the cylinder to the area of the steam passage, (art. 154.) The temperature for condensation which affords the greatest useful effect will be found by (art. 166.) If the area of the cylinder in feet be multiplied by half the velocity in feet, and that product by 1·23 added to 1·4 divided by the diameter in feet, (art. 163,) the result divided by 1480 will give the cubic feet of water required for steam per minute. If from 1220 the temperature of condensation be deducted, and the result divided by the difference between the temperature of the cold water, and the temperature of condensation, the quotient will be the number of times the quantity of water required for injection must be greater than that required for steam, (art. 284;) in general it will be about twelve times the quantity, but it had better be a little in defect than excess. The aperture for the injection must be such that the above quantity of water will be injected during the time of the stroke. The moving force in the first instant is only that due to the height of the cistern, and therefore in order that the injection be sufficiently powerful at first, the head should be about three times the height of the cylinder; and making the jet apertures square, the area should be the 850th part of the area of the cylinder, or its side should be $\frac{1}{7\cdot7}$ of the diameter of the cylinder. The conducting pipe should be about four times the diameter of the jet.

396.—*To determine the power of an atmospheric engine.* The moving force is the pressure of the atmosphere, from which the whole of the friction, and the force of the uncondensed steam is to be deducted.

The moving force is the pressure of one atmosphere, or - - - - - 1·00
The loss of force measured in atmospheres consists of,
First, the uncondensed steam corresponding to the temperature
     of condensation (usually about 160°) - - - - - = ·34
Second, the force to expel it, and the air from the cylinder
     (art. 154.) - - - - - - - - - - - - - - = ·007
Third, the friction of the piston (art. 474.) - - - - - - = ·050
Fourth, the force required to open and close the valves, raise
     injection water, and overcome the friction of the
     axes - - - - - - - - - - - - - - - - - = ·093

                                                        0·49

The portion of the pressure of the atmosphere equal the effective pressure is - - - - - - - - - - - - - 0·51
or 5·9 pounds per circular inch.

397.—RULE, for the power of the common atmospheric engine. Multiply 5·9 times the square of the diameter of the cylinder in inches, by half the velocity of the piston in feet per minute, and the product is the effective power of the engine in pounds raised one foot high per minute.

To find the horses' power divide by 33000.

Example. Let the diameter of the cylinder be seventy-two inches, and the length of the stroke nine feet, making nine strokes per minute: in this case half the velocity is $9 \times 9 = 81$ feet per minute, consequently $5\cdot9 \times 72^2 \times 81 = 2,477,433\cdot6$ pounds raised one foot per minute, or

$$\frac{2,477,433\cdot6}{33,000} =$$

seventy-five horses' power. This example is the size of the Chase Water engine, designed by Smeaton, (see art. 24, p. 23.) His estimate of the equivalent horses' power differs from this chiefly through using a different measure of that power; but he also estimated on condensing at a lower temperature than 160°: and to estimate correctly, the proper force of the uncondensed steam should be inserted in the causes of loss of force for each particular case.

398.—The engine may be regulated by cutting off the steam before the piston has arrived at the top, and cutting off the injection sooner; further means of regulation are described in Sect. VIII. By cutting off the steam it acts expansively, and a less quantity produces the effect. Water for the top of the piston, and for the supply of the boiler should be raised from the hot well. The quantity of water required to supply the boiler being ascertained in cubic feet per minute, (art. 395,) the fuel will be known by referring to (art. 190,) and the size of the boiler by (art. 225, or 229.) In the case of the Chase Water engine,

$$\frac{6^2 \times \cdot 7854 \times 81}{1480} \left(1\cdot23 + \frac{1\cdot4}{6} \right) =$$

2·4 feet of water per minute, or 144 cubic feet per hour: and 8·22 pounds of caking coal convert one cubic foot of water into steam, therefore the quantity per hour will be 1183·7 pounds, or

$$\frac{1183\cdot7}{75} =$$

sixteen pounds per hour for each horse power. The boiler may be either rectangular, or cylindrical, with the steam limited to one pound on the circular inch.

399.—The atmospheric engine is applicable for raising water in most cases where coals are abundant; the engine is simple in construction and in operation, and does not require that accuracy of workmanship which is necessary for an engine acting by steam pressure. On a small scale it has less advantage, for when the cylinder is not more than about two feet in diameter, the consumption of fuel becomes great in proportion to the effect; the drainage of coal mines, and raising water to supply towns, and for irrigation where fuel is cheap, are its proper objects.

400.—*Atmospheric engines with a separate condenser.* The manner in which an engine of this kind may be constructed, is shewn in Plate VI. Fig. 2. Where C is the cylinder with its piston P; the steam comes from the boiler by the pipe S, and by a slide B is let into the cylinder at D, or kept out. A is a pump with a solid piston, to receive the condensed steam, air, and water, and expel it; the injection is made into the pipe E; and I, is the injection cock; F, is a cock to let out any air that may collect below the piston *p* when the engine is at rest. To begin the operation the slide B, must be raised above S, and steam admitted till all the air be blown out at the valve Q; the pistons being at the top in both the cylinder and pump; then shut off the steam by the slide B, and open the injection; then in consequence of the condensation produced by the jet both the pistons will descend, and during the first descent the cock F should be open, but afterwards closed; the injection being stopped and the slide B moved to close the passage to the condenser, on opening that for the steam the pistons will again ascend, and the air and water of condensation will be expelled at the valve Q; the alternate opening and closing of the passages and the injection cock are required to continue the action.

401.—This engine may be regulated by closing the valve B, at any period of the ascent, and the cock I, at any period of the descent; and as the application will be limited to raising water, the velocity in feet per minute should be ninety-eight times the square root of the length of the stroke, (art. 342;) the length of the cylinder twice its diameter; the area of the steam passages to the area of the cylinder, as the velocity in feet per minute is to 4800, (art. 154.) The air pump should be one-fourteenth of the capacity of the cylinder, (art. 349, note;) or making the stroke of the air pump half that of the steam piston, the diameter of the pump should be three-eighths of the diameter of the cylinder. The quantity of steam is found by multiplying the area of the cylinder in feet, by half the velocity in feet, with the addition of one-fifth for loss by cooling, (art. 161,) and waste; the result divided by 1480 will give the quantity of water per minute required to supply the boiler, and twenty-four times that quantity will be that required for injection, (art. 284.) The diameter of the aperture for the injection should be one thirty-sixth of the diameter of the cylinder, and the injection pipe one-ninth.

402.—*The power of this atmospheric engine* will be the difference between the pressure of the atmosphere on the piston, and the retarding force multiplied by half the velocity.

The pressure of the atmosphere being - - - - - - - - -		1·000
The retarding forces are,		
First, the resistance of the uncondensed steam, temperature 125°	= ·134	
Second, the force to expel it through the passages, (art. 154.) -	·007	
Third, the loss by cooling in the cylinder, &c. (art. 161.) - -	·067	
Fourth, the friction of the piston, (art. 474.) - - - - - -	·050	
Fifth, the force required to open the valve, raise water for injection, and overcome the friction of the axes - -	·100	
Sixth, the force required to work the air pump, (art. 354.) -	·100	
Sum of the retarding forces - - - - - - - - - - - - -		0·458
Portion of the pressure of the atmosphere equal to the effective pressure		·542

This is equivalent to 6·25 pounds on a circular inch; the excess of force of the steam in the boiler, is a full compensation for some other causes of loss of power.

403.—RULE. Multiply 6·25 times the square of the diameter of the piston, by half the velocity in feet per minute, and the product is the effective power in pounds raised one foot high per minute.

Divide by 33000, and the quotient will be the number of horses' power.

Example. If the diameter of a cylinder be thirty-two inches, and half the velocity be 110 feet per minute; then $6\cdot25 \times 32^2 \times 110 = 704000$ pounds raised one foot per minute, or

$$\frac{704000}{33000} =$$

twenty-one horses' power.

404.—As the quantity of water required for the boiler is found by (art. 401,) the quantity of fuel is easily found from (art. 190.) In the example of the preceding article we have

$$\frac{1\frac{1}{3} \times 2\frac{2}{3} \times 2\frac{2}{3} \times \cdot7854 \times 110}{1480} =$$

·49 feet of water per minute, or 29·4 feet per hour; consequently $29\cdot4 \times 8\cdot22 = 246$ pounds of caking coal per hour; or

$$\frac{246}{21} =$$

11·7 pounds per horse power. For the proportions of the boiler, (see Sect. III.) and the beams and other parts for strength, (see Sect. VIII.)

405.—For raising water this species of atmospheric engine is admirably adapted; it can be constructed without difficulty by ordinary workmen, and for water works, drainage, irrigation, canals, and other cases where water is required in considerable quantities, it is an economical mode of obtaining power.

Steam Pressure Engines.

406.—*Boulton and Watt's single engine.* The essential parts and operation of a single engine having been described, (art. 386,) we have only to shew the construction as it regards effect. (Fig. 4. Plate V.) shews a section of the cylinder C, condenser B, and air pump A, of a single engine, arranged as is most convenient for exhibiting the parts. The steam enters from the boiler to the cylinder by the pipe S, through the valve *o;* and presses down the piston P, which is supposed to be taken at the time of its descent; the steam below it goes into the condenser, and is condensed by the jet which plays into it. The air pump bucket *p* is descending in the air and vapour which the pump had received from the condenser during the previous ascent. When the piston is at the bottom of the cylinder, a motion is given to the rod O which shuts the valves-*a* and *c,* and opens the valve *b;* there is then a communication open by the pipe E, between the top and bottom of the cylinder, and the pressure of the counter weight must be sufficient to overcome the friction of the piston, and expel the steam from the upper to the lower side of the piston; the action of the counter weight has also to expel the air and water of condensation through the valve Q by means of the air pump. The mode I have shewn of placing the valves and moving them by a single motion is not Messrs. Boulton and Watt's, but is one intended to render the motion of the steam from the upper to the under side of the cylinder more quick, by the pipe E being exhausted: the motion of the valves is simple and easily balanced. The valves of Messrs. Boulton and Watt are similar to Fig. 5, but they move them independently of one another, and this ought to be the case for an engine to work expansively, unless a separate valve acted on by a regulator be used to cut off the steam, (see Sect VIII.) An elevation of Boulton and Watt's single engine is represented in Plate XII. as applied to raising water.

407.—*The proportions of the parts.* The length of the cylinder should be twice its diameter, (art. 329.) The velocity of the piston in feet per minute should be ninety-eight times the square root of the length of the stroke, (art. 342.) The area of the steam passages should be equal to the area of the cylinder, multiplied by the velocity of the piston in feet per minute, and divided by 4800, (art. 154.) The air pump should be one-eighth of the capacity of the cylinder, or half the diameter and half the length of the stroke of the cylinder, (art. 351,) and the condenser should be of the same capacity. The quantity of steam will be found by multiplying the area of the cylinder in feet by half the velocity in feet; with an addition of one-tenth for cooling (art. 160,) and waste; and this divided by the column of the steam corresponding to its force in the boiler, (art. 121,) gives the quantity of water required for steam per minute, from whence the proportions of the boiler may be determined, (see Sect. III. art. 224, and 227.) At the common pressure of two pounds per circular inch on the valve, the divisor will be 1497. The quantity of injection water should be twenty-four times that required for steam, (art. 284;) and the diameter of the injection pipe one thirty-sixth of the diameter of the cylinder. The valves in the air pump bucket should be as large as they can be made, and the discharge and foot valves not less than the same area. For the proportions of the beams and other parts for strength, (see Sect. VII;) and the modes of regulation and management (see Sect. VIII.)

408.—The power of the single engine may be ascertained as follows:—

The effective pressure on the piston is less than the difference between the force of the steam in the boiler, and the resistance of the uncondensed steam. Let the force in the boiler be denoted by   1·000

First, by the force producing the motion of the steam into the cylinder, (art. 154.) - - - - - - - - - -   ·007

Second, by the cooling in the cylinder, (art. 160,) and pipes, (art. 148.) - - - - - - - - - - - - - -   ·038

Third, by the friction of the piston and loss by escape (art. 474.)   ·05

Fourth, by the force necessary to expel the steam through the passages - - - - - - - - - - - - -   ·007

Fifth, by the force required to open and close the valves, raise injection water, and the friction of the axes - - -   ·100

Sixth, by the steam being cut off before the end of the stroke   ·100

Seventh, by the power required to work the air pump (art. 354.)   ·100

————

·402

————

·598

The force of the steam in the boiler is commonly thirty-five inches of mercury, that of the uncondensed steam, (temp. 120°,) is 3·7 inches, hence, $35 \times ·598 = 20·93$ inches, and $20·93 - 3·7 = 17·25$, or 6·66 pounds is the mean effective pressure on the piston; and when the steam in the boiler is of any other force, the mean effective pressure may be determined in the same manner.

409.—RULE. Multiply the mean effective pressure on the piston by the square of its diameter in inches, and by half the velocity in feet per minute, and the product is the effective power in pounds raised one foot high per minute.

Divide by 33000 and the result is the number of horses' power.

Example. Let the force of the steam in the boiler be thirty-five inches of mercury, the diameter of the cylinder forty-eight inches, and half the velocity 135 feet per minute. Then the mean pressure is 6·66 pounds, and $6·66 \times 48^2 \times 135 = 2,071,526$ pounds raised one foot, or

$$\frac{2,071,526}{33,000} =$$

sixty-three horses' power.

The water required would be

$$\frac{1 \cdot 1 \times 4^2 \times \cdot 7854 \times 135}{1497} =$$

1·27 cubic feet per minute, or 76·2 cubic feet per hour; and by (art. 190,) 76·2 × 8·22 = 626·4 pounds of caking coal,* or

$$\frac{626 \cdot 4}{63} =$$

9·94 pounds of coals per hour for each horse power.

410.—The application of the single engine is limited by the nature of its action to raising water or other works admitting of an inefficient returning stroke, but for these purposes it has great advantages. I would suggest as an improvement, that the condensation should be effected as described for the atmospheric engine, (art. 400,) and that it should always act more or less by expansion; the full effect of expansion cannot however be obtained unless the action be equalized by a proper arrangement of the pressures and counter weight.

411.—*Single engine acting expansively.* When the single engine acts expansively, it is necessary to determine the point of the stroke at which the steam should be cut off. Now the pressure on the piston should never be less than the mean moving force, otherwise it would be overpowered, and the column of water would descend again. Consequently we may adopt this analogy. As the whole force of the steam in the boiler is to one, so is half the greatest effective force on the piston added to the resistance from friction, &c. to the portion of the stroke at which the steam should be cut off. Thus, if the force in the boiler be thirty-five inches of mercury, and the resistance of the uncondensed steam 3·7 inches, then 3·7 + (35 × ·402) = 17·77 inches, the loss of power from friction, &c. (art. 408,) and consequently

$$\frac{35 - 17 \cdot 77}{2} + 17 \cdot 77 = 26 \cdot 38 =$$

pressure on the piston at the end of the stroke; therefore 35 : 26·38 : : 1 : ·75 = ¾ of the

---

* This is equivalent to raising 17,600,000 pounds one foot by a bushel of coals, or 198,000 pounds by one pound of coal.

stroke. The steam will obviously act expansively in its ascent in the same proportion, whence a less counter weight is necessary.

412.—To find the mean pressure on the piston in an expansive engine. Let the portion of the stroke made when the steam is cut off be

$$\frac{1}{n.}$$

Then the nth part of the whole force in pounds per circular inch, of the steam in the boiler, multiplied by the 2·3 times the common logarithm of n, added to ·3, is the mean moving force or pressure; which is to be used in the rule, (art. 409,) for finding the power, and also for adjusting the load.

Example. Suppose the steam to be cut off at three-fourths of the stroke, then

$$\frac{1}{n}$$

= three-fourths, or $n = 1·33$, and its logarithm is $0·125156$; the whole force being thirty-five inches of mercury, or 13·5 pounds per circular inch, we have

$$\frac{3 \times 13·5 \times \overline{(2·3 \times ·125156 + ·3)}}{4} = 13·5 \times ·441 =$$

5·95 pounds per circular inch for the mean pressure.

413.—The velocity should be found by the rule (art. 343,) and the quantity of steam will be as much less than that required for an engine working at full pressure as the portion of the stroke at which the steam is cut off is less than the whole stroke; and in other respects the quantity of water, fuel, water for condensation &c. should be determined by the rules in (art. 407.*) The counter weight will be less in the same ratio, as the pressure on the piston is less than it is in a common engine. Owing to a larger sized engine being required, the expansive method is not valued as it ought to be, except when the force of the steam in the boiler is increased, and this I would recommend to the extent of two atmospheres, but not higher.

414.—*The double engine of Boulton and Watt.* It has been already shewn in what the double engine differs from a single one, (art. 389.) The parts are shewn in Fig. 1. Plate V. where C is the cylinder; the steam enters at S, and passes into the upper part of the cylinder at F, or into the lower part at D, as in Fig. 3; Fig. 1. shewing the piston in

---

* Taking the example of (art. 409,) we find 22,000,000 pounds may be raised one foot by a bushel, or nearly 250,000 pounds by one pound of coal; and I do not think more has been actually done with low pressure steam by a single engine.

the state of descending and Fig. 3, as ascending. From the lower part of the cylinder in Fig. 1, the steam escapes through D, into the condenser B, where it is condensed by a jet of cold water, which plays into it constantly; and the uncondensed gases and water pass through the valve G, during the ascending stroke, and during the descending one they pass from the lower to the upper side of the pump bucket, through its valves, and are drawn up by the ascending stroke, and expelled at the valve Q into the hot well. When the steam piston P ascends, the steam from the upper part of the cylinder, passes through F down the pipe E to the condenser. The steam passages D, and F, are opened and closed by a *D-slide*, so called from its plan resembling the letter D; it is moved by the rod O, by tappets or other methods, (see Sect. VII.) where the different methods are described. In small engines the steam passages are frequently opened and closed by cocks, in larger ones by valves, or slides, the species of which and the pistons and other parts are described in Sect. VII.

415.—*The proportions of the parts for a double engine acting with full pressure.* When the case to which the engine is applied will admit of it, the length of the cylinder should be twice its diameter, (art. 329.) The velocity of the piston in feet per minute, should be found by multiplying the square root of the length of the stroke by 103 for machinery, or by 98 for raising water, (art. 337 and 342.) The area of the steam passages should be equal to the area of the cylinder multiplied by the velocity of the piston in feet per minute, and divided by 4800, (art. 154.) The air pump should be one-eighth of the capacity of the cylinder, or half the diameter, and half the length of the stroke of the cylinders, (art. 351,) and the condenser should be of the same capacity. The quantity of steam will be found by multiplying the area of the cylinder in feet by the velocity in feet, with an addition of one-tenth for cooling and waste; and this divided by the volume of the steam corresponding to its force in the boiler, (art. 121,) gives the quantity of water required for steam per minute, from whence the proportions of the boiler may be determined, (see Sect. III. art. 224, and 227;) at the common pressure of two pounds per circular inch on the valve, the divisor will be 1497. The quantity of injection water should be twenty-four times that required for steam, (art. 284,) and the diameter of the injection pipe one thirty-sixth of the diameter of the cylinder. The valves in the air pump bucket should be as large as they can be made, and the discharge and foot valves not less than the same area. For the proportions of the beams and other parts for strength, (see Sect. VII;) and the modes of regulation and management (see Sect. VIII.)

416.—*To determine the power of a double acting engine.* Let the force of
the steam in the boiler be denoted by - - -   - -   1·000
Then besides the loss from uncondensed steam there is loss,
First, by the force producing the motion of the steam into the
cylinder, (art. 154.) - - - - - - - - - -   ·007
Second, by the cooling in the cylinder, (art. 157,) and pipes,
(art. 148.) - - - - - - - - - - - - -   ·016
Third, by the friction of the piston and loss (art. 474.)   - -   ·125
Fourth, by the force necessary to expel the steam through the
passages, (art. 154.) - - - - - - - - - -   ·007
Fifth, by the force required to open and close the valves, raise
injection water, and the friction of the axes - - -   ·063
Sixth, by the steam being cut off before the end of the stroke   ·100
Seventh, by the power required to work the air pump, (art.
354.) - - - - - - - - - - - - - - -   ·050

                                                                ·368

                                                                ·632

The force of the steam being generally thirty-five inches of mercury in the boiler, the temperature of the uncondensed steam 120°, and its force 3·7 inches; hence, (35 × ·632) — 3·7 = 18·42 inches, or 7·1 lbs. per circular inch for the mean effective pressure on the piston.*

417.—RULE.   Multiply the mean effective pressure on the piston by the square of its diameter in inches, and that product by the velocity in feet per minute, the result will be the effective power in pounds raised one foot high per minute.

To find the horses' power divide the result by 33,000.

Example.   The diameter of the cylinder of a double engine being twenty-four inches, the length of the stroke five feet, the number of strokes per minute twenty-one and a half, and the force of the steam in the boiler thirty-five inches of mercury, or five inches above the pressure of the atmosphere, required its power.

The velocity is $2 \times 5 \times 21\frac{1}{2} = 215$ per minute, and the mean effective pressure on the piston will be 7·1 lbs. per circular inch; therefore, $7\cdot1 \times 24^2 \times 215 = 879,264$ lbs. raised one foot high per minute, or

---

* This is 9·05 lbs. per square inch.

$$\frac{879264}{33000} =$$

26·64 horses' power. The nominal power of this engine would be only twenty horses' power by Boulton and Watt's mode of calculation, but it will be found that the nominal and real power nearly agree when the steam acts expansively, (art. 422.)

The water required for the above engine, (art. 415,) will be

$$\frac{1\cdot1 \times 2^2 \times \cdot 7854 \times 215}{1497} =$$

·5 cubic feet per minute, or thirty cubic feet per hour; and (art. 190,) 30 × 8·22 = 246·6 lbs. of caking coal, or

$$\frac{246\cdot6}{26\cdot64} =$$

9·2 lbs. of coal per hour for each horse power.*

When an engine is of less than ten horses' power, the consumption of fuel will be greater per horse power about in the ratio given in (art. 221.)

418.—This engine is applicable to every purpose for which a stationary engine is adapted, and it is only in cases where water is procured with difficulty that it is not applied. It has also been lately brought into use as a moving agent in steam vessels. (See Sect. X.) When the steam acts expansively the power is obtained with a smaller quantity of fuel, and to save fuel is the great object in every application of steam power.

419.—*Double engine acting expansively.* The motion of a double engine acting expansively ought to be equalized by a fly or some other method, (see Sect. VIII.) otherwise the effect cannot be perfectly obtained. To determine the point of the stroke at which the steam should be cut off, we have this proportion.

As the whole force of the steam in the boiler is to 1, so is ·368 times that force, (art. 416.) added to the resistance of the uncondensed steam, to the part of the stroke to be made before the steam be cut off.

Thus, if the force in the boiler be thirty-five inches of mercury, and the resistance of the uncondensed vapour 3·7 inches, we have

$$35 : (35 \times \cdot 368) + 3\cdot7 ; : 1 : \cdot473 = \frac{1}{2\cdot1}$$

of the stroke.

---

* Mr. Watt states to the effect that 8·7 lbs. is the quantity equivalent to a horse power, but no doubt he means when working expansively. Notes on Robison, Vol. II. p. 145.

420.—To find the mean pressure on the piston of an expansive engine, the part of the stroke at which the steam is cut off being

$$\frac{1}{n},$$

divide 2·3 times the common logarithm of n by n, and multiply the quotient by the whole force of the steam in the boiler in pounds per circular inch, the result will be the mean moving force on the piston on a circular inch.

Example. Suppose the steam to be cut off at

$$\frac{1}{2 \cdot 1}$$

of the stroke, then $n = 2 \cdot 1$, and the logarithm of 2·1 is ·322219; consequently,

$$\frac{2.3 \times 322219}{2 \cdot 1} = \cdot 354;$$

and as the pressure corresponding to this point of cutting off the steam is thirty-five inches, or 13·5 pounds per circular inch, we have $13 \cdot 5 \times \cdot 354 = 4 \cdot 8$ pounds per circular inch, the mean pressure.

421.—The velocity should be found by (art. 336, or 343,) and the quantity of steam will be

$$\frac{1}{n}$$

part of that required when the engine works at full pressure; therefore the water for steam, the fuel, injection water, will be less in the same proportion in regard to the dimensions of the cylinder, but the passages, pumps, boiler, and other proportions should be found by the rules in (art. 415,) in order that the engine may work either at full pressure or expansively as circumstances may render desirable.

422.—Taking the dimensions and force of steam of the engine given as an example, in (art. 417,) its power as an expansive engine would be $4 \cdot 8 \times 24^2 \times 215 = 594,432$ pounds raised one foot high per minute, or

$$\frac{594432}{33000} =$$

eighteen horses' power. At the full pressure, the fuel was 246·6 pounds; in this case it is

$$\frac{246 \cdot 6}{2 \cdot 1} =$$

117 pounds,* hence,

---

* This is the same as raising 27,000,000 pounds one foot high by a bushel of coals.

$$\frac{117}{18} =$$

6·5 pounds per horse power; the advantage is therefore as 6·5 : 9·2, or as 10 : 14.* For small engines this quantity requires to be increased in the ratio given in the table, (art. 221.)

423.—The mode of cutting off the steam by giving two movements to the slide during the stroke is shewn in Plate V.; Fig. 2, shews the position of the slide when the piston is descending and the steam cut off, with the passage D to the condenser still open. Slides have the defect of requiring a separate passage to introduce the steam to expel the air from the engine at the time of starting, technically called "blowing through :" but in other respects they seem to afford the most simple and durable means of opening and closing the passages.

Combined cylinder Engines.

424.—In Hornblower's engine with two cylinders the steam acts at full pressure in the one, and expansively in the other; as a single engine it is decidedly inferior to Boulton and Watt's construction in every respect, except that of the moving force being more nearly uniform, for there is the additional friction of the small piston, and it is a singular fact, that a single engine of this kind is more complex than a double one. As mine engines they appear to be nearly abandoned, and therefore it is not necessary to occupy space in describing a species which will be sufficiently understood by imagining two single engines acting on one beam, the one of which works at full pressure, and the steam which propels it acts expansively in the other cylinder during the next stroke. In both cylinders the steam has to change from the upper to the lower sides of the piston during the ascent. The ratio of the size of the expansion cylinder to the other should be determined by the same rule as for double engines of this kind, (art. 426,) and in other respects the proportions should be as for single engines.

425.—*The double engine with combined cylinders.* This engine will be understood most easily with a simple mode of letting on and off the steam. Let C be the small cylinder, Plate VI. Fig. 3, and D the large one, and S the place where the steam enters the pipes. The steam enters the small cylinder at *a* when the piston descends, and the portion below its piston passes through *b*, and rising in the passage *c*, enters the large cylinder

---

* If we take the mean between 6·5 and 9·2 or 7·85 it is what we may expect to be the ordinary consumption of an engine with a variable resistance, when of the best kind.

at *d*, while the steam passes to the condenser through *e*. When the motion is reversed by the slide being moved till the parts are on the other side of the passages, then similar motions take place in the reverse directions, and the vapour passes through *f* down a pipe to the condenser. Thus the whole apparatus is reduced to a slide box, the rod of which has only one motion for each stroke, and though it is here shewn between the cylinders for convenience, it may be placed in the angle they form when close to each other.

426.—*The proportions of combined engines.* The smaller cylinder should have the same proportions as for a noncondensing engine working with steam of the same force, (art. 366,) and the loss of force must be the same, that is, 0·4 of the force of the steam in the boiler.

The loss of force at the piston of the large cylinder, when its power is 1, will be

> First, by the cooling in the cylinder and pipes - - - - - ·016
> Second, by the friction of the piston - - - - - - - - ·125
> Third, by the force necessary to expel the steam through the
>   passages - - - - - - - - - - - - - - ·007
> Fourth, by the power required to work the air pump - - - ·050
>   ———
>   ·198

Consequently, ·6 × ·198 = ·1188 = the portion of the whole power, which added to the loss in the small cylinder, the total loss is ·1188 + ·4 = ·5188, or ·52 nearly. Hence, if *f* denote the whole force of the steam in the boiler, 3·7 the resistance of the uncondensed steam, and *n*, the times the capacity of the large cylinder is to exceed the small one, we have

$$\frac{f}{\cdot 52\,f + 3\cdot 7} = n.$$

If for example the force of the steam in the boiler be 120 inches of mercury, then

$$\frac{120}{(\cdot 52 \times 120) + 3\,7} =$$

1·82 = *n*, that is, the large cylinder should be 1·82 times the capacity of the small one; if it be larger a loss of effect must necessarily ensue.

427.—The power of a combined cylinder engine is easily ascertained from the investigation, (art. 382,) by substituting the proper constant numbers. The resulting rule for the mean pressure, supposing it to be collected on the surface of the small piston, is 2·3 times the common logarithm of the number of times the large cylinder is greater than the smaller one, multiplied by the force of the steam in the boiler on a circular inch. Thus if the force be 120 inches of mercury, then the capacity of the large cylinder should be 1·82 times the small one; therefore 2·3 × log. 1·82 = ·575: and as each inch of mercury is equivalent to ·385 pounds on a circular inch, 120 × ·385 = 46·2 pounds, and 46·2 × ·575 = 26·56 pounds on a circular inch, for the mean pressure collected at the small piston.

428.—RULE. The mean pressure being found as above, let it be multiplied by the square of the diameter of the small cylinder in inches, and by the velocity of the small piston in feet per minute, the result will be the power in pounds raised one foot per minute.

Divide by 33000 for the horses' power.

Example. If the force of the steam be 120 inches of mercury, the diameter of the small cylinder eleven inches, and the velocity of its piston 160 feet per minute, then the mean pressure is 26·56 pounds; and 26·56 × 11² × 160 = 514200 pounds raised one foot per minute, or

$$\frac{514200}{33000} =$$

fifteen and a half horses' power.

429.—The quantity of steam required per minute will be equal to the area of the small cylinder in feet multiplied by the velocity; and the quantity of water will be found by dividing by the volume the steam of a cubic foot of water occupies, when of the force it is in the boiler, allowing one-tenth for waste. In the above example it is

$$\frac{1·1 \times ·66 \times 160}{479} =$$

·242 cubic feet of water per minute, or 14·52 feet per hour. The fuel will therefore be 14·52 × 8·22 = 119·35 pounds of caking coal per hour, or

$$\frac{119·35}{15\frac{1}{2}} =$$

7·7 pounds of coal per hour for each horse power. Comparing this with (art. 422,) we find there is no advantage in using two cylinders as regards economy of fuel.

430.—The effects that may be obtained by engines of different species, have now been reduced for the first time, to definite measures, and their proportions referred to scientific principles. I have in these two sections endeavoured to render assistance to the practical engineer in as condensed and easy a form as possible, and yet with the minute circumstances in detail which are susceptible of variation by improvement in action or construction. He will see that the sum of the particulars must be near the truth, and the circumstances which increase or diminish any one of them must be apparent, or easily known by a reference to the article where it is investigated; and if he will be careful to distinguish actual practice from pretension, he will find that science and practice go hand in hand, the one supporting the conclusions of the other. I think I assert an undeniable proposition when I state, that the ultimate bearing of practice is towards that which is most economically adapted for its object; and the proper use of science is to assist in arriving at right conclusions with the least expense in trials. But at the same time that economy of power is considered, I think appropriate forms, good proportions, and excellent workmanship should be attended to in all machinery, and in many instances it is desirable that they should be beautiful; for a beautiful machine will be so attended to as to produce economy where an inferior one would perish by neglect.

SECTION VII.

OF THE PROPORTIONS, AND THE CONSTRUCTION, OF THE PARTS OF STEAM ENGINES.

431.—The steam engine has hitherto been studied as a whole, but in order to become more perfectly acquainted with its nature, we must dissect it, and study it in parts. This forms the object of the present section. Some of these parts have to be considered only as far as strength is concerned, as beams, shafts, cross-heads, &c.; others in respect to the motions they are to produce, as the parallel motion, the eccentric motion, &c.; others depend on the combination of moveable parts with accuracy of workmanship, as pistons, valves, &c. besides the modes of constructing joints &c. According to the dependence of these parts on one another, it seems desirable to treat them in the order of valves, pistons, stuffing boxes, hand gear, piston guides, parallel motion, strength of parts (as beams, cranks, wheel-arms, gudgeons, and teeth of wheels,—cross-heads and frames,—shafts and journals,—piston rods, connecting rods, and parallel motion rods,—cylinders, pipes, and boilers,) and joining pipes.

Of Cocks and Valves.

432.—Under the head cocks and valves may be included all those methods which may be found useful for opening and closing passages for steam. It is of some advantage, in discussing their respective merits, to class them, and the most simple method of doing

this seems to be by the motion which opens them. Following this method our arrangement is as under.

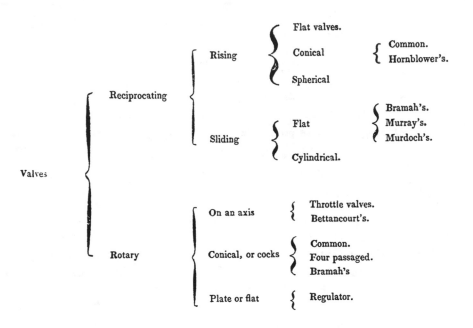

433.—The office of a valve or cock being to open, or to close a passage in the most
perfect manner, and either instantly or progressively, as may be proper for the object in
view, it is evident that those which offer the least obstruction to the passage, and are
opened with the least force are the best. Hence, the convenience of considering them in
succession in this place, and referring them to their respective uses.

434.—All the species of valves become more difficult to manage in proportion as the
apertures become larger. The area of the passage when open should be rather more
than equivalent to that of the narrowest part of the pipe, and by comparing these areas
the proportions of the valve boxes or apertures will be easily found for each species.

435.—When valves, cocks, or sliders are to be moved to admit steam to a steam
engine, the motion should be as quick as circumstances will permit, so that the passages
may be wholly opened, or wholly closed at the proper time with the least delay; for it
may be easily shewn that a considerable loss of effect arises from valves opening or shutting with a slow motion.

Rising Valves.

436.—The common *clack valve* is one of the most simple; in its common form it is a plate of leather a little larger than the valve aperture with a part of it fixed in the joint as a hinge. The leather is strengthened by a metal plate on each side, the lower one less, and the upper one larger than the valve aperture. It must open to an angle of about 30°, to allow a free passage equal to its aperture, and the box should be one and a half times the diameter of the valve aperture.

Its chief application in the steam engine is for the valve between the condenser and air pump, called the foot valve, and for the blow valve, but on account of the heat of the water it is necessary to use metal instead of leather, and to grind the parts to fit.

The foot valve G, Fig. 1, and Fig. 4, Plate V. is sometimes suspended by a hinge joint to the upper side of the passage, and falls against an inclined seat, the inclination being as much as to cause the weight of the valve to close it and not more.

437.—A *double clack valve*, consists of two semicircular valves, and is used for pump buckets; the construction of this is similar to the single clack valve, and the valves must rise to the same angle. They have the advantage of being more convenient than single ones for large pistons. The air pump bucket in a steam engine is furnished with metal valves of this species. See Plate V. *p*, Fig. 1, and 4.

To afford a greater quantity of passage with less resistance to open the valves, a kind of pyramidical valve consisting of four triangular pieces is sometimes used, but the construction is complex, and without corresponding advantages.

438.—*A flat metal plate* has frequently been recommended for a valve, particularly for a safety valve; it requires a guide sufficient to keep it in its proper seat; which may be most effectively done by a spindle sliding in holes in cross bars, above and below the valve. The diameter of the box should be to that of the valve as 3 : 2; and the parts should be ground together with emery, till they fit steam-tight. Its advantage as a safety valve is supposed to consist in its being less liable to stick fast, and with this opinion I perfectly agree; in other respects it differs little from the conical valve.

439.—The *conical steam valve* is a plate of metal, with its edge bevelled to fit into a conical seat, it is sometimes called a puppet or T-valve. The steam valves of Watt's engines were at first made of this kind. In this valve the diameter of the box should be to the greater diameter of the valve as 3 : 2, and it should rise not less than one-fourth of its greatest diameter when quite open; but both these proportions must be increased if the valve be out of the centre of the box. These valves and seats are often made of brass, but gun metal is better, the plate and the seat for it being of the same metal. They are turned as nearly to fit as possible, and afterwards the one is ground into the other, till it accurately fits the seat, with fine emery powder.

The best angle for the valve to fit in its seat is forty-five degrees, for then the pressure is balanced by the reaction of the sides. With less taper the valve has a tendency to set fast, with greater it occupies more space. When the conical valve exceeds five or six inches in diameter, it requires great power to open it against the pressure of the steam, and therefore is inconvenient. Mr. Watt applied a piston to the stem of the valve, fitted to a cylinder of the same diameter as the valve, on the opposite side of the passage, and the steam acting on the valve and piston equally, the difficulty of raising it was much reduced.

When the valve is to be self acting, that is, to move as soon as its narrower surface is exposed to a given pressure, then the weight of the valve must be equal to the square of the diameter multiplied by the pressure in pounds in a circular inch.

440.—A valve is sometimes made with the seat a portion of a sphere, and the valve either a portion or a complete sphere to fit it. This species, under the name of a *cup valve*, has been strongly recommended for safety valves, and by suspending the weight below the valve it is expected it will in a steam vessel be constantly in motion, so as to prevent sticking. See U, Plate XVII. Fig. 1. In other respects the cup valve seems to be inferior to the conical valve.

441 —*Hornblower's valve.* A common valve must often have to be opened against a pressure depending on its surface; to avoid this, a valve on a different principle was invented by Hornblower. This valve, Fig. 4, Plate VI. is inclosed in a box, and consists of a short cylinder resting on two conical seats, one on the exterior of the cylinder, the other is an interior seat at the bottom of it. The valve is raised or depressed by the usual methods applied to the cross bar at the top, and it is guided by the rod which slides in a socket in the lower seat. If there be strong steam on the upper side of the valve, and light vapour below, the pressure tending to keep the valve close is exerted only on the horizontal areas of the two seats, instead of being distributed over the whole surface of the valve.*

This reduction of pressing surface is obviously considerable in large valves. The principal passage for the steam is very direct, and at the lower seat the steam in its passage going chiefly down through the body of the valve, it is interrupted only by the cross bar at the top.

442.—*Improved form for Hornblower's valve.* The obvious difficulty of the valve,

* Professor Robison saw the theoretical advantages of this construction, but why has the account he gave of it been omitted in the reprint of his works?

is to make it fit steam-tight on two seats, but if we make the outside of the cylinder to slide in a stuffing box, or in an elastic packing of metal, (see V, Fig. 1, Plate VI.) that difficulty is removed, and the largest valves may be made with no other resistance to being opened, than the pressure on the seat, and the friction of the surface of the cylinder. It is simply the common conical valve inverted, and that which formed the seat in the common valve moves instead of the plate; and should obviously slide in a steam-tight case.

Sliding Valves.

443.—The sluice is the oldest form of this valve, but its advantages for any other than rough work in wood do not appear to have been understood: indeed it was not to be expected that metallic surfaces would slide on each other so closely as to be tight and durable, unless very truly worked, and of a hard metal.

Mr. Watt endeavoured to employ them at first but did not succeed, and it was not till more accurate methods of workmanship were introduced about thirty years ago, that the slide valve appeared.

444.—*Bramah's slide valve.* This slide valve is extensively used for pipes of water works, breweries, gas works, and various other purposes, and is exceedingly well adapted for steam passages. It consists of a box with a slider at right angles to the passage, moved by a rod passing through a stuffing box.

The slider is ground to fit accurately against the circumference of the passage with one surface, and is held close by a spring; it is moved by a handle for small apertures, and for larger ones by a rack and pinion.

445.—The first idea of employing slides for more than one aperture appears to have been to the air pump by Lavoisier or some of his associates, on which Dr. Robison has remarked, that a sliding plate performs the office of four cocks in a very beautiful and simple manner; he adds, however, " that the best workmen in London thought they would be difficult to execute." The same principle was applied to the steam engine by Murray in 1799, a sliding box answering the purpose of opening and closing four steam passages, to use Dr. Robison's words " in a beautiful and simple manner."*

446.—*Murray's slide.* The apertures all terminate in a steam-tight case, and within this a smaller box slides up and down, so as alternately to open and close the passages. A section of it is shewn in the annexed Plate VI. Fig. 5. The sliding part is moved by the rod *o* passing through a stuffing box. The steam from the boiler enters at S, and passes through *a* to the top of the cylinder, when the slide is down, while the passage *c*

---

* Art. Pneumatics. Robison's Mech. Phil.

F F

to the condenser is open through the interior portion of the slide; in like manner when the slider is up, the passage b for the steam to the bottom of the cylinder is open, and the passage a from the top to c the condenser is open.

A small reciprocating motion is obviously sufficient for the motion of the slide: its friction from the pressure of the steam against the box is considerable; but in order to reduce it, the rubbing surfaces should not be too small, and the harder they are the better; for steam boats gun metal is used, but where salt water is not to be employed, the sliding parts which apply together may be made of steel, and hardened; they then act and wear extremely well.

447.—*Murdoch's slides.* In slides formed in the preceding manner there is a loss of steam, in consequence of the apertures being opened and closed at some distance from where the steam enters the cylinder. This has been avoided in Messrs. Boulton and Watt's engines, where they have used similar slides invented by Murdoch, in which the strong steam is in the place assigned by Murray to the weak: and in engines with a long stroke, they make the two sliders separate, and move them by a rod of communication; because it would be more difficult to fit a long slide so that there would be a certainty of its rubbing surfaces being in complete contact, as the least deviation of these sliders, whether at the top or bottom of the cylinder, would cause a great leakage. Maudslay also, in his later boat engines, has adopted the same arrangement of slides as Boulton and Watt. See Fig. 2, Plate IV.

448.—Slides are getting into considerable repute for many purposes, and even in appearance the intricacy of a double engine is much diminished by using them. The contrivance of the slide to shut off the steam at any portion of the stroke is a point of some importance. Mr. Millington justly esteems the want of the power to do so a defect, and says it is common to the slide and four-passaged cocks;* but this objection may be removed in both cases by increasing the quantity of motion of the sliding surfaces one-half. For this purpose the slide should be the depth of the aperture shorter than will cover both the apertures to the cylinder, (see Fig. 1, 2, and 3, Plate V.) and it should be moved twice during the stroke by an adjustable tappet: the first motion shuts off the steam, as in Fig. 2; the second opens the passage to the condenser, and admits the steam at the other end. In this case let F and D represent the passages to the cylinder, S the place where the steam enters, and E the passage to the condenser. Suppose the steam to have been admitted to the upper part of the cylinder by the passage F, Fig. 1, and the slide to have been moved its first motion in Fig. 2, so as to cover F, and still leave D open to the condenser; then, at the next movement, Fig. 3, the slide will be at the bottom and admit

* Epitome of Natural Philosophy, p. 313.

steam at D, and F will be open to the condenser. The steam should encircle the pipe E; it then does not increase the friction materially by its pressure.

449.—The chief object of attention in setting out a slide, is to shorten its motion as much as possible, so as not to reduce the area of the passages. The area of the rubbing surface can scarcely be estimated at less than eight times that of the passages, which will be about one twenty-fifth of the area of the cylinder, (art. 154,) hence, eight twenty-fifths = the pressure; and taking the maximum pressure to be double the mean pressure, and the friction being supposed one-eighth of the pressure, it will be two twenty-fifths of the moving force, and it will be, in a short cylinder in action, about one-fifth of the length of the stroke; whence the loss amounts to about one sixty-second of the power of the engine. In long cylinders the ratio will be less.

450.—The *cylindrical slide* of metal, like a piston in a tube, was applied by Edelcrantz to the safety valve, but such a slide would obviously either be subject to stick fast, or allow steam to escape, as it would bear neither wear nor corrosion. Woolf's slide for regulating the quantity of steam passing an aperture is of the same kind, and seems to have no useful application whatever.* The attempt has been made to apply the metallic piston as a slide, and there is no doubt that it may be used both for that purpose and for the back of a flat slide; the object must be to construct it so as to be tight, and wear equally when applied in a cylinder. The advantages of such a slide I have endeavoured to shew in Plates IV. and VI.†

Rotary Valves.

451.—Axis valves are the most simple of the valves moved by rotary motion. A valve of this kind consists of a plate of metal fixed on an axis in the passage; the axis crosses the centre of the plate, and is made to pass through an air-tight aperture to the outside. They are extremely useful where perfect tightness is not required, as in the throttle valve, for dampers and the like. Belidor applied an axis valve to pump work, by putting the axis a

* Philosophical Magazine, Vol. XVII. p. 164.

† In Fig. 4, Plate IV. I have shewn a mode of construction for the piston slide, which would possess some advantages. A ring, cylindrical on the outside and conical in the inside, may be cut into two or more parts, with lap joints, and these parts may be expanded by the pressure of the steam on a conical part made to fit the interior of the ring; on the opposite side there should be a plate ground to fit the surface of the ring, and between this plate and the bottom of the cone, an elastic packing of hemp should be inserted; and the whole held together by nuts upon the piston rod. The steam apertures should be divided so that no single aperture should exceed one-eighth of the circumference.

little to one side of the centre; it then, however, becomes so very difficult to fit, that its use has not been continued; and this difficulty must always exist in a valve with two seats, otherwise it is easy to simplify Belidor's valve.*

452.—A species of slide to revolve on an axis was designed by Bettancourt for a double engine; such a slide would not, however, keep in order for any length of time, and does not appear to have been used.†

453.—*Cocks* are so well known as to need no description; and on a small scale they are certainly the best adapted for opening and closing pipes of any thing that has yet been proposed. They do not answer so well when they are in constant action; but even then it is doubtful whether or not they are inferior to other methods, and much depends on their being properly constructed. For a single or common cock the plug should be nearly cylindrical, where it has to be exposed to much pressure. The common reduction of the diameter is about one-sixth of the length.

454.—For various purposes a double passage cock is useful, and in some cases one with a triple passage may be required; but the one most commonly applied to the steam passages of steam engines is of the kind called the four-way cock, and is in fact a rotary slide. Of these we have to consider two kinds: the common one the application of which was suggested by Leupold, (art. 12,) and applied by Trevithick; and Bramah's improved one.

455.—*A four-way cock*, by its motion round its axis, opens a communication alternately from the boiler and condenser, to the top and bottom of the cylinder of a steam engine, Fig. 1, Plate IV. The simplicity of its action in some degree compensates for its friction, but there is the disadvantage of part of the steam being lost in the pipes at each stroke. Its form should be nearly cylindrical, otherwise its friction and tendency to wear unequally will be increased. When it is ground to fit truly, the pressure of the steam tends to keep the surfaces in contact, and to wear the cavity into an elliptical shape; hence, it is soon necessary to grind it to fit again.

456.—The cock applied in this manner does not admit of the steam being cut off at any portion of the stroke without the use of other valves. But by dividing the spaces, so that the solid part on each side of the aperture by which the steam passes to the condenser is double the aperture, the cock may be moved at twice, so as to cut off the steam at the first movement, and leave the passage to the condenser open till the second. See Fig. 6 and 7, Plate VI. The cock must move back and forward in this case, but it will be obvious that the disposition of the surfaces is such as will prevent the wear being so destructive as it is in the common form.

* Architect. Hydraulique, Tom II. p. 220.   † Prony, Architect. Hydraulique, Tom I. p. 572.

457.—*Bramah's four-way cock.* In the common one the pressure being wholly against the side of the conical plug, its wear is unequal and friction considerable: to remove these, the conic frustrum is formed on a cylindric axis, and the steam is admitted upon its larger end, by which the pressure on the seat is nearly equalized; and by turning in the same direction constantly the wear is equalized, notwithstanding the inequality of pressure.

These cocks, with some deviations, have been very much employed by Mr. Maudslay in small engines; and an example of their application to his portable engine is shewn in Plate XV, and the parts to a larger scale in Plate VIII. In the plan, C is the cylinder, I the four-way cock, and E the pipe by which the steam enters. The cock is represented with all the apertures shut; but the figure above the plan is a section through the cock. The steam enters at E, flows over the top of the cock, and by an aperture G in the top it passes either to the top or to the bottom of the cylinder, according as the aperture in the side of the cone is turned to the one or the other of these passages.

By comparing the effect of turning the cock to the right or left from the position it has on the plan, the manner of opening and closing the passages will be obvious. The higher passage leads to the condenser, (marked F in the two sections,) the middle one to the top, and the lower one A, to the bottom of the cylinder. If the cock be turned to the right, so that the opening in the triangular aperture through which the steam descends from the top is opposite the middle passage, then the steam will pass to the top, and the condensed vapour will have a passage open from the bottom to the condenser, through the body of the cone. If the cock be turned to the left, the centre of the triangular passage will be opposite the passage to the bottom of the cylinder, and the steam will pass in that direction, and a passage from the top to the condenser will be open through the body of the cock. In this cock the motion is back and forwards.

The escape of steam at the lower part of the cone is prevented by a packing of hemp round the cylindric part, and the cylindric part of the top is pressed by a spiral spring, with an oil cup H, and screw above it to act on the spring if occasion requires.

The pressure and friction of this cock will not be greater than that of a slide, if both be equally well executed. The loss of steam in the passages is an objection, and the steam cannot be shut off without closing the passage to the condenser; this, however, is in some degree compensated for by the application of Field's valve. See Plate XV.

458.—*Four-way cock to cut off the steam at any portion of the stroke.* The mode by which this may be done, is to make the cock so much larger that there will be the breadth of two apertures between the middle and each adjoining passage. The diameter will be increased only in the ratio of 10 to 8; the rubbing surfaces will remain nearly the same, and the cone will be more equally pressed into its socket.

459.—*Double passaged cocks.* By far the most simple method in practice would be to use two double passaged cocks; the apparent simplicity of one cock involves more trouble and care, and after all is not so good as two small ones. Two are as easily moved as one when the movements are simultaneous, and more conveniently managed where the steam is to be cut off.

460.—*Plate* or *flat valves*. The general nature of these valves may readily be conceived by imagining two flat plates to be ground to fit one another, and one to turn on an axis passing through the other plate; the plates being both pierced with apertures which coincide in one position of the moveable plate, and are all closed in another position. Valves of this kind, made of hard steel, were resorted to by Perkins for high pressure steam. When accurately made and applied so that the pressure is tolerably equal on the moving plate they might be useful. They admit of reducing the quantity of motion to open them in a considerable degree, but not without dividing the passage into small apertures.

461.—*Regulator*. The steam valve is called a regulator in the atmospheric engine; it is a kind of rotary plate valve, but it is formed wholly on one side of the axis, and hence, is more difficult to make work air-tight. Its construction, as designed by Smeaton,

FIG. 17.

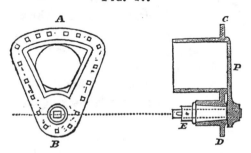

is shewn in the annexed figure; where A B is the under side of the aperture, and C D a section, with the plate P which covers it, and which is turned by a handle applied at E.

Of Pistons.

462.—The great desideratum in a piston is that it should admit of no leakage; and have as little friction as is consistent with this indispensable quality.

Pistons may be rendered tight by an elastic packing of vegetable, or animal matter; but the latter kind of packing cannot be used for steam, on account of the heat destroying it.

Pistons may also be made wholly of metal, constructed so as to admit of a certain degree of elasticity.

After considering some particulars common to all pistons, we will treat of pistons as below, dividing them into two classes.

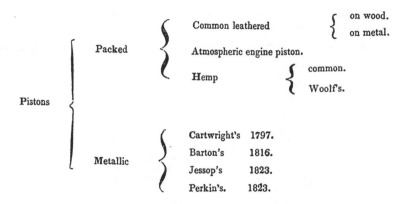

Pistons
- Packed
  - Common leathered
    - on wood.
    - on metal.
  - Atmospheric engine piston.
  - Hemp
    - common.
    - Woolf's.
- Metallic
  - Cartwright's 1797.
  - Barton's 1816.
  - Jessop's 1823.
  - Perkin's. 1823.

463.—When a piston rod is to be pushed, as well as drawn, unless it be of a certain thickness in proportion to the diameter, it is liable to stick if there be the slightest inequality in the friction, or in the centring of the piston rod. If it were a thin plate, nothing but its connection to the rod would prevent it turning with the slightest inequality of its friction, on being pushed; and as we make it thicker, the thickness interferes more and more with any tendency to turn. The proportions which will secure us from the risk of this evil are not difficult to ascertain.

Let the pressure on the piston A B move the rod C D. Then in order that the piston may move steadily, its friction at the circumference multiplied by half the diameter of the piston, should be equivalent to the pressure producing that friction, multiplied by half the thickness of the piston; consequently, the thickness should be to the diameter, as the friction is to the pressure of the rubbing surfaces.

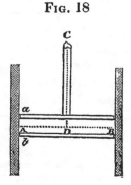

FIG. 18

The friction of brass on iron is at an average one-eighth of the pressure; hence, the thickness of metallic pistons should be not less than one-eighth of the diameter.

The friction of hemp packing on iron is about one-sixth of the pressure, hence, the thickness of the packing should be one-sixth of the diameter. Practice is extremely variable on this point, but the mean appears to be not far distant from the rule. For leather on iron the friction is greater, the average approaching to one-fifth of the pressure.* When the pressure on the piston draws the rod a thickness not more than four-tenths, that which is required when the rod is pushed will be sufficient.

It will be evident enough that the central part of the thickness of the piston adds little to the steadiness of its motion, though it increases the friction; hence, that construction of a piston has the advantage which renders the upper and lower part *a b* tight, without putting a like stress on the intermediate ones at A.

464.—The common piston is a double cone of wood, Fig. 19, having two bands of strong

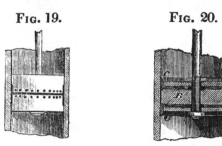

FIG. 19.   FIG. 20.

leather fastened round it with nails or hoops. The joints in the leather are not seamed, but closed as accurately as possible, and not put opposite to one another.

465.—If the parts be made of metal, a cylinder of brass should be turned to fit the barrel or the cylinder it is to move in, Fig. 20, so that it will slide freely without sensible resistance. Then an upper and lower plate, made of sufficient thickness for the piston to be of the depth the diameter requires, confines two cupped leathers, C C, with the edges cut to an angle of about 45°.

Both these pistons have the advantage of the friction being at the upper and lower edges; and bevelling the edges of the leather causes the force of the fluid to spread it against the surface of the pump barrel. This mode of bevelling the leathers seems to have been first used by Mr. Smeaton for a fire engine bucket; and the general principle of construction was first applied in his air pump in 1752. Mr. Bramah applied it to the various parts of his presses, and found its advantage in the application of high pressures.

---

* Belidor, who seems to have first applied the solid piston, gives the proportions so that the thickness is nearly equal to the diameter; the friction, with such a proportion, must be greatly increased, as it must be in every part air-tight. His plates in another place shew the thickness somewhat less than one-third. Archi. Hydraulique, Vol. II. p. 117 et 223.

466.—The atmospheric engine piston consists of a plate of cast iron, about one-eighth of an inch less in diameter than the cylinder, and about one inch and a half thick, formed with a rim about four inches from the edge. A flat ring corresponding to the part beyond the rim is fitted upon it, and both have holes for bolts to screw them together, which is done after a packing of soft hemp or gasket, saturated with tallow, has been inserted. In order to render it more tight a portion of water is kept constantly on the upper side of the piston.

Smeaton had a superior method of constructing the piston for atmospheric engines, which rendered the loss by the condensation of steam much less. The construction of that for the Chase Water engine, with a seventy-two inch cylinder, being given, will shew this method. The bottom of the piston was made of wooden planks fastened by bolts to the piston plate, with rings on the under side of the planks to receive the heads of the bolts. The advantage of wood for this purpose, in cases where the injection is made in the cylinder, is obvious. See Fig. 1, Plate VI.

The plank bottom, of elm or beech, was about two inches and a quarter thick when worked, and was formed by two planks halved together, in the form of a cross, and grooved on the edges with a three quarter inch groove, to receive the ends of the pieces to fill the corners between the cross; put in so that the pieces may have the grain radiating from the centre: a few rivets to hold the cross planks together were inserted where they were halved into each other, at their intersection, and the whole being hooped with a good iron hoop half an inch thick, and two and a quarter broad, it bound all tight together. The outside diameter of the hoop was a quarter of an inch less than the cylinder. The flat iron rings for the under surface of the piston, should be let in flush with the surface of the wood, and the bolt heads counter sunk; the planking was screwed on with a double thickness of flannel and tar between it and the iron piston plate, and any irregular hollows filled up with additional thicknesses of flannel and tar, so as to exclude the air between the plate and the wood. The bolts were carefully secured so as to make a water-tight joint from above. The plank was covered on the lower side by a lining of deal boards, shot clear of sap, and three quarters of an inch thick, nailed to the planks, with a single thickness of flannel and tar between so as to exclude the air; after this the lower surface of the lining was made perfectly flat and smooth.

467.—*The hemp packed piston* is now most commonly employed for steam engines, and the usual mode of construction is as follows. The bottom of the piston *b*, Plate VII. Fig. 1, is fitted as accurately to the cylinder as it can be done, to leave it at full liberty to rise and fall through the whole length. The part of the piston immediately above this is from one to two inches, according to the size of the engine, less all round than the cylinder, to leave a circular space into which unspun long hemp, or soft rope prepared for the purpose, and called gasket, is wound as evenly and compactly as possible, to form the packing. This packing is compressed together by a plate or cover C, which is put over the top of the piston, having a projecting ring to fit over the lower part, and complete the upper side of the space for the packing, the pressure being produced by screws S, S, &c. Both the upper and lower part of the space round the piston, to contain the packing, is a little curved, that the pressure produced by the screws on the packing may force it against the inside surface of the cylinder, into as close contact as possible.

The screws being tightened when the piston is in the cylinder, the particular form of the piston has the effect of squeezing out the packing, and causing it to press forcibly against the inside of the cylinder at its upper and lower edges. When the packing wears so as to become too small by use, these screws, which are more or less in number according to the size of the piston, are always resorted to for tightening it, as long as they are capable of acting, and when this is no longer the case, the piston top must be removed, and an additional quantity of new packing introduced. The piston rod is generally attached to the bottom part of the piston, by passing it upwards into a conical hole made to receive it, to which the bottom of the rod is exactly fitted, and a screw nut, or a wedge, between the top and bottom is inserted which effectually secures it.

The piston is kept supplied with melted tallow by means of a funnel on the top of the cylinder lid, provided with a cock to prevent the escape of steam.

468.—*Woolf's piston.* In the usual method, whenever the piston, by continued working, becomes too small and occasions a waste of steam, it is necessary to take off the top of the cylinder, in order to get at the screws, even when fresh packing is not wanted. This being laborious work, is therefore generally avoided by the person who attends the engine, as long as it can possibly be made to work without taking this trouble; and the neglect occasions a great and unnecessary waste of steam, and consequently of fuel in proportion.

The object of Mr. Woolf's improvement is to enable the engine man to tighten the piston, without the necessity of taking off the cover of the cylinder, except when new packing becomes necessary. He accomplishes this by the following methods.

To the head of each of the screws a small toothed wheel is fixed, so that it may be turned, and therefore tightened, by means of a central toothed wheel which works upon the piston rod as an axis; if one of the small wheels be turned, it turns the central wheel, and the latter turns the others. The one which is to be turned by the handle is furnished with a projecting square head, which rises up into a recess in the cover of the cylinder. This recess is surmounted by a plate fixed on with screws called a cap or bonnet, that being easily taken off, or put on again in its place.

The other method is similar in principle but different in construction. Instead of having several screws all worked down by one motion, there is in this but one screw, and that one is a part of the piston rod, Plate VII. Fig. 2; on this is placed a wheel *d* of a convenient diameter, the hole in the centre of which is a female screw cut to work into that of the piston rod. The wheel is turned round so as to tighten the piston by means of a pinion *a*, provided with a square projecting head for that purpose, rising into a recess in the cylinder cover of the kind already described, and the cover or top plate is prevented from turning with the wheel by means of the pins *e e*, called steady pins.

Metallic Pistons.

469.—*Cartwright's piston.* The idea of employing metal instead of elastic vegetable matter, to render the pistons of steam engines tight, was one part of the patent obtained by Cartwright in 1797. It consisted in using six or more solid masses of metal in the place of the usual packing; these masses being segments of rings, *a a*, Fig. 3, Plate VII. made to fit the internal surface of the cylinder, with a second series *b b*, crossing the joints of the other, and both series were pressed against each other and the cylinder by V-springs; and by having two sets, with the joinings of the rings in the one set, opposite the solid parts of the rings of the other set, the escape of steam at the joints was to be prevented. The upper and lower parts were connected by plates to which the piston rod was joined. (See the section, Fig. 3.)

The two exterior rings of brass were made of the full size of the cylinder, and cut into several segments, as shewn at *a a a*, and laid one above the other so as to cross the joints. The joints in the under rings are shewn by dotted lines in the figure, and in like manner are disposed the two interior rings, both being confined to their places by a top and bottom plate to which the piston rod is fixed. The segments are pushed away from the centre by steel springs, of the form of the letter V.

Pistons on Cartwright's plan have not been quite successful in practice, when the cylinders have not been truly bored; and the causes were pointed out very clearly by Mr. W. Nicholson, soon after the invention was brought before the public.* The pieces forming the piston having a determinate curvature, and being too strong to be sensibly flexible, cannot be expected to accommodate themselves to any irregularity in the cylinder in different parts of its length, as is done by the elastic stuffing of hemp. And there is reason to doubt in applying them whether the pressure of the rings or pieces together, has not been too powerful for the springs to perform their office when applied in this manner.

As to the actual difference between the friction of metal, and hemp against metal,

---

* Philosophical Journal.

when the pistons are equally steam-tight, it is undoubtedly in favour of metallic pistons, (art. 463.)

470.—*Barton's piston.* A piston considered superior to Cartwright's, was made by Mr. Barton, Plate VII. Fig. 4. It consists of one thick ring E, of brass or cast iron, made very nearly to fit the cylinder, and then cut into three or more equal segments; the equal triangles remaining are used as wedges to expand the segments of rings into a larger circle. The segments, and small triangles or wedges, are secured between a top and bottom plate, as in the piston last described, with spiral springs to press the triangles outwards from the piston rod, making them act as wedges to press the segments against the inside of the cylinder, and as these wear by use, the points of the wedges themselves protrude, and being formed of the same metal, still make part of the piston. A piston of this kind, and a true cylinder, has been known to work for some years without requiring any other attention than keeping it properly greased, but it is easy to prove that the wedges and segments do not expand equally, hence, in this state it was not applicable to high pressures; besides, the imperfection of Cartwright's piston still remained. It has however been recently much improved by Barton, and therefore I propose to describe it more fully in its improved state.

The piston is represented by a plan and section, Fig. 4. It is composed of a solid cylindrical cast iron body A, having a conical hole, B, to receive the enlarged end of the piston rod G, to which it is secured by a cross pin D, passing through both. A space or groove is formed round the body of the piston to receive four brass, cast iron, or cast steel hardened and tempered, segments, marked E, which are spread asunder by four triangular wedges G, of the same metal as the segments, acted on by eight spiral springs of tempered steel. These springs are inserted in cylindrical cavities at both ends, in order to render them secure from bending, and yet allow them to play freely. With the same view each spring has a cylindrical pin of steel within it, a little shorter than the spring. In pistons made for high pressure steam there are three grooves, formed round the exterior part of the segments, as in Fig. 5, the middle one *a* designed to hold oil, or grease, to lubricate the rubbing surfaces. The upper and lower grooves *b*, *b*, are for hoops of tempered steel having a forked loose joint, (as Fig. 6,) at one point in each. These hoops are nicely fitted to the grooves; and when the piston is placed in the cylinders their jointed ends meet. Each hoop is prevented from turning round in the groove, by a pin or stud, in order that the two hoops may not have their joints opposite to each other. These hoops, or rather springs, form an important addition, and assist greatly in preventing the leakage which otherwise would take place through the unequal expansion of the segments and wedges. For the point of the wedge will move outwards over *n m*, while the segments move only over *n o*, and consequently would wear the cylinder into grooves, were it not rounded off, and the hoops added, to prevent the escape of steam.

But by combining hardness and elasticity, Barton has done much to render these pistons tight and durable : they still however depend chiefly on the skill of the workman; when they are done well by a person who understands them, they undoubtedly answer effectively.

471.—To avoid the effect which the unequal expansion of the parts of Barton's piston produces, I would recommend the construction shewn by Fig. 7, where the wedge-formed pieces do not extend to the surface of the cylinder; and to prevent there being an aperture at each joint, two series of segments and wedges should be used, as shewn in the section : the joints of the lower series are shewn by dotted lines in the plan.

472.—It is of importance to remark, that the metallic packing is pressed so as to be steam-tight by the steam itself; and it is essential to their perfect operation that the steam has egress to the cavities in the piston, and that the parts fit perfectly against each other in all the horizontal joints.   Let strong steam be on the upper side A, of the piston, Fig. 7, and the lower side B, be open to the condenser; then the steam enters at the joints $e\ e$, presses the segments close on the lower plates, and fills the interior so as with the assistance of the springs to press the segments outward against the cylinder.   Also when the lower side is open to the steam, and the upper one to the condenser, the steam enters at f, f, pressing the segments close against the cylinder and upper plate.   If this were not so the springs could not possibly press with sufficient force to keep the joints steam-tight; for a fluid cannot be confined by a force less than its own elastic force, and hence, the pressure producing friction is always greater than the pressure of the steam on the rubbing surface, by that due to the pressure of the springs.*

473.—*Jessop's piston.*   A completely different method of applying metal to render pistons steam-tight was invented by Mr. Jessop, and secured by patent in 1823.   It consists of an expanding coil of metal, which binds round the piston body in a spiral form.   Fig. 8, Plate VII. shews a section of a piston of this kind, where A A is the elastic spiral of metal, which when at liberty and removed from the piston, assumes the form shewn in Fig. 9. To form a piston of this kind a bed of hemp packing, B B, is first prepared, which answers the double purpose of preventing steam passing at the joints, and of supplying a means of pressing the springs against the surface of the cylinder.   A small addition of hemp packing is at times necessary to make up for the wear.

The action of the steam in keeping this piston tight, is by pressure on the top and bottom

---

* So little is known by many mechanicians of the nature of the action of pistons, that it is not unusual to hear them express an opinion on the friction of a piston from the force which it requires to move it in an open cylinder; and on a level with it is the method of estimating the friction of an engine by the power it requires to move it when it is doing no work.   The true state of the fact is that the friction is as the stress on the parts, and this stress bears a relation nearly in proportion to the work done.

plates, as in the common hemp packed pistons. The pressure and wear of these pistons will be more equable than in the other metallic kinds; when they are equally well made; and they have been as successful in practice.*

474.—*Of the friction of pistons.* The rubbing surface of a piston must be pressed against the cylinder with a force at least equal to the pressure of the steam it confines, otherwise the surfaces would separate and the steam escape. Now it has been shewn (art. 463,) that the thickness of the rubbing surface should be equal to that portion of the diameter which expresses the friction; therefore, let r be the friction when the pressure is unity, $t =$ the thickness, $d =$ the diameter, and $p =$ pressure of the steam; then, $3 \cdot 1416 \; t \, d \, p \; r =$ the friction; or since $t = r \, d$, it is $3 \cdot 1416 \; p \; d^2 \; r^2 =$ the friction; to which one-tenth may be added for that of the piston rod.

The moving force is

$$\frac{3 \cdot 1416 \, p \, d^2}{4},$$

consequently that part of the moving force equal to the friction is

$$\frac{4 \cdot 4 \times 3 \cdot 1416 \; p \, d^2 \; r^2}{3 \cdot 1416 \; p \, d^2} = 4 \cdot 4 \; r^2$$

In double engines with metallic pistons, $4 \cdot 4 \; r^2 =$

$$\frac{4 \cdot 4}{8 \times 8} =$$

$\cdot 069$ of the power.

In double engines with hemp packed pistons $4 \cdot 4 \; r^2 =$

$$\frac{4 \cdot 4}{6 \times 6} =$$

$\cdot 1222$ of the power.

In single engines with hemp packed pistons $\cdot 4 \times 4 \cdot 4 \; r^2 =$

$$\frac{176}{6 \times 6} =$$

$\cdot 049$ of the power.

In high pressure engines the friction is supposed to be in the same ratio, but the loss of steam past the piston being in respect to the power in the inverse ratio of the diameter of the piston, I have assumed the friction and loss to be two-tenths of the power, and that because it corresponded with observation in two cases where I had a tolerably certain

---

* An arrangement of the parts of a metallic piston was one of the objects of a patent obtained by Perkins; but it is so obviously inferior to those already described, that it will be sufficient to refer to it. See Repertory of Patents, Vol. I. p. 224.

means of comparing the power and effect. Calculation gives the loss a little more; see note to (art. 384.)

Piston rod Collars, or Stuffing Boxes.

475.—The piston rod collar, or stuffing box, is a contrivance for rendering the place where a smooth rod or plunger passes into a vessel air-tight. This mode of giving motion without admitting air must have been long in use; we meet with it in various works without an allusion to the time of its invention. It is so similar to the construction of a piston that a separate detail seems scarcely to be necessary. As in the piston, so in this the effect is produced by elasticity; and leather, hemp, cotton, cork, and metal have been used for the purpose.

Where the heat of steam is not likely to be injurious, leather is generally employed. It was first applied in discs, cut to fit the rod, and pressed together by screws. The next was cupped leathers, and the first instance of their application seems to have been at the York Buildings water works,* and they were used by Smeaton, for his air pump; he also applied them to the piston rods of the blowing machines at Carron; and describes how to form the cups by stamping them into a cylinder of the size of the rod they are intended for.† What renders Smeaton's stuffing box for the blowing cylinders more curious is, that he uses a block of hard wood for the rod to pass through, and the rods it seems were draw-filed. The application of cupped leathers to the plungers of the hydrostatic press by Mr. Bramah, put them to the test on a large scale, under immense pressures.

476.—The stuffing box with the hemp packing is made to fit tight round the piston rod in a manner nearly similar to the piston. A collar with a hole through it, just sufficient to give easy passage to the rod is screwed down, to confine the packing, and cause it to press against the rod; it is cup-formed at the top to contain tallow to grease the rod. See Plate IV. and V.

477.—Metallic packing was tried for piston rods by Cartwright; and has since been much improved by Barton. It is however a part of so much less importance than the piston, that it will not be very often thought prudent to be at the expense, though the ingenuity of the contrivance renders it desirable to describe it. Barton's metallic substitute for stuffing boxes is shewn in the annexed figures; where D is the piston rod, E the

---

* Architecture Hydraulique, Vol. II. p. 62. Description of the Pumps of York Buildings Water Works, London.

† Phil. Transactions, Vol. XLVII. p. 415. Reports, Vol. I. p. 360.

box with a ledge for the cast iron plate F to rest on, and G another above it to receive
the cast iron plate H. The cover I of the box is secured by screws in the usual manner,

FIG. 21.

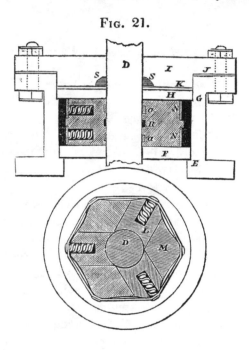

with plates of lead in the joints J, and K, for the purpose of making the joints closer.
The three principal metal blocks L embrace the piston rod D, and three wedging blocks
M fill up the spaces between them. Two thin hoops N N of tempered steel, firmly
riveted together at their ends, surround the outside of the blocks, binding upon the
rounded exterior angles of the blocks, and these angles are left on in the middle to keep
the hoops in their places. At each of the exterior angles of the blocks L, there are two
spiral springs, fitted to cylindrical holes, and also provided with cylindrical pins, as those
of the piston. By these and the elastic hoops the blocks L are strongly pressed towards
the piston rod. Two other hoops *a a* of elastic steel, cut across, are inserted in two
grooves to be in contact with the rod, and serve to close the joints more perfectly; they
are fixed in a similar manner to the rings round the piston before described. The
middle groove R, is formed between the two others to receive grease, and a circular
cavity S S, is also made around the hole in the cover of the cylinder for the same
purpose.*

* Gill's Technical Repository, Vol. IV. p. 242.

In constructing this collar the blocks L should be parallel, otherwise the wear will be irregular, and the springs will soon be ineffective; the blocks M should not wear unless the piston rod wears; and I do not expect that it will be steam-tight, if it be not assisted by a hemp packing behind the hoops N N.

Modes of opening Valves, Cocks, and Slides.

478.—The motion may be given either from the reciprocating or from the rotary parts of an engine. In engines which have no rotary parts, motion is communicated to the valves by a rod or beam, called a *plug tree*, attached to the engine beam near to the end moved by the piston rod. This plug tree is provided with certain adjustible projections called *tappets*, which strike the levers or handles of the valves, and thus open and shut them at the proper intervals as the beam ascends or descends. These handles turn on axes, and act as levers to move the valves, slides, or cocks. The most important point is to render the action certain, for the effect of the engine depends on the passages being opened and closed at the proper times. When valves are employed they are generally opened by weights. See Plate IX. Fig. 3. A weight w, sufficient to overcome the friction and open the valve, acts by a short arm a on the axis, which requires to be turned to move the valve; the weight is kept suspended by a spring catch b while the valve is close, and when the catch is disengaged by the handle c, being moved by the tappet d, the valve opens. If the valve be large it requires a considerable weight w to open it against the pressure of the steam; and in that case either the valve described in (art. 442,) or Watt's mode of relieving the pressure may be adopted. It will naturally be inquired, why weights are raised to open the valves instead of using the direct power of the beam. The only reason assigned for so doing, is, that a weight opens a valve more rapidly, and the loss by closing them slowly was not quite so readily detected; though the absolute loss is about the same, and the practice is becoming more common to open them by direct action.

The descent of the weight which opens a valve is regulated by an ingenious method: it either descends into, or forces a piston into a vessel of water, (see C Fig. 3, Plate IX,) while the aperture by which the water escapes from under it, may be increased or diminished at pleasure; the weight therefore acts with its full force to open the valve, but as soon as it begins to move it is retarded by the water, till it be finally stopped. During the ascent, a valve opens inwardly at the bottom of the vessel, and therefore the engine has no more than the weight to raise again.

In engines for raising water this mode of opening valves has always been followed. The difficulty of opening large valves was probably the cause of its introduction, and the ingenuity of its mechanism has preserved it in use; but I think there will be an advantage both in simplicity and effect, to let the motion of the plug tree act directly on the valves, as shewn in (art. 482;) the tappet by which the steam is shut off should be capable of considerable range, whether for adjusting by hand or by a self-acting apparatus. (See art. 554.)

479.—In an engine having a fly, it is esteemed better to apply an eccentric wheel within a hoop upon the fly-wheel shaft, and this by its revolution alternately pushes and draws a rod connected to the hoop, and thus gives motion to the valves, cocks, or slides. Such an apparatus is shewn at Fig. 2, in Plate XV. in which N is a cross section of the fly-wheel shaft, and k the eccentric wheel fixed upon and revolving with it; a circular hoop of metal encompasses the eccentric wheel in such a manner as to permit its turning round, and from this hoop the arm i projects, and it is braced to increase its strength. It terminates in an arm upon a centre, which by a second arm gives motion to the rod l; and causes another axis to move, which by a pair of bevelled wheels moves the cock of the engine partly round upon its axis n, and back again. The advantage of an eccentric wheel is the easy changes of motion it makes; for being constantly moving it gives no stroke at the times of change; and in large engines part of the weight of the eccentric apparatus is balanced by a weight, so that there is only a slight pressure on the shaft. See Plate XIX.

Let r be the radius of the eccentric circle, and d the distance of its centre from the centre of motion; then $r + d - (r - d)$ will be the extent of the movement, $= 2 d$, or twice the eccentricity; and in any other position the place counted from the centre will be $(d \cos. a)$ where a is the angle between the centres, whose cosine is equal to the horizontal distance. When they are in a vertical line, $a = 90$, and cos. $a = 0$, and the distance is o, and this corresponds to the termination of the stroke. Now we know from the nature of the circle that the cosines increase rapidly at first in departing from the angle of 90°; but at one-sixth of the stroke counted from either end of it, a valve, a slide, or a cock can be only half way opened, and unless its motion be greater than that required to open it, the time it will be about fully open will be only one-ninth part of the stroke.

480.—Eccentric rollers to raise the valve rods have the same defect; but the application is ingenious. Conceive the shaft Y, Fig. I, Plate VIII. to be kept in motion by the crank shaft of a double engine, causing the shaft Z to revolve by means of the wheels 7, 8; then if on Z two eccentric wheels 4, 4, be fixed, under two rods which slide vertically in guides, (see z, z, Fig. 2,) and provided with friction rollers 3, 3, the revolution of the shaft Z will alternately raise and depress the rods, which, by the arms 9, 10, 11, 12, raise and depress the valves by their stems. The lever or handle 13 is used to open or close the valves by hand in setting to work, &c. It will be remarked that this construction does not admit of cutting off the steam without also shutting the condenser.

481.—As far as regards opening and closing the passages more rapidly, a good improvement has been made on the eccentric motion, by altering the form of the portion fixed on the shaft so as to act more nearly as a tooth or cam, and by placing adjustible spanners on the eccentric rod; but why not at once form it as a tooth, or a series of teeth in the best manner to produce the movements required? Suppose the object be to cut off the steam at some part of the stroke by a slide or cock, then there must be two motions, the one double the length of the other. Let A B, Fig. 1, and 2, Plate IX. be the first, and B G the second, and from the centre D describe circles through these points; set off A E for the time to be expended in closing the passage to the condenser, and A F for the time of opening the passage for the steam; then, that the action may be easy, the curve H G should be drawn, so that each of its parts may be a parabola, the one with its vertex at H, that of the other, at G.* To produce the second motion another wheel should be placed on the same axis, behind the first one, with the curve I K. If these curves have corresponding ones, and act on connected rollers, the motion will be certain, and the range confined, and the motions of the engine may be reversed in the case of boat or carriage engines; for the position of the slide being changed by hand, the pressure of the steam will impel the crank shaft in the contrary direction, and the toothed wheel will move the slide or cock in the proper directions.

In order that the steam may be cut off at any period of the stroke, according to the resistance or the work on the engine, the wheel with the curve I K may be made to slide round on its axis, and the curve I K may be placed so that the period of cutting off the steam may be varied from N to O.

482.—If valves are to be opened, the weight of the valves and rods is generally sufficient to close them, hence, the rods do not require to be connected so as both to push and draw, but on the other hand a separate rod for each valve is required for a valve engine to work expansively,† and the toothed wheels or cams to move the rods will be placed with most advantage under the rods, as on the axis Z, Plate VIII. Fig. 1, and 2.

---

* The best curve for generating motion from rest is the common parabola. See Emerson's Mechanics, 4to Ed prop. 91. case 3.

† From the nature of the motions of the valves, slides and cocks being incompatible with the employment of the expanding force of steam in the engines of most makers, we infer that, Boulton and Watt's excepted, very few have availed themselves of this great source of economy. The proprietors of engines are too anxious about the power that an engine of a given sized cylinder possesses, forgetting that if an engine work with a minimum quantity of fuel, it must have a larger cylinder to do the same work. In estimating the comparative economy of engines, nominal power should not be considered, but the effect produced by each pound of fuel.

483.—To apply the same principle to a reciprocating engine; let A B, Fig. 4, 5, Plate IX. be the plug tree, with the curve C D to act on the roller at C, which as soon as the plug tree descends to C, begins to cause the roller frame to slide, and turn the axis E so as to depress the slide rod by the arm F. The steam will be shut off by H I on the descending stroke, and by K L in an ascending one.

484.—To regulate the period of cutting off the stroke, the portions containing the curves I H, and K L, may be made in two parts, to slide side by side by means of a screw : and if the rod having the screw upon it slides in a wheel acted upon by either a governor or other regulator, the engine will regulate itself. (See art. 554.)

485.—In all cases an axis to be alternately moved in opposite directions should be balanced, and the stress of all heavy parts should be relieved by counter balancing them by weights acting on levers. The hand gear should be a power proportioned to the force required to move the slides, cocks, or valves, (see art. 449.)

Piston Guides.

486.—The motion of the piston rod should be in a straight line in the direction of its length, and when the point it acts upon describes the part of a circle, the construction must be such that each may be confined to its proper motion; and yet the piston rod must produce the circular motion with as little oblique action as possible.

The most simple method is to confine the piston rod to its direction by means of a guide or guides, and to let it act on the part which moves in a circular direction by means of a connecting rod. To reduce the friction of the guides, rollers may be added. A very simple and efficient combination of this kind is shewn in Plate XV. Fig. 1. A wheel or roller F is fixed on the piston rod D, and is confined to a vertical motion by the guides G G, and the motion is transmitted to the crank I by a connecting rod H H. When the fly is of sufficient power the whole loss of force in this combination is simply the friction produced by oblique action, and is less in proportion as the connecting rod is longer; provided the stress from weight be not materially increased.*

---

* The whole increase of stress required for converting a reciprocating into a rotary motion cannot double the friction on the crank axis in any case, and as double this friction never amounts to a tenth part of the power of an engine, there is no reason to hope for an equal degree either of economy or simplicity, by using the rotary action of steam. (See art. 313—317, and the table, art. 487.

Cranks.

487.—The crank is one of the best contrivances for changing a reciprocating into a rotary motion. There are three different cases:

First, the moving force may be uniform and in a straight line,
Second, the moving force may be uniform and in a curved line,
Third, in either case the force may be variable.

A crank increases the velocity of the moving force, and in the usual construction, in the ratio of the circumference of a circle to twice its diameter; but this ratio is susceptible of variation, as is also the action of the power. This will be evident from the annexed figure; as if A B be the motion of the piston rod, the crank may be any where in the lune repre-

Fig. 22.

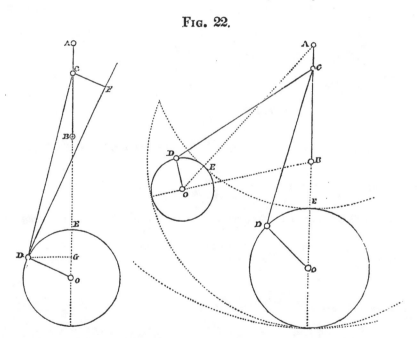

sented in dotted lines. If we sum up the forces acting in the circle, we find them exactly equal to the mechanical power in the straight line, the additional friction excepted.

The following table is calculated for a uniform force acting in a straight line; the moving force in the straight line is supposed to be 1, and the table shews the pressure it produces in the direction of a tangent to the circle, at the quarters and at every thirty degrees of its path. It will enable the reader to judge of the effect of a variable force; and where the acting point describes a curve, it is less regular but not so different nor so important as to require investigation : the last column is added to shew the additional stress on the axis above that which would take place, if the axis were turned by toothed wheels.*

* When the moving force reciprocates in the straight line A B, and the end D of a connecting rod is moved round in a circle. Let a be the angle the connecting rod forms with the direction A B of the motion, and c be the angle or arc described counted from E. The force, in the direction of the connecting rod, is P sec. a, where P is the force when the rod is vertical. Also,

$$1 : \sin. (c + a) :: P \sec. a : P \sec. a \times \sin. (c + a) =$$

the force in the direction F D of a tangent to the circle. But, sin. $(c + a) = \sin. c. \cos. a + \sin. a. \cos. c$, (see Gregory's Trigonometry, p. 42,) consequently,

$$P (\sin c. \cos. a + \sin. a. \cos. c) \sec. a =$$

the force. Also, since when the connecting rod is n times the length of the crank, sin. $c = n$ sin. a. we have

$$P \sin. c \left(\frac{\cos. c}{\sqrt{n^2 - \sin.^2 c}} + 1 \right) =$$

the force at any angle c.

The additional stress on the axis or shaft, and consequently the friction, is as

$$P \left(\cos. c \mp \frac{\sin.^2 c}{\sqrt{n^2 - \sin.^2 c}} \right).$$

The lower sign to be used after the rod becomes a tangent to the circle; the additional friction is therefore never greater than

$$\frac{P d}{16 r};$$

where d is the diameter of the shaft, and r the radius of the crank, both in inches; the friction being one-eighth of the pressure.

By construction of the figure the above ratios may be found; for if C G be the pressure, F D is the force in the circle, and C F the stress on the axis.

Table of the Variation of rotary Force, when a Crank is impelled by a constant Force.

Portion of the stroke described, the whole being 1	Portion of circle described, in degrees from the beginning.	Proportion of the length of the connecting rod to the length of the crank, the crank being 1						The stress on the axis, when the connecting rod is 6 times the length of the crank.
		2	3	4	5	6	7	
0	0°	0	0	0	0	0	0	1·0
0.067	30	·72	·65	·61	·59	·57	·56	·825
0·146	45	·97	·87	·83	·80	·78	·77	·624
0·25	60	1·10	1·01	·98	·95	·94	·93	·375
0·5	90	1·00	1·00	1·00	1·00	1·00	1·00	·169
0·75	120	·62	·75	·75	·78	·79	·80	·625
0·854	135	·43	·57	·57	·60	·62	·63	·790
0·933	150	·27	·39	·39	·42	·43	·44	·907
1·000	180	0	0	0	0	0	0	1·00

The length of the crank is supposed to be 1; and the table applies to any other length of crank, when the connecting rod is 2, 3, 4, 5, 6, or 7 times its length; the columns below these numbers shew the force corresponding to the positions indicated in the first and second column.

Parallel Motion.

488.—The next method to be described for communicating motion from a piston rod to a beam, is that called the *parallel motion.* It was first discovered by Mr. Watt, who gave a slight notice of his first attempts on the subject in Robison's Mechanical Philosophy; but its theory has been most generally investigated by Prony. I shall by confining myself to practical conditions, however, be able to treat it more briefly; and shew the best proportions for practice.

There are two cases, which for simplicity I propose to investigate separately; but they are generally both in use in the same engine.

489.—*First case.* If each of two bars A B, C D, Plate X. Fig. 4, has an axis at one end, round which it moves, and the other end be connected with a third bar, B D, by moveable joints; then, there is a point E in the middle bar which will nearly describe a straight line. The rectilinear movement of the air-pump rod, in a steam engine, is often obtained by this method, and as the motion is not perfectly rectilinear, it is most desirable to determine the point which renders it most nearly so.

490.—In any regulating apparatus of this kind, it is of considerable importance that the strains on the parts should not change their directions during the stroke; and this condition being premised, we shall have less difficulty in forming them to act with regularity and certainty. The beam A B, and the radius bar C D, will be both nearly in a horizontal position at the middle of the length of the stroke; and in order that the strains may not change their directions so as to jolt or force their axes to and fro, the connecting bar B D should not pass the vertical at either termination of the stroke: and to limit it to this condition, we will suppose the bar, as shewn by the dotted lines, to be exactly vertical, or coinciding with the direction of the rod it guides at each end of the stroke.*

When A B, is equal D C, the point E is in the middle of the length of the bar B D.

491.—Rule. With any other proportion between the lengths of the bars A B, and D C; for instance if A B : C D : : n : m. Then from the number n subtract half the square root of four times its square, less one, for a first number. Also from the number m subtract half the square root of four times its square, less one, for a second number. Divide the first number by the first added to the second, and the quotient multiplied by the length, B D, of the link or bar, will give the distance of the point E from B.

Example. Let A B be to C D as $2 : 3$; then $2 \times 2 \times 4 = 16$, and $16 - 1 = 15$, of

---

* Let A B and C D be the bars, and B D the connecting rod, and E the point to which the piston rod is to be attached; $b\,d$ being the direction it is to move in. Put A B $= n\,s$, D C $= m\,s$, B D $= l$, and the length of the stroke of the piston rod s, which is equal to the chord of the arc described by the bar A B. Make the versed sine of that arc x, and the versed sine of the arc described by the end D of the radius rod $= v$. Then a B is the sum of those versed tines $= x + v$; and $x + v : x : : l :$ B E $=$

$$\frac{l\,x}{x + v}.$$

But, by the properties of the circle, we have $s\left(m - \sqrt{m^2 - \cdot 25}\right) = v$, and $s\left(n - \sqrt{n^2 - \cdot 25}\right) = x$; consequently,

$$\mathrm{B\,E} = \frac{l\left(n - \sqrt{n^2 - \cdot 25}\right)}{\left(m - \sqrt{m^2 - \cdot 25}\right) + \left(n - \sqrt{n^2 - \cdot 25}\right)}.$$

When $m = n$, that is, when A B $=$ D C, then B E $= \frac{1}{2}\,l$.

which the square root is 3·873 nearly, and its half is 1·9365; and 2 — 1·9365 is ·0635 for the first number.   Next 3 × 3 × 4 = 36; and 36 — 1 = 35, of which the square root is 5·916, and its half is 2·958; and 3 — 2958 = ·042, therefore

$$\frac{\cdot 0635}{\cdot 0635 + \cdot 042} =$$

·602 nearly.

Hence, the length of the link or bar B D, multiplied by the decimal ·602 is the distance of the point E from B, when A B is to C D, as 2 is to 3.   Or if the point E be given, then B E divided by ·602 is equal B D, the length of the bar, link, or point of connection D from B.   The parallel motion of the engine in Plate XI. and that of Plate XIX. are examples.

492.—*Second case.*—In this case to a bar, which moves on an axis at A, Fig. 5, Plate X. conceive three shorter bars to be added to the end, so as to form with a part of the bar the parallelogram, B D C F; and let another bar D C, which moves on a centre at its extremity C, be attached to the lower angle of the parallelogram D, which is most distant from the centre C, round which the bar moves.   Then the piston rod being attached to the other lower angle G of the parallelogram, its motion will be nearly rectilineal in the direction G H.

The like reason of rendering the stress in the same direction during the whole of the stroke, would determine me to prefer the construction which renders the links B D, F G, vertical at both extremities of the stroke; this is not however the usual mode, for the line of motion of the piston rod is commonly made to divide the arc described by the end of the beam into two equal parts.   The very small difference from strictly rectilinear motion is rather increased by this mode, and it occurs at two points in the stroke instead of one, and causes irregular action, but the difference in this respect does not affect the investigation of the rules, for the length of the radius bars, &c.

In any case, except when the radius bar D C, and parallel bar D G, are of the same length, the deviation is increased by increasing the quantity of angular motion of the beam.   Hence, beams having short parallel bars, should be limited in the extent of angular movements; indeed the motion should not in general exceed twenty degrees, and this is very nearly the case when the distance of the end F of the beam from its centre of motion A, is to the length of the stroke I H, as 3 : 2; and then the radius bar may be found as follows.

493.—RULE. To find the length of the radius bar, when the length of the beam from the centre of motion is to half the length of the stroke, as 3 is to 2. From three times half the length of the stroke, subtract twice the length of the parallel bar, and multiply the difference by the half length of the stroke. Divide the product by ·343146 times the length of the parallel bar, and the quotient

I I

added to the length of the parallel bar will be the length of the radius bar.*

When twice the parallel bar D G, is equal three times half the length of the stroke, the radius bar and parallel bar should be of equal length.

Example. Let the length of the stroke I H, be eight feet, its half is four feet, and let the length of the parallel bar D G be three feet; then $\overline{3 \times 4} - \overline{2 \times 3} = 6$; and $6 \times 4 = 24$, which divided by $\cdot343146 \times 3 = 1\cdot029438$ is

$$\frac{24}{1\cdot029438} =$$

---

* Put $b =$ the length of the beam from the centre of motion, A F.

    $c =$ the length of the parallel bar D G.

    $r =$ the length of the radius bar D C.

    $v =$ the versed sine of the angle described by the radius bar.

    $a =$ half the angle described by the beam.

Assume it possible that the radius bar may be horizontal when the beam is horizontal; this connot be strictly but is very nearly true. Then, $(b - c)$ sin. $a = \sqrt{2\,r\,v - v^2} =$ half the chord of the arc described by the end D of the radius bar. But $v = c\,(1 - \cos. a)$; and substituting this value of v in the equation, it becomes

$$(b - c)^2 \text{ sin. }^2 a = 2\,r\,c\,(1 - \cos. a) - c\underline{2}\,(1 - \cos. a);$$

and by reduction,

$$\frac{(b^2 - 2\,b\,c)\,(1 - \cos.^2 a)}{2\,c\,(1 - \cos. a)} + c = r.$$

When $b = 2\,c$, then the first member of the equation disappears, and it becomes $c = r$; and this is the only case in which the length does not vary with the increase of the angle. By substituting for $(1 - \cos.^2 a)$ we have

$$\frac{(b - 2\,c)\,s^2}{2\,b\,c\,(1 - \cos. a)} + c = r;$$

which is a convenient formula when the angle is fixed; but when it is not, we have

$$\frac{(b - 2\,c)s^2}{2\,c\,(b - \sqrt{b^2 - s^2})} + c = r,$$

the length of the radius bar, where $b =$ the length of the beam, $c =$ the length of the parallel bar, and $s =$ half the length of the stroke.

If the beam from the centre of motion to the point F be one and a half times the length of the stroke; then,

$$\frac{(3\,s - 2\,c)s}{\cdot343146\,c} + c = r.$$

23·3137; add to this the length of the parallel bar three feet, we have 23·3137 + 3 = 26·3137 feet, for the radius bar D C.

I have taken a short parallel bar in the example, to shew the great length of radius bar, required in such a case.

The length of the links D B, G F, are from four to five-tenths of the length of the stroke, depending on convenience, and space; but the longer they can be made, the less oblique strain will take place during the motion. The vertical distance, between the centres of motion of the beam and the radius bar, should be exactly equal to the length of the links.

494.—RULE II. To find the length of the radius bar when there is no assigned proportion between the length of the stroke, and the radius of the beam. First, from the length of the radius of the beam, substract twice the length of the parallel bar, and multiply the difference by the square of half the length of the stroke.

Secondly, Find the square root of the difference between the square of the length of the radius of the beam, and the square of the half length of the stroke, and substract this root from the length of the radius of the beam, and multiply the difference by twice the length of the parallel bar. Use this product for a divisor, and the number found by the first operation as a dividend, and the quotient, added to the length of the parallel bar, will be the length of the radius bar.

Example. Let the radius of the beam A F, be twelve feet, the length of the stroke six feet, and the length of the parallel bar D G, five feet. Then the first operation is 12 — $\overline{2 \times 5}$ = 2, which multiplied by the square of half the length of the stroke is, 3 × 3 × 2 = 18.

By the second operation, the square of 12, less the square of 6, is, 144 — 36 = 108, of which the square root is 10·3923, and 12 — 10·3923 = 1·6077; this being multiplied by twice 5, is 10 × 1·6077 = 16·077.

Hence,

$$\frac{18}{16·077} + 5 =$$

6·12 nearly.

495.—When the proportions to obtain parallel motion have been found by the preceding rules, the point for the air-pump rod in the link D B, is easily found by drawing a line from G to A, and then the rod must be attached to the point of intersection. Its distance from the point B may also be found by the proportion, as

$$A F : F G :: A B : B E = \frac{A B \times F G}{A F}$$

Thus if A F be twelve feet; F C three feet; and A B seven feet; then

$$\frac{7 \times 3}{12} =$$

1·75 = B E.

In like manner, in any complex case, as in Woolf's engine with two cylinders, the points of connection for the piston rods must all be in the line A G, as is shewn in the examples in the plate; or the point for the air-pump rod being found by the rule, (art. 491,) the point for the piston rod may be ascertained by drawing a line through the points A, E, Fig. 4, Plate X. till it cuts the line in which the piston rod is to move at G; then draw G F parallel to the link B D, and G H parallel to the beam, and B F G H are the moveable joints of the parallelogram, and G the point to which the piston rod should be connected. The construction of the parallel motion adopted for steam boat engines, Fig. 1, Plate X. may be most conveniently solved by the rule for the first case, as will be evident from the figure, and the concluding sentence of the example, (art. 491.)

Of the Strength of the Parts of Steam Engines.

496.—In considering this important branch of my subject, I propose to follow the most simple methods I can devise, and those most readily applied in practice. The foundation of the inquiry must be the power of the steam in the boiler; or rather the greatest power it can possibly acquire without escaping at the safety valve. Now since there is always a risk of the safety valve not being in perfect order, we may in a great degree provide against such a risk, by taking the load on the valve at double the actual load upon it. Thus, if the load on the valve be eight pounds on a circular inch, consider it sixteen pounds; and sixteen pounds added to 11·5 pounds, the pressure of the atmosphere, will give 27·5 pounds for the strength of the steam, or the pressure which must cause the machine to move backwards.

497.—In the case of steam boats, a greater degree of surplus of strength, ought to be provided, because accidents at sea are attended with more serious consequences. And I would recommend all good machinery to be regulated by the following rule. It is to add the load per circular inch on the safety valve to the pressure of the atmosphere, and to take double this quantity, as the utmost force of the steam. That is, if the load on the safety valve be eight pounds on a circular inch, let this be added to 11·5, the pressure of atmosphere, the sum is 19·5, and double this is thirty-nine pounds, for the possible pressure per circular inch on the piston.

If the parts be formed to resist this pressure, then, in the case of the machinery being impelled backwards by an excess of resistance, they will not be injured by it, except where the momentum of a heavy fly wheel, renders it necessary to provide a resistance to impulsive force.

498.—The datum for the resistance of the material, must be the strain it will bear without a permanent derangement of its parts, and this strain is about one-third of its cohesive force.*

499.—In respect to the effect of the friction of an engine, it ought to be added to the power in estimating the strength, because when the resistance is capable of reversing the motion of the engine, it also must have to overcome the friction of the intermediate parts; but when the force of the steam is considered double its whole pressure, as limited by the safety valve, the friction may be neglected.

500.—The stress on any of the moving parts of a steam engine may be most easily found by comparing the number of revolutions or vibrations it makes for each double stroke of the piston; the stress is inversely as the number of revolutions or vibrations multiplied by the diameter of the circle, or the chord of the arc described by the point where the force acts; thus if a wheel be four feet in diameter, and makes three revolutions while the piston makes one stroke, and the length of the stroke be five feet; then $4 \times 3 : 5 :: $ pressure on the piston : stress on the teeth of the wheel, equal five-twelfths of the pressure on the piston. The stress thus found is to be considered as a weight applied at the point to which the motion belongs.

In like manner, the period of the motion of the working point of any machine may be considered unity, and by comparing the chords of the arcs described in the same time, and the revolutions in the same period, the stress may be found in terms of the force required to overcome the resistance at the working point.

501.—The method to be followed in determining the strength, is, when there is only one working point to proceed from the engine, taking its power as the measure of the stress at every point, and to make the part so that it shall be sufficiently strong to bear a reversion of the motion. But if the power of the engine be divided among various trains of machinery, then its power should be the measure of strength, only to the point where the trains branch, and for each separate train the greatest possible stress at the working point, should be made the measure of the strength of its parts.

The advantage of reasoning by general formula is so great that it will be adopted, and the rules as they arise given in words at length with examples.

502.—Let D be the diameter of the piston in inches, L the length of its stroke

---

* See Practical Essay on the Strength of Cast Iron, &c.   Sect. V. Second edition.

in feet and P the double of the whole elastic force of the steam in the boiler, in pounds per circular inch.

Also let l be the length from the centre of motion to the centre of stress in feet, $d =$ the depth or diameter, and $b =$ the breadth in inches ; $f =$ the cohesive force of a square inch at the point of alteration, and R = the radius of any wheel.

The force on the piston is $D^2 P$ in pounds.

503 —*Strength of rods where the strain is wholly tensile.* There is in every case of this kind a possibility of the strain deviating one-sixth of the diameter of the rod from the axis, and when it does so the resistance is

$$\frac{d^2 f}{2 \times 1 \cdot 27} = \frac{d^2 f}{2 \cdot 5}$$

nearly ; consequently,

$$D^2 P = \frac{d^2 f}{2 \cdot 5}, \text{ or } D \left(\frac{2 \cdot 5 \, P}{f} \right)^{\frac{1}{2}} = d.$$

For malleable iron $f = 17,800$, consequently,

$$\frac{D}{84} \sqrt{P} = d.$$

This rule applies to rods subject to a tensile strain only, such are piston rods of single acting engines; pump rods.

For head links it becomes

$$\frac{D^2}{84} \sqrt{P} = b \, t =$$

the breadth multiplied by the thickness in inches.

504.—RULE. Multiply the diameter of the steam piston in inches, by the square root of twice the elastic force of the steam in the boiler, in pounds per circular inch, and the product divided 84 is the diameter of the rod in inches.

Example. If the force of the steam be sixteen pounds per circular inch, and the diameter of the cylinder fifty-four inches, then, the square root of 32 is 5·657 and

$$\frac{54 \times 5 \cdot 657}{84} =$$

3·6 inches, the diameter required.

For the atmospheric pressure it is one-sixteenth of the diameter.

505.—*Of the strength of rods alternately extended and compressed.* In the compression of rods the force increases with the flexure, but if the length never exceed about thirty-six times the diameter, its error will be very small to assume that degree of flexure, and by taking in addition the greatest possible deviation, from misfitting, which is half the diameter of the rod,* with this simplification we have

$$\mathrm{D}^2\,\mathrm{P} = \frac{d^2\,f}{8\cdot75} \text{ nearly, or } \mathrm{D}\left(\frac{8\cdot75\,\mathrm{P}}{f}\right)^{\frac{1}{2}} = d.$$

For malleable iron $f = 17,800$, and

$$\frac{\mathrm{D}}{45}\,\sqrt{\mathrm{P}} = d.$$

For cast iron $f = 15,300$, and

$$\frac{\mathrm{D}}{42}\,\sqrt{\mathrm{P}} = d.$$

For tempered steel $f = 45,000$, and

$$\frac{\mathrm{D}}{72}\,\sqrt{\mathrm{P}} = d.$$

This rule applies to piston rods of double engines, parallel motion rods, air pump and force pump rods, and the like; and if P be increased in the ratio of the radius to the sine of the greatest angle a connecting rod makes with the direction, it applies to connecting rods.

506.—RULE. Multiply the diameter of the piston in inches by the square root of twice the pressure of the steam on a circular inch, and dividing the product by 45, for wrought iron, gives the diameter in inches.

For cast iron, divide by 42 instead of 45, for steel by 72.

Example 1. The force of the steam being sixteen pounds per circular inch, and the diameter of the piston eighty inches, that of the piston rod should be, for wrought iron,

$$\frac{80 \times \sqrt{32}}{45} =$$

9·8 inches.

Example 2. The force of the steam being four atmospheres = forty-six pounds per circular inch, and the diameter of the cylinder eleven inches, the diameter of a piston rod of wrought iron should be

$$\frac{11 \times \sqrt{92}}{45} =$$

2·34 inches.

If the rod be of steel, then

---

* Practical Essay on Cast Iron, art. 246.

$$\frac{11 \times \sqrt{92}}{72} =$$

1·46 inches.

Example 3. The force of the steam being sixteen pounds per circular inch, and the diameter of the piston twenty-four inches, the diameter of a cast iron connecting rod should not be less than

$$\frac{24 \times \sqrt{32}}{42} =$$

three inches and a quarter.

The middle is commonly expanded into a form of greater lateral strength, and in all cases should be of larger diameter than the ends, in the proportion of about one-tenth.

507.—For air pump rods, the pressure of the atmosphere and the diameter of the pump must be taken, instead of the force of the steam and the diameter of the cylinder. Parallel motion rods should be three-sevenths of the diameter of the piston rod, except in the case of that for steam boat engines, when there is lateral stress. Connecting rods for giving motion from the cross-head to beams, or to cranks, should be seven-tenths of the diameter of the piston rod.

508.—*Of the strength of arms of beams, cranks, &c.* It may be assumed as a principle that a beam of uniform thickness should not be of less thickness than one-sixteenth of its depth, otherwise it is liable to overturn; besides, in cast iron it is not safe to trust the strength of a casting, which is not a sixteenth part of its depth in thickness. Now for the case, when the velocity is the same as that of the piston $D^2 P l = 212 b d^2$,* and when $16 b = d$, and $12 l = n D$, it becomes for cast iron

$$D \left(\frac{1·34 P n}{212}\right)^{\frac{1}{3}} = d.$$

That is, when $D =$ the diameter of the piston in inches, and $d =$ the depth of the beam in inches, and the breadth one-sixteenth of that depth, n the number of times the diameter is contained in the length from the centre of motion to the point where the force is applied, and P double the force of the steam in the boiler, in pounds per circular inch. The depth at the end should be half the depth at the centre of motion, and the breadth uniform, and an access of strength may be given by forming the section, so as to increase the thickness at the edges to one-ninth of the depth, or till the parts between be reduced to the thickness of one-sixteenth of their width.

For wrought iron put 240 instead of 212, and for wood sixty-four instead of 212.

---

* Practical Essay on Strength of Iron, art. 116.

509.—Example 1. *Beams.*—An engine beam is three times the diameter of the cylinder, from the centre to the point where the piston rod acts on it, the force of the steam in the boiler is fourteen pounds per circular inch, its double is twenty-eight, and the diameter of the piston is twenty-four inches. In this case

$$D \left(\frac{1\cdot34 \; P \; n}{212} \right)^{\frac{1}{3}} = D \left(\frac{1\cdot34 \times 28 \times 3}{212} \right)^{\frac{1}{3}} =$$

·81 D = d, or d = 19·4 inches, and the mean breadth is 1·22 inches, and the breadth at top and bottom 2·16 inches.

If of wrought iron with the same proportions, ·78 D = d, and the breadth one-sixteenth of the depth.

Of wood with the same proportions, ·78 D = d, but the breadth one-fourth of the depth.

510.—*Cranks.* A crank should embrace a shaft, so that its depth at the shaft should be 1·5 times the diameter of the shaft; hence, if S D be the diameter of the shaft, the depth of the crank must be 1·5 S D, but since (art. 508,) $D^2 \, P \, l = 212, \, b \; d^2$, we have

$$\frac{P \, l}{2\cdot25 \, S^2 \times 212} = \frac{P \, l}{477 S^2} = b.$$

Example 2. A crank shaft is equal in diameter to ·31 times the diameter of the cylinder, and the force of the steam in the boiler being fourteen pounds per circular inch, or P = twenty-eight pounds, required the breadth of the crank at the shaft, its radius being 2·5 feet. In this case

$$\frac{P \, l}{477 \, S^2} = \frac{28 \times l}{477 \times \cdot31^2} =$$

·6 l = b in inches; and as l = 2·5 feet, it is ·6 × 2·5 = 1·5 inches, and the depth is 1·5 × ·31 × 30 = fourteen inches.

511.—*Wheel arms.* The arms of wheels may be considered in respect to strength only; and if the rim be of equal strength, then a wheel should in all cases have six arms, when it is of sufficient magnitude to require its strength to be found by rule. With this condition we have 2 D^2 P R = 212 × 6 $b \; d^2$, or D^2 P R = 3 × 212 $b \; d^2$. If we consider the arms to be one-third of the breadth of the wheel, and allow as excess of strength, that which is added to give it lateral strength, then

$$\frac{D^2 \, P \, R}{212} = b \, d^2, \text{ or } D \sqrt{\frac{P \, R}{212 \, b}} = d.$$

When R the radius is = 1, and P = 28 = twice the force of the steam in the boiler, then

$$\frac{D}{\sqrt{7\,b}} = d,$$

as inserted along with the proportions of teeth in the table, (art. 513.)

512.—The teeth of wheels will be most conveniently given in a tabular form, with a correction for curvature in determining their breadth, which is not included in the formula in my treatise on cast iron. The first column shews the stress on the teeth in pounds; the second the horses' power nearly equivalent, when the velocity is three feet per second; the third column the pitch; the fourth the thickness; and the fifth the breadth of the teeth : the sixth column shews the greatest depth of the middle of the arm at the base in the direction of the wheels' motion, when that part is one-third of the breadth of the teeth, and the radius is one foot; hence, being multiplied by the square root of any other radius in feet, it will be the depth for it : the seventh column shews the breadth of the rib which strengthens the arm; and the eighth the diameter of a cylinder, when the force of the steam is thirty-five inches of mercury in the boiler, and the teeth move at the same velocity as the piston. For any other velocity the stress will be found by (art. 500.)

513.—*A Table of the Strength, &c. of Teeth and Arms for Wheel Work.*

Stress in pounds.	Horses' power at 3 feet per second.	Teeth of Wheels.			Wheel with 6 arms.		Diameter of the cylinder for low pressure steam, teeth moving at the same velocity as the piston, in inches.
		Pitch in inches.	Thickness in inches.	Breadth in inches.	Depth of arm for 1 foot radius, in inches.	Breadth of rib in inches.	
22	¼	·25	·119	·75	·87	0·25	2·
85	½	·5	·238	1·25	1·24	0·42	3·7
191	1	·75	·357	1·75	1·67	0·6	5·5
337	2	1·00	·475	2·5	1·76	0·8	7·4
520	3	1·25	·59	3·0	2·	1·	9·2
800	4	1·50	·73	4·0	2·2	1·3	11·3
1040	5	1·75	·835	4·25	2·4	1·4	12·9
1370	7	2·00	·955	5·0	2·5	1·7	14·8
1720	9	2·25	1·07	5·5	2·7	1·8	16·6
2100	10½	2·5	1·19	6·0	2·85	2 0	18·4
2560	13	2·75	1·31	6·75	3·0	2·2	20·3
3000	15	3·0	1·43	7·25	3·2	2·4	22·2
3600	18	3·25	1·55	8·00	3·3	2·6	24·
4150	21	3·5	1·67	8·5	3·4	2·8	26·
4800	24	3·75	1·79	9·25	3·5	2·9	28·
5700	27½	4·00	1·91	10·25	3·6	3·4	29·5
6300	31½	4·25	2·025	10·5	3·7	3·5	31·5
6900	34½	4·5	2·15	11·0	3·8	3·7	33·3
7700	38½	4·75	2·27	11·75	3·9	3·9	35·
8500	42½	5·00	2·39	12·25	4·0	4·0	37·

514.—The strength of beam gudgeons may be determined by the rule, $P D^2 = 854 \, d^2$.* It reduces to

Essay on Strength of Iron, art. 139.

$$\frac{D \sqrt{P}}{30} = d,$$

and the length should be not less than eight-tenths of the diameter. For low pressure steam, twice its force is twenty-eight pounds per circular inch, or $P = 28$, and therefore in that case, one-sixth of the diameter of the cylinder should be that of the gudgeon. For pins for connecting rods where the bearing is double, the stress is reduced one-half, and

$$\frac{D \sqrt{P}}{43} = d.$$

Or in the case of low pressure steam

$$\frac{D}{8} = d.$$

515.—*The strength of shafts.* The shafts are supposed to be supported so as to render the lateral stress as small as possible, then the resistance to twisting alone has to be considered, and as no part of the shaft should be less than the bearings or journals, therefore allowing one-sixth for wear $R D^2 P = 960 d^3$;* when the shaft revolves in the same time, the piston makes a double stroke, and if the radius $R = n D$, we have

$$D \left(\frac{n P}{960}\right)^{\frac{1}{3}} = d,$$

the diameter in inches.

If it revolve N times while the piston makes a double stroke, then (art. 500,) we have

$$D \left(\frac{n P}{960 N}\right)^{\frac{1}{3}} = d.$$

For wrought iron the divisor should be 1080 instead of 960.

Example. What should be the diameter of a shaft of cast iron, when the crank arm is equal to the diameter of the cylinder, double the force of the steam in the boiler twenty-eight pounds per circular inch, the piston thirty inches in diameter, and one revolution of the shaft made in the same time as a double stroke? In this case n and N are each equal one, hence,

$$D \left(\frac{P}{960}\right)^{\frac{1}{3}} = D \left(\frac{28}{960}\right)^{\frac{1}{3}} =$$

$0.31 D = d$ in inches, and $0.31 \times 30 = 9.3$ inches.

Essay on Strength of Iron, art. 224, R being in this case in inches.

Of the strength of Pipes and working Cylinders.

516.—The thickness for pipes and cylinders of solid metal is more frequently determined by the condition, that the castings may be sound and perfect, than by a regard to strength, yet it is necessary to shew the proportions essential for strength, that a mistake in this respect may not occur.

The data required are the tensile strain a square inch of the metal will bear without permanent alteration, at the proposed temperature, the pressure of the steam on a circular inch, including such allowance as is proper for the risk of increase, and the diameter of the cylinder. I advise to take double the whole force of the steam when it escapes at the safety valve of the boiler.

We may safely consider the cylinder to be of equal resistance throughout its length; and hence, if we take the stress upon an inch of that length, that stress will be equal to the diameter in inches, multiplied by the greatest possible force on a square inch, and the resistance will be twice the thickness of the cylinder, by one-fourth of the limit of tensile strain of the metal, the tension being considered to be unequal on the resisting part. Therefore we have this rule.

517.—RULE. For the thickness of solid metal, pipes or cylinders to bear a given stress, the whole being of an equal temperature.

Multiply the 2·54 times the internal diameter of the cylinder, by the greatest force of steam on a circular inch; divide by the tensile force the metal will bear without alteration, the result is the thickness in inches.

Example. To determine the thickness of a cast iron cylinder, sixty inches diameter, for a pressure not exceeding 3·2 pounds per circular inch, in addition to the atmospheric pressure. In this case twice the force is thirty pounds on the circular inch, and the resistance of cast iron is 15,000 pounds per square inch; hence,

$$\frac{2\cdot54 \times 60 \times 30}{15,000} =$$

0·305 inches.

518.—Were there the direct force alone to consider, we see that a very thin cylinder or pipe is sufficient, but the pressure is often aided by a powerful strain from unequal expansion. If e be the extension the metal will bear without alteration, and t its thickness, d being the diameter of the pipe, we have

$$\frac{t}{2} : e : : d : \frac{2\,e\,d}{t} =$$

the greatest quantity which the expansion of one side of the pipe should exceed the other $=$ $h\,a$, when $h =$ the excess of heat, and $a =$ the expansion for one degree.   In cast iron

$$e = \frac{1}{1200}, \text{ and } a = \frac{1}{162000};$$

hence,

$$\frac{270\,d}{t} = h =$$

the increase which would strain the metal as far as it would bear without permanent derangement.

519.—Here we suppose the heat to be confined to a single point, but generally or rather in all cases, a considerable portion of surface is directly affected by the heat; in this case a near approximation will be to double the effect of expansion, or make

$$\frac{135\,d}{t} = h.$$

In the case of pipes and cylinders, the greatest difference of temperature will never exceed 300°; and then

$$\frac{300\,t}{135\,d} = \frac{2\cdot2\,t}{d},$$

the force of cohesion the cylinder loses by unequal expansion.

If this be added to the former, we must have

$$\frac{2\cdot54\,d\,p}{15000} + \frac{2\cdot2\,t}{d} = t; \text{ or } \frac{d\,p}{6000}\left(\frac{d}{d-2\cdot2}\right) = t.$$

The effect of irregular expansion is sensible only in small cylinders; in the case of a cylinder of sixty inches diameter we found its thickness ·305 inches, it became only ·315 when corrected for expansion.

For pipes of less than five inches diameter, the equation will be

$$\frac{d\,p}{6000 \times (1 - 0\cdot116\,d)} = t.$$

For working cylinders both wear and other causes of pressure exist; the latter will require at least that the thickness should be double, and for wear half an inch may be added, as about the proper quantity of allowance.

520 —RULE. For the thickness of a working cylinder. Multiply four times the elastic force of the steam in pounds per circular inch by the diameter in inches, and divide by 6000. The result multiplied by the quotient arising from dividing the diameter by the diameter, less 2·2, is the thickness for strength, to which half an inch may be added for wear.

Example 1. A cylinder twenty-four inches in diameter is to be made of cast iron, for steam not exceeding three pounds and a half per circular inch on the safety valve, or 11·5 + 3·5 = 15·4 elastic force, required its thickness.

Hence,

$$\frac{15 \times 4 \times 24}{6000} =$$

24; and

$$\frac{24}{24 - 2\cdot2} \times \cdot24 =$$

255, which added to half an inch, is 765 inches, the thickness required.

Example 2. A cast iron cylinder for a high pressure engine being nine inches in diameter, and for steam fifty pounds per circular inch elastic force, required the thickness.

In this case

$$\frac{4 \times 50 \times 9}{6000} =$$

·3 ; and

$$\frac{9}{9 - 2\cdot2} \times \cdot3 =$$

4, to this adding ·5 for wear, it is 0·9 inches.

521.—*Of the strength of flat plates to bear the pressure of steam, or other elastic fluids.* The strength of a plate is limited by the curvature it takes by the strain; and when the length and breadth are equal, the resistance is the same in both directions, but in any other case the two flexures do not correspond, and the resistance depends chiefly on the curvature in the shortest direction of support.

From the laws of deflexion it will be

$$l^2 : b^2 :: 1 : \frac{b^2}{l^2} =$$

the resistance in the longitudinal direction; and

$$1 + \frac{b^2}{l^2} =$$

the multiplier which multiplied by the resistance in the shorter direction gives the whole resistance.

When a plate is fixed at the edges, the flexure lessens the quantity of strain on the resisting part, but only in a small degree, and in bending to the new position, the inner part of the matter must be partially compressed, and the resistance to tension will extend only to a little more than half the thickness, and varies as the distance from the neutral line; hence, it is only one-fourth tf, when one-fourth of the thickness is an inch, t being the whole thickness, and f the cohesive force of a square inch; and allowing for rivetted plates

$$\frac{tf}{4} \times \frac{2}{3} = \frac{tf}{6} =$$

the resistance in one direction; and

$$\left(1 + \left(\frac{b}{l}\right)^2\right)\frac{tf}{6} =$$

the whole resistance.

The stress is as the force on a given portion of the curve, resolved into its tendency to split the material. If z be a part of the curve, and r the radius of curvature,* we have

---

* But the curvature is limited by the stretching and bending in the shorter direction, and if we suppose it to be wholly by bending, we have

$$\epsilon : 1 :: \frac{t}{2} : r = \frac{t}{2\,\epsilon};$$

therefore in this case

$$\frac{1\cdot27\,p\,t}{2\,\epsilon} = \left(1 + \left(\frac{b}{l}\right)^2\right)\frac{tf}{6},$$

or

$$p = \frac{\left(1 + \left(\frac{b}{e}\right)^2\right)f\,\epsilon}{3\cdot8}$$

Hence, we find that the resistance of a plate is quite independent of its thickness when it bends in this manner, but that the pressure is limited. For wrought iron, $f = 17800$, and

$$\epsilon = \frac{1}{1400},$$

hence,

$$\frac{f\,\epsilon}{3\cdot81} =$$

3·33 : and therefore

$z : r :: 1\cdot27\ p\ z : 1\cdot27\ p\ r =$ the stress, when p is the pressure on a circular inch; hence,

$$1\cdot27\ p\ r = \frac{t\ f}{6}\left(1 + \left(\frac{b}{l}\right)^2\right)$$

or

$$\frac{7\cdot62\ p\ r}{f\left(1 + \frac{b^2}{l^2}\right)} = t.$$

In square or circular plates it becomes

$$\frac{3\cdot81\ p\ r}{f} = t.$$

For wrought iron, $0\cdot006\ r = t$ when low pressure steam is to be confined, and both r and t are in inches, when the length is great compared with the breadth, or the bounding edges are not properly confined in one direction, then put the diameter instead of the radius of curvature.

Of the Excess of Strength to render Boilers safe.

522.—The pressure tending to separate a boiler is about proportional to the load on the safety valve; that, to crush it together is equal to the pressure of the atmosphere. In the latter case it cannot exceed that pressure, in the former a considerable excess may take place if any derangement happens to the valves, and it is to provide against accident, in the event of the valves being out of order, that an excess of strength in the boiler is necessary.

$$3\cdot33\left(1 + \left(\frac{b}{l}\right)^2\right) =$$

the greatest stress in pounds on a circular inch that a plate will bear. When the plate is either square or circular, it becomes 6·85 pounds per circular inch, = 8·5 pounds per square inch. When of other proportions as in the equation, and when the length is very considerable it becomes simply 3·33 pounds per circular inch, or 4·25 pounds per square inch.

Copper bears about the same strain.

This is important, as it shews us that flat surfaces cannot be used with safety to confine high pressure steam.

It is clearly a matter of opinion, founded on the experience of past accidents, as to the degree of excess required, and it has been almost universally allowed, that three times the pressure on the valve in the working state, should be borne by the boiler without injury.

This degree of excess of power seems to be fully sufficient for the ordinary low pressure steam boilers; indeed, I should think twice a proper allowance, and were it always provided, there would be little chance of accident if the valves be properly constructed and attended to.

It becomes insufficient in high pressure boilers, because a common low pressure boiler contains about ten times the volume of steam required for one stroke of the engine, consequently the time of twenty strokes must elapse before the density of the steam could accumulate to three times its working density, supposing the engine to be stopped, and the valve out of order. But if the boiler contains only as much steam as is required for one stroke, the force will be increased to three times, in the time the engine would have made two strokes. This rapidity of the increase of force does not leave the necessary time to examine, nor even to open the valves in this extreme case, and the hazard must be in consequence greater. In all cases the time of accumulating power should not be shorter than it is in the common boiler. Besides, in working an engine where the excess of force increases so fast, the loss of steam would be considerable from any variation of the heat of the fire, even were the valve to act properly, and therefore there is a temptation to load the valve beyond its regular weight. To render the security on the stoppage of the engine equal in all cases, the excess of strength should be inversely as the space allowed for steam.

It is still more important to consider the subject, in relation to the danger arising from unequal action of the fire, and for this the excess of strength should be inversely as the whole contents of the boiler expressed in units of the power.

Thus taking the horse power as the measure, if one boiler contains twenty cubic feet for each horse power, and another only ten, the boiler with only ten feet of space should be of twice the strength. For equal powers require equal fires, and the effect of excess of fire in raising the temperature, and force of the steam, is inversely as the quantity of matter acted upon; hence, the risk of the dangerous increase of strength is inversely as the quantity of water and steam the boiler contains.

523.—The proportion for excess of strength, I shall therefore consider to be two times that which is proper for the working pressure, when the boiler contains twenty cubic feet for each horse power, and containing any other quantity as n cubic feet per horse power, it will be

$$n : 20 :: 2 : \frac{40}{n}$$

The effect of unequal expansion, of improper form and flexure, and of wear must be included in the calculation of the strength, for these are not allowances for risk, but actually necessary for security.

Boilers may fail from strains produced by other causes besides the force of the steam, and these may be noticed to guard against the circumstance which produces them.

If a boiler flue rise from the fire, and then *descend* again before it enters the chimney, it will in particular states of the fire be liable to fill with inflammable gas, which takes fire and explodes. The effect of such an explosion in the flues of a boiler, must cause an impulsive strain on the boiler, under which it may fail.

The danger may be avoided by making the flues lead off to the chimney without depression, and constructing the damper so that it cannot be perfectly closed, and it should either rise so as to close the upper part of the aperture last, or move horizontally.

Hydrogen gas may be, and frequently is formed in steam boilers, through the water being in contact with a part of the boiler which is red hot, and it seems to be regularly produced during the formation of steam at very high temperatures. And though it appears to me that it would not add to the risk of an explosion happening, it undoubtedly would render it more destructive if it should take place.*

Boilers formed of Plates.

524.—Having determined the resistance of plates of any curvature, it is easy to apply these rules to rectangular boilers; remarking that it is indifferent, whether the curve be convex or concave to the pressure, provided it have either abutments as an arch, or forms a complete circle. I doubt the efficacy of the usual abutments, and I think the fact that boilers fail round the seats, is greatly owing to the strain and motion of the parts at every change of force or temperature.

* In a letter I received from Mr. W. Williams of Cyrfartha Iron Works, he attributes the destructive effects of an accident in that neighbourhood, to an accumulation of hydrogen inflaming, when the boiler burst. The boiler was constructed of the old spherical form, twenty feet in diameter, the thickness of the plates when new was, top plates a full quarter of an inch, bottom plates half an inch, load on the safety valve seven pounds per circular inch. Many lives were lost by this explosion, and the boiler was thrown to a distance of 150 feet, to a place thirty feet above the level of its former seat. The upper plates were undoubtedly too weak.

A rectangular boiler may be considered as a cylinder, taking the greatest diagonal line of its section for the diameter, and its strength will be (by art. 521,)

$$\frac{3 \cdot 81 p d}{f} = t;$$

or putting the value of $f = 17,800$ pounds for wrought iron

$$\frac{p \, d}{4660} = t.$$

The excess of strength for risk being

$$\frac{40}{n}$$

(see art. 523,) we have

$$\frac{p \, d}{120 \, n} = t.$$

And for copper $f = 11,000$, and

$$\frac{p \, d}{72 \, n} = t.$$

525.—RULE. For the upper plates of long rectangular and cylindric boilers. Multiply the load in pounds per circular inch on the safety valve, by the greatest diagonal of the section of the boiler in inches, and divide the product by 120 times the cubic contents of the boiler per horse power, the result is the thickness in inches. For copper divide by 72 instead of 120.

The bottom plates should be as much thicker as will compensate for wear; usually one and a half times the thickness of the top ones.

Example 1. In a rectangular boiler the greatest diagonal being eight feet, and consequently equivalent to a radius of curvature of ninety-six inches, the load on the valve three pounds and a half per circular inch, and the space for steam for each horse power sixteen feet, required the thickness for the top plates of wrought iron.

In this case

$$\frac{3 \cdot 5 \times 96}{120 \times 16} =$$

$0 \cdot 173$ inches.

The bottom plates may be $1 \cdot 5 \times \cdot 173 = \cdot 39$ inches.

This nearly corresponds with the practice of the best makers.

526.—Example 2. If the boiler be a long cylinder of which the diameter is sixty inches, and the pressure on the safety valve thirty pounds, the boiler containing twenty feet for each horse power of the engine; then

$$\frac{30 \times 60}{120 \times 20} =$$

0·75 inches.

In practice boilers of this kind are made barely equivalent to the working pressure, can we wonder that they sometimes fail ?

The same rule applies to internal flues with addition for the effect of the fire.

527.—Of spherical boilers. A spherical boiler has its dimensions equal, and consequently, (art. 521,) its strength is

$$\frac{3\cdot81\ p\ d}{2\ f} = t.$$

Hence for wrought iron

$$\frac{p\ d}{240\ n} = t;$$

and for copper

$$\frac{p\ d}{144\ n} = t.$$

RULE, for spherical boilers. Multiply the diameter in inches by the pressure on the valve in pounds per circular inch, and divide by 240 times the cubic contents of boiler for each horse power for malleable iron, or for copper by 144 times instead of 240.

Example. A boiler is of a spherical form, twenty feet in diameter, with twenty cubic feet to a horse power, and the load on the valve seven pounds per circular inch, what should be its thickness. The diameter is 240 inches, therefore

$$\frac{7 \times 240}{240 \times 20} =$$

·350 inches, or a little less than three-eighths of an inch. (See note to art. 523.)

When cylindric boilers have spherical ends, the radius of curvature may be equal to the diameter of the cylinder, and they will be equally strong with the same thickness of metal : and flat segments are more convenient in construction, and occupy less space to get the same effect.

Cast Iron Boilers.

528.—The preceding rules apply only to boilers of ductile metals, and in forming one for brittle ones, the effect of unequal expansion must be considered. For cylindric boilers the equation is

262 OF THE PARTS OF [SECT. VII.

$$\frac{2\cdot54 \; p \, d}{f} \times \frac{40}{n} = \frac{p \, d}{150 \, n} = t,$$

(art. 517 and 523;) and the mode of allowing for expansion was shewn in treating of cylinders, (art. 519.)

In boilers composed of solid tubes it is possible that in a tube having a diameter of eight inches or more, the excess of temperature on one of its sides may be 1000°, and then

$$\frac{7\cdot4 \; t}{d} =$$

the loss of force indicating that in those circumstances a tube would ultimately burst of whatever strength it were made; if $7\cdot4 \, t$ were made greater than d, for whenever the quotient is unity or more than unity, the unequal expansion alone is beyond the power of the material.

This explains the known fact that such tubes break without apparent defect, or the use of steam stronger than usual.

From these principles we derive the following rule, for cast iron boilers; d being the diameter, and p the elastic force of the steam,

$$\frac{d \; p}{150 \; n} =$$

the thickness for strength, which added to a thickness equivalent to the loss of force it may sustain by unequal expansion, is, for boiler cylinders above eight inches diameter,

$$\frac{d \; p}{150 \; n} + \frac{7\cdot4 \; t}{d} = t;$$

whence

$$\frac{d^2 \; p}{150 \; n \, (\, d - 7\cdot4 \,)} = t.$$

For boiler tubes, or cylinders under eight inches in diameter,

$$\frac{d \; p}{150 \; n \, (\, 1 - 0\cdot116 d \,)} = t.$$

In either case there is a risk of failure, when the diameter is less than $7\cdot4$ inches in the first, and when $0\cdot116$ times the diameter is greater than one in the second case. If the thickness be much more than the rule gives, the risk from unequal expansion increases; if it be less the joint effect of pressure and inequality may cause failure.

529 —RULE. For the strength of cast iron tubes exceeding eight inches in diameter. Multiply the square of the diameter by the pressure on the safety valve, in pounds on a circular inch, and divide the product by 150 times the cubic feet of space in the boiler

per horse power, multiplied by the difference between the diameter and 7·4 inches, the result is the thickness in inches, which should be increased for wear and tear in proportion to the degree of durability required.

Example 1. The internal diameter of the tube being ten inches, the cubic feet of boiler per horse power ten, and the load on the valve thirty-six pounds on the circular inch, what should be its thickness. In this case

$$\frac{10 \times 10 \times 36}{150 \times 10\ (\ 10 - 7\cdot4\ )} =$$

92 inches.

Example 2. The internal diameter of a cast iron cylinder for a boiler being three feet, and the force of the steam to be confined to five atmospheres, fifty-eight pounds per circular inch on the valve, what should be its thickness, the space of boiler for each horse power being sixteen feet.

In this case

$$\frac{36 \times 36 \times 58}{150 \times 16 \times (\ 36 - 7\cdot4\ )} =$$

1·1 inches.

Of Joining Pipes and other Parts of Engines.

530.—Joints are generally connected by screw bolts passing through flanches; between these flanches an elastic material of a durable nature is |inserted, or a compound, called a cement, which unites and forms one mass with the joined surfaces.

Iron cement is the most valuable of the latter kind, it may be compounded as follows : To two ounces of sal-ammoniac, add one ounce of flowers of sulphur, and sixteen ounces of clean cast iron filings or borings, mix all well together by rubbing them in a mortar, and keep the powder dry. When the cement is wanted for use, take one part of the above powder and twenty parts of clean iron borings or filings, and blend them intimately, by grinding them in a mortar. Wet the compound with water, and when brought to a convenient consistence, apply it to the joints and then screw them together. A considerable degree of action and reaction takes place among the ingredients, and between them and the iron surfaces, which causes the whole to unite as one mass; the surfaces of the flanches become joined by a species of pyrites, all the parts of which cohere strongly together. Mr. Watt found that the cement is improved by adding some fine sand from the grindstone trough.

531.—For some purposes it is more convenient to join the parts with white lead paint mixed with a portion of red lead to a proper consistence, and applied on each side of a piece of thick canvas, flannel, or plaited hemp, previously shaped to fit the parts, and then interposed between them before they be screwed together. It makes a close and durable joint, and is generally used for those joints which have occasionally to be opened, and for those which must be separated repeatedly before a proper adjustment is obtained; and when this is the case the white lead ought to be predominant in the mixture, as it dries much slower than the red.

532.—There is another cement often used by coppersmiths, to lay over the rivets and edges of the sheets of copper in large boilers, to serve as an additional security to the joinings, and to secure cocks, &c. from leaking; it is made by mixing pounded quick lime with serum of blood or white of egg; it is made into a paste, and must be applied as soon as it is made, for it speedily gets so hard that it becomes unfit for use. The properties of this cement have been long known to chemists, and it may be found useful for many purposes, to which it has never been yet applied. It is cheap and very durable.

533.—Steam-tight joints may also be formed by fitting the parts very accurately to a conical aperture, and screwing them close together with bolts of a less expansible metal; and the same method may be followed where the pressure of the steam tends to close the joint.

When two flat surfaces are to be joined they may be made to fit together very accurately, and a single ring of fine copper wire inserted in between them, before screwing them close. The pressure of the screws partially flattens the wire, and makes it fit so accurately as to prevent the escape of even very high pressure steam.

SECTION VIII.

OF EQUALIZING THE ACTION, REGULATING THE POWER, MEASURING THE USEFUL EFFECT, AND MANAGING THE STEAM ENGINE.

534.—The action of a steam engine is variable; consequently, when an equable motion is necessary, its action must be equalized. It may also be employed in one hour to overcome a small resistance, and in another to overcome a considerable one; therefore, the means of regulating the power to the work should be provided: we have also to consider certain methods which may be made subservient to ascertaining the useful effect of an engine after it is erected, or in the language of technical men, its *performance*; and, lastly, the mode of managing the generation of steam, and the working of an engine.

Of Equalizing the Action of Steam Engines.

535.—An equable motion is desirable in almost every kind of machine, it being strained much more by an irregular desultory one, as well as the fabric that supports it, than when the motion is equable. The strength of the machine must be adapted to the greatest strains that occur, but the quantity of work done is equivalent to the mean action only, and more is not performed by a desultory motion, than by one at a mean rate and uniform. There are two modes used for equalizing the action of an engine, which we propose to describe. The one is by the *fly wheel*, the other by a *counter weight*.

536.—*Of the fly wheel*. A fly wheel is a wheel with a heavy rim which absorbs the surplus force at one part of the action, to distribute it again when the action is deficient; it has been aptly compared by Professor Leslie, to "a reservoir which collects the intermitting currents, and sends forth a regular stream."* To equalize a motion which is subject to variation at each reciprocation in the steam engine, the fly is used. Its heavy mass of matter must be so shaped, as to balance itself in any position on an axis connected with the machinery, and turning round with a part of it.

The proportions of the fly wheel must be derived from the laws of rotary motion. They are not often stated very clearly, and rather in too comparative a form for the purpose of application; Dr. Jackson's equation† is derived most in unison with my own methods, and adding the time, the radius corresponding to the angular velocity of the exterior ring of the wheel, and comparing with the force of gravity to obtain the co-efficient, it is

$$\frac{32\ \mathrm{P}\ d\ r\ t}{b\ x^2} = n\ v.$$

In this equation P is the mean quantity the moving force varies in its intensity in excess above the resistance, and t the time in which that variation takes place; v the velocity, and $n\ v$ the greatest variation of velocity; d the leverage the force P acts with, and r the radius corresponding to the velocity v; and b the weight of the fly acting at the distance x from the axis.

It is obvious that the mass of the fly must be sufficient to receive the excess of force during the time it acts, and afford it again to the machine in an equal lapse of time; and so that the velocity shall not vary more than the nth part. The only point therefore, which depends on practical experience, is what variation of velocity may be allowed. On this point however there is no difficulty, as the practice of different makers is so different, as to shew that it may be taken with considerable latitude.

The weight of the rim may always be considered to be collected at the extremity of the radius; and then $x = r$, and the equation becomes

$$\frac{32\ \mathrm{P}\ d\ t}{b\ r} = n\ v.$$

The effect of the arms of the wheel may be neglected, as it is a problem which neither requires nor admits of a very refined solution, in consequence of the uncertainty regarding the precise variation of the intensity of the moving force; hence, it ought not to be rendered complicated.

537.—From this equation it appears, that when the weight or the diameter of the rim is considerable, and still more when both are so, it may acquire a great momentum with

---

* Natural Philosophy, Vol. I. p. 152.   † Theoretical Mechanics, art. 400—403.

but little increase of angular velocity, or lose a considerable momentum with little diminution of that velocity. It thus becomes a receptacle for the surplus energy of the power, when it acts with most intensity, or when the resistance is least, and preserves it for future demand.

By either a diminution of resistance, or an increase of power, the machine would otherwise be considerably accelerated; the excess of motive force is however, in a great measure, expended upon the fly, in which it generates a proportional momentum with little increase of velocity: again, when the resistance is increased, or the moving power diminished, the machinery would be very sensibly retarded, if the momentum accumulated in the fly did not continue the motion with little diminution of its own velocity; and other things being the same, the shorter the interval of reciprocation, or of unequal resistance, the less will be the change of velocity.

The greater the angular velocity of the axis of the fly is, the greater will be its dominion or equalizing power, all other things being equal, for the variation of velocity is inversely as the velocity of the rim.

Every part of a machine which has either a continuous or pendulous motion, particularly when it is massive, will obviously act as a fly in equalizing the motion of the machine.

The greater part of these remarks have been made in a less general form by Dr. Robison,* and Dr. Jackson;† but they also state that when a more perfect equalizer is wanted, we should increase the power of the fly wheel by enlarging the diameter rather than the mass, because we thus produce the same effect with less weight, consequently with less transverse strain upon the axle and supports, and less friction.

This must however be carried only to small extent, for a mass of matter with an immense velocity, sustained by arms which must be completely incapable of resisting its impulse, becomes a very dangerous appendage to a machine. Arms of cast iron could not resist a sudden check with a rim moving at the velocity of eighteen feet per second, and equal to the weight of the arms,‡ consequently, such wheels should be of limited diameter.

538.—When it is necessary to exceed a velocity of twelve feet per second at the rim, malleable iron arms should always be used, and a velocity of thirty-three feet per second at the rim is about the extreme limit for a fly, even where the ring is of malleable iron. For cast iron rims, with arms of malleable iron, I should not think a velocity exceeding eighteen feet per second safe.

---

* Mechanical Philosophy, Vol. II. p. 250.   † Theoretical Mechanics, p. 227.
‡ Essay on the Strength of Cast Iron, art. 261.

With these explanations we may now proceed to form rules from the equation. The equation is

$$\frac{32\,\mathrm{P}\,d\,t}{r\,n\,v} = b,$$

but the dimensions of the rim will be more convenient than the weight; and the weight $b = 2 \times 3{\cdot}1416\,r\,a \times 3{\cdot}2$ pounds for cast iron $= 20\,r\,a$, consequently,

$$\frac{1{\cdot}6\,\mathrm{P}\,d\,t}{r^{2}\,n\,v} = a =$$

the area of the section of the rim in inches. If the fly wheel shaft makes N revolutions per minute, then in the time t it will make

$$\frac{t\,\mathrm{N}}{60};$$

and as $v =$

$$\frac{6{\cdot}2832\,t\,\mathrm{N}\,r}{60},$$

we have

$$\frac{15\,\mathrm{P}\,d}{r^{3}\,n\,\mathrm{N}} = a.$$

539.—The next point to be considered, is the degree of equalization a machine requires. Its own parts have much effect, and the species of parts which act as flies, are most numerous in machines which require the equalizing power of the fly the most. At a mean, perhaps a variation of one-tenth is nearly corresponding with practice, and with this condition the rule is

$$\frac{150\,\mathrm{P}\,d}{r^{3}\,\mathrm{N}} = a.$$

540.—*Case I. A double engine with a crank.* In this case the variation is from the full force of the steam to nothing at each quarter of the stroke; hence, the mean excess is one-fourth of the greatest force P on the piston, and the rule in the nearest simple expression is

$$\frac{40\,\mathrm{P}\,d}{r^{3}\,\mathrm{N}} = a.$$

Rule. Multiply forty times the pressure on the piston in pounds, by the radius of the crank in feet, and divide this product by the cube of the radius of the fly wheel in feet, and by the number of its revolutions per minute, the result is the area of the rim of the fly in inches.

The number of horses' power, multiplied by 200, will be the greatest pressure on the piston, nearly.

Example. The pressure on the piston of an engine being 4000 pounds, the radius of the crank 2·5 feet, and the revolutions per minute twenty-two, required the section of the rim of a fly of nine feet radius. In this case

$$\frac{40 \times 4000 \times 2·5}{9 \times 22} = \frac{400000}{16038} =$$

twenty-five inches nearly, for the area of the section of the rim.

541.—*Case II.* *In a single engine with a crank* the mean excess is half the moving force; hence,

$$\frac{80 \, P \, d}{r^3 \, N} = a,$$

or the rim of the fly wheel should be double that required for a double engine with the same sized cylinder, or of twice the power.*

542.—*Counter weights.* If the beam of a single engine be balanced when at rest, that weight which it is necessary to add or subtract, to cause the piston to rise at the proper speed, is called the counter weight. The excess of force of the steam overcomes the friction of the parts, and the additional weight ought to be sufficient to cause it to rise and acquire double the velocity of the engine, if it freely accelerated during the whole stroke. If W be the whole weight of matter moved, $w =$ the counter weight, and $l =$ the length of the stroke, then

$$\frac{64 \, l \, w}{W + w} = 4 \, v^2,$$

or

$$w = \frac{v^2 \, W}{16 \, l - v^2,}$$

But (art. 342,) $v^2 = 2·66 \, l$; v being here in seconds: hence, $w = 0·2$ W; consequently, the counter weight should with these proportions be one-fifth of the mass of matter it has

---

* In single acting atmospheric engines a weight has been applied to the fly wheel, such that its force to turn the shaft should be exactly half the force of the steam to turn it, and placed so as to rise while the piston was descending, and descend during the rise of the piston. To find the weight, we have

$$\frac{P \, d}{2 \, r} = w;$$

when w is the weight and P the mean pressure on the piston. It is supposed to be applied to the rim of the fly, and the section of the continued rim should be the same as for a double engine of the same power. This mode is described in Fenwick's Essays on Practical Mechanics, p. 39. Woolf proposed to equalize the motion of an engine by a piston working in a cylinder; this however has no other effect than a weight, while the friction and expense of construction are considerably increased. See Nich. Journal, Vol. VI. p. 218, and Vol. VII. p. 134.

to move, supposing the whole to be collected at the ends of the beam, and it is most easily found by trial. The resistance of the water in the pumps will reduce the accelerated to an uniform motion of half the final velocity it would have acquired with no such resistance.*

Of regulating the Power of Engines.

543.—An engine is frequently applied where the work to be done is not constantly the same; and when the machinery of a part of it is suddenly stopped, or suddenly set on, if the moving power were to remain the same, an alteration of the velocity must take place, it must move faster or slower. This change of velocity would in most cases be very hurtful to the work, and cause considerable loss; besides, there is always a velocity at which a machine will act with greater advantage than at any other; therefore the change of velocity arising from the above cause, is in all cases a disadvantage, and in all delicate operations exceedingly injurious. In a cotton mill, for example, where the power moved the spindles with a given speed, if so much of the work were at once thrown off as to increase the velocity in a considerable degree, a loss of work would immediately take place, and an increase of waste from the breaking of the threads; on the other hand, there would be much loss of the time of the attendants, if the machinery moved too slow.

An equally bad effect is observed in raising water, and other species of work.

544.—*The throttle valve.* The power of a steam engine is usually regulated by increasing or diminishing the steam passage, and this is generally performed by admitting the steam into the cylinder, more or less freely, by means of what is called a throttle valve; this valve is formed of a circular plate of metal, *a*, Fig. 1, Plate VIII. having a spindle fixed across its diameter. The plate is accurately fitted to an aperture in a metal ring of some thickness, through which the spindle is fitted steam-tight, and the ring is fixed between the flanches of that joint of the steam pipe which is next to the cylinder. A square part is formed on one end of the spindle to receive an arm or lever *b*, by which the valve may be turned in either direction.

---

* Smeaton arranged his engines to make the returning stroke in less time than the acting one, (Reports, Vol. II. p. 360.) Watt states it to be generally agreed that the reverse should be the case. (Robison's Mech. Phil. Vol. II. p. 99.) The reasons for making them equal are stated in (art. 340.)

545.—For many purposes engines are thus regulated by hand at the pleasure of the attendant; but where a regular velocity is required, means must be applied to open and shut it, without any attention on the part of the person who has the care of the engine. For this purpose Mr. Watt, after trying various methods, fixed upon the conical pendulum which he called a governor. (See art. 550.)

An axis valve of this kind has much advantage over a valve of any other form for a circular pipe, because it contracts the aperture without being difficult to move, or presenting more than the necessary obstruction. But it is by no means an economical mode of varying the power of the steam engine.

546.—*To regulate by working more or less by expansion.* This may be done by adjusting the motion of the steam valves, so that they may be closed at an earlier or later period of the stroke, according as the engine has less or more work upon it This method is confined chiefly to regulating by hand, (see art. 481, and Plate IX.) The self-acting regulator in use applies with good effect only to valve engines, as neither the common slide nor cock can be adjusted otherwise than to close the passage to the condenser. (See art. 448, and 456.)

547.—*Field's valve.* An ingenious mode of cutting off the steam at any period of the stroke, has however been discovered by Mr. Joshua Field. It consists of a valve placed in the situation usually assigned to the throttle valve, that is, near to the place where the steam is admitted to the cylinder. This valve is to be opened at once, at the commencement of the stroke, so as to afford full passage to the steam, and shut at once, after a certain part of the stroke is made, that the rest of it may be completed by the expansive power of the steam. This may be done by causing the valve to open by a tooth or cam on a cylinder, on one of the revolving shafts formed to raise the valve, and keep it open till the shaft has made part of its revolution, and then shut it. If the toothed cylinder be made to slide on the shaft, and the form of the tooth be such, that as the cylinder is moved in one direction the valve will shut sooner, and in the other direction later, there is then the means of regulating the period the valve shall be open, and consequently of regulating the power of the engine. This may either be done by hand, or by causing the cylinder having the tooth to slide by the governor. Its application to Maudslay's portable engine, where it is moved by the governor, is shewn in Plate XV. It was there first applied by way of experiment, which will account for the indirect passages for the steam, and for retaining the throttle valve; the saving of power, according to the experiment, amounted to about ten per cent.

548.—When atmospheric engines condensing in the cylinder have to work under loads inferior to their whole power, they are regulated by lessening the quantity of injection, or by shutting the injection cock sooner. But in almost all engines employed for raising water which are regulated by hand, it is necessary to provide the means of warning the attendant of the power being in excess.

549.—*Spring beams.* In engines with fly wheels no precaution is necessary to limit the motion of the beam, because this is most effectually done by the length of the crank, while the fly continues the rotary motion so as to prevent strain on the crank shaft; but in engines where a crank is not used, as in engines for pumping, a very strong piece of timber is bolted across the top of the beam at each end, as shewn in Plate XII. each of which strikes against two wooden springs, one placed on each side of the beam, on the two longitudinal beams which support the axis of the engine.beam, and which are on this account called the spring beams of the engine. To prevent noise the springs are covered with cork at the place where they receive the stroke, and when they are bent beyond a certain degree they cause a bell to ring, which gives the attendant notice that the engine requires regulation.

The Conical Pendulum or Governor.

550.—If two or more balls be suspended from a revolving axis so as to revolve with it, the balls will rise when the velocity is increased, and fall when it is diminished; and by connecting arms to the rods by which the balls are suspended, their rising or falling may be made to move a lever so as to open or close a valve, or the like, on any change taking place in the velocity of the machinery; and hence, it is employed to render an engine the regulator of its own power to the effect it is to produce.

In the construction of this apparatus, there is to consider the place of the balls corresponding to the mean velocity, the range of motion, and the weight and velocity of the balls.

Different modes of combining the parts are used by different engineers; one of these is shewn in Plate VIII. Fig. 1, where g is the revolving axis, f the point of suspension, $j\,j$ the balls, $e\,e$ the rods by which the balls are suspended. These rods are connected to the rods $i\,i$, and by that means raise or depress the sliding piece h, and with it the lever l which acts on the throttle valve. The parts marked $k\,k$, are two rests to receive the balls when the engine is not in motion.

551.—The vertical distance between the point of suspension and the plane in which the centre of the balls revolve, is the same as the length of a pendulum which makes one vibration, forward and back again, in the same time the balls make one revolution. The usual velocity for the the axis is thirty revolutions per second, and therefore the height should be the same as the length of the seconds pendulum, that is, 39·14 inches. To find the height for any other number of revolutions per minute, divide 35,226 by the square of the number; thus, for twenty revolutions, 20 × 20 = 400, and

$$\frac{35226}{400} =$$

88·065 inches, the height required.

552.—The range may be settled from considering the greatest change of velocity the machinery may acquire without injurious effect on the work; and with this range the governor ought to be capable of completely cutting off the acting power. Now the greatest variation should not generally exceed one-tenth of the velocity, that is, one-twentieth on either side of the mean; and the range of the plane of revolution will, in that case, be nearly one-fifth of the height of the point of suspension above the planes of revolution at the mean velocity.* Thus, if the mean height be 39·14 inches, then one-tenth on each side will be

$$39\text{·}14 + 3\text{·}914 \quad 43\text{·}054$$
$$39\text{·}14 - 3\text{·}914 = 35\text{·}226$$

One-fifth of 39·14   =   7·728, the range.

Where a throttle valve is acted on by a governor, the steam passage should be fully

---

* For if v be the mean velocity, and it increases to $v + nv = v(1 + n)$; then the height of the plane of revolution will be altered from h to

$$\frac{h}{(1 + n)^2};$$

or in the ratio of $(1 + n)^2 : 1$, consequently, the change in the velocity will be to the change in the height of the plane of revolution, as $1 + n : (1 + n)^2$, or as the increments are as $n : 2n + n^2$, or as $1 : 2 + n$; and when n is a small friction, it is nearly as $1 : 2$.

From want of attention to this point, the governor has been supposed to be deficient in sensibility to the changes of velocity in a nice machine; and M. Preus has proposed to use a small pump to raise water to a cistern, from whence it escapes by an aperture which can be regulated at pleasure. When the engine moves at a greater speed than the proposed one, the water rises in the cistern, and raises a float which closes the throttle valve. See Phil. Mag. Vol. LXII. p. 298. It is obvious that it cannot exceed the governor in sensibility, while it will require considerable attention to keep it in order.

N N

open at the usual velocity of the engine, and contracted only when it exceeds that velocity, otherwise the steam must be always throttled, except when the engine is working against an unusual resistance.

553.—The balls are made from thirty to eighty pounds each; their effect, however, depends considerably on the angles formed by the combination of bars. In the form in Fig. 1, Plate VIII. the force is small, but the quantity of motion is considerable; while that in Fig. 3, Plate XV. has more force and less motion. The angle the ball rods make with the axis, should be about thirty degrees when they are at rest; and provided the range be sufficient, the angle the connecting rods make with the axis, may be made acute with advantage in point of power.*

554.—*The regulator.* The velocity of an engine for raising water may be regulated by a small cylinder provided with a piston, and fixed on a pipe from the air vessel of the main; which, when the engine goes too quick, forces water into the lower part of the small cylinder and raises its piston. The piston is loaded with a weight corresponding to the proper velocity of the engine, and therefore it is only when it goes too rapidly, that, the friction increasing in the main pipes, the pressure in the air vessel increases also, and this pressure being also communicated by the small pipe to the regulating cylinder, causes its loaded piston to rise, and the motion is communicated by a wire to the throttle valve, so as to close it and diminish the supply of steam; or, on the other hand, if the engine works too slow, the pressure in the air vessel diminishes, and the loaded piston descends and opens the throttle valve.

In order to prevent the motion of the piston being too great, the load is divided into links like a chain, and as the piston rises more links are raised, consequently the load increases; and also as it descends, the links, by resting on the ground, diminish the load. A spring might be applied to produce a similar effect.

555.—In some cases the further improvement has been adopted, of using this method to adjust the tappets which shut off the steam. For this object, the motion of the small piston is communicated to a wheel which turns a pair of bevelled wheels, the one of which is on the square part of a screw rod attached to the plug tree; and whenever the motion is too rapid the rod is turned, and moves the tappet so as to cut off the steam sooner, and the reverse. The square part of the rod slides in the wheel upon it without change, except when that wheel is moved by the regulator piston.

556.—*Of the cataract.* The power of an engine for raising water may also be regulated by increasing or diminishing the interval between its strokes; this is done by caus-

---

* Several trials have been made to apply the governor to boat engines, but it appears to me that the changes are too sudden for this mode of regulation.

ing the tappets to disengage a loaded piston, which descends in a small air vessel, expelling the air from it by a pipe, which can be regulated by a cock at pleasure; the valves are not free to open till this piston be at the end of its stroke. The air vessel is a cylinder of from five to six inches diameter and twenty inches in length, open at the top with a valve opening inwards at the bottom, that it may ascend without unnecessary resistance. It is provided with a pipe from the bottom, of sufficient diameter to allow the air to escape when the engine is at full speed, which has a cock to regulate the time of discharge. It is also fitted with an air-tight piston, the rod of which is connected with the apparatus which opens the valves. Two air vessels are required for a double engine.

Of the Methods of ascertaining the State and effective Power of a Steam Engine.

557.—Certain instruments have been invented which are of great use in ascertaining the state of an engine; and these ought to be kept in good order, so as to be capable of affording the required proof at any time. Mr. Watt has most justly remarked, " It is the interest however of every owner of an engine to see that they, as well as all other parts of the engine, are kept in order.*

The instruments consist of a steam gauge, the condenser gauge, and the indicator.

558.—*Steam gauge.* The steam gauge, Plate VIII. Fig. 1, (18,) is a short bent tube of iron, nearly half an inch in diameter, with one end fixed into the boiler or the steam pipe, and open to it; with a portion of mercury in the bent part of the tube. The part joined to the boiler or steam pipe is freely open to the steam, which, pressing on the surface of the mercury in the pipe, raises it in the other leg of the tube, which is open to the air at the upper end, and the height it is raised is measured on a scale (20) by a slender stem from a light float on the surface of the mercury; which therefore shews the elastic power of the steam over that of the atmosphere. The scale should be adjusted by allowing the air free access to the mercury on both sides.

The scale is commonly divided into inches and parts; each inch corresponds to two inches of mercury, and to a pressure of 0·775 lbs. on a circular inch, and to 0·98 lbs. on a square inch. If each of the divisions of the scale be made 1·3 inches, and these each divided into ten equal parts, the pressure in pounds and tenths on a circular inch will be shewn by the gauge. Some divide the scale into half inches, then each division represents an inch of mercury.

* Robison's Mechan. Phil. Vol. II. p. 156.

Sometimes a cock (19) is placed between the mercury and the steam, so as to use it or not at pleasure.

To render the divisions of the gauge larger, Mr. Watt made his guage pipe of glass, to terminate in a cistern of mercury inclosed in an iron box. The action is then like a common barometer; the steam having free access to the surface of the mercury in the cistern.

559.—*Condenser gauge.* This is sometimes called the barometer gauge, from its resemblance to a barometer. It is made of iron tube in the form of an inverted syphon, Plate VIII. Fig. 1, (21) with one leg about half the length of the other. To the upper end of the longer leg (24) a pipe is joined which communicates with the condenser, and has a stop cock (22) to open or close it. When a proper quantity of mercury is poured into the short leg of the syphon, and it is open to the atmosphere at both ends, it naturally stands level in the two legs. A light float with a slender stem is placed in the short leg, and a scale (25) attached, which is usually divided into half inches; and as by the exhaustion in the condenser, the mercury rises as much in the long leg as it falls in the short one, these divisions will be equivalent to inches on the common barometer.

The condenser gauge should indicate the state of the vapour in the condenser, to be capable of sustaining from two to three inches of mercury. While it does not exceed three inches the condensation may be esteemed very good; and about two inches is the best I have seen obtained in practice.

The difference between the elastic force of the vapour in the condenser, and the elastic force of the steam in the boiler, as shewn by the gauge, added to the height of the barometer at the time, gives the relative force of the steam to move the engine, but many deductions have to take place before we have the real moving force; nevertheless they shew the state of two very important parts of the engine. (See Sect. V. and VI.)

560.—*The indicator.* The force of the steam and the state of exhaustion in the cylinder, at the different periods of the stroke of the engine, cannot be ascertained by the condenser gauge, and for that purpose it was necessary to form an instrument less subject to vibration; the instrument in use is call the *indicator*, and is found to answer the end tolerably well. It consists of a cylinder about one inch and three quarters in diameter, and eight inches long, exceedingly truly bored, with a solid piston accurately fitted to it, so as to slide easy by the help of some oil; the stem of the piston is guided in the direction of the axis of the cylinder, so that it may not be subject to jam or cause friction in any part of its motion. The bottom of this cylinder has a cock and small pipe joined to it; a flat pillar D (Plate XVI. Fig. 1,) is screwed to the cylinder of the indicator C, and supporting the frame E E, which is twelve inches by seven inches, with the upper and under rail grooved to retain the sliding board K.

The piston rod G is about five-eighths of an inch in diameter, and sixteen inches long;

and H is the guide for it screwed to the pillar D, at about six inches above the top of the cylinder C.

A spiral spring I is attached to the piston at F, and to the guide at H. It should be about seven inches long when at rest, and of such a strength as to allow the piston F to descend to within about an inch of the bottom of the cylinder C, when it is loaded with fifteen pounds upon every square inch of its area; and the spring should admit of being compressed one inch and a half.

The board or pannel K slides in the grooves of the frame E E, and should be seven inches square; and a small brass slider L should be set at any height on the piston rod G, by means of a screw. A short pencil is inserted in the other end, with a weak spring to push it against the surface of the board K, which is caused to slide by a weight N, attached to a line passing over a pully; the opposite line O being attached to any convenient part of the parallel motion of the engine, so as to cause the board K to traverse a space of about fourteen inches and a half during each half stroke of the engine.

Operation. By opening the stop cock B, a direct communication is made between the cylinder of the steam engine, and the cylinder of the indicator. When the force of the steam in the cylinder is greater than the pressure of the atmosphere, the piston F will rise; when the force of steam is less than atmospheric pressure, it will sink. The indicator will consequently rise when the upper steam valve opens, and will be at a height proportional to the force of the steam in the cylinder during the stroke of the engine; and when the eduction valve opens, it will sink, and by the rapidity and quantity of its descent, denote the state of the vapour in the condenser. During the motion of the piston F, the sliding board will move horizontally, and the pencil in the socket in L will trace on the board K, or on a paper on its surface, a figure P, Q, R, S, resembling those shewn on an enlarged scale in the annexed figures. Of this figure, P Q is described during

FIG. 23.

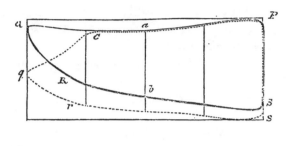

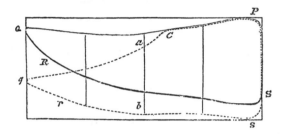

the descent of the steam piston, at Q, the condensation takes place, and the indicator is forced down by the pressure of the atmosphere, till it be balanced by the resistance of the spring, and the vapour in the cylinder. While the engine makes the ascending part of the stroke, the line R S is described, and the line S P is described during a fresh admission of steam by the upper valve.

The area P Q R S, is proportional to the force of steam on the piston during the stroke. But the steam is not exerting the greatest power in proportion to the quantity of fuel, when this area is the greatest, for when the steam acts by expansion, the area described will resemble the figure P Q R S, the steam being cut off at C, and more power will be exerted by a given quantity of steam. In the same engine doing different quantities of work the figures shew two cases; the black lines represent the power when the throttle valve is used for regulation, and the dotted lines when the engine is regulated by cutting off the steam.

561.—If p be the number of pounds on a circular inch of the indicator piston, which causes it to descend d inches, and let m be the length of the line $a\,b$, measured in inches on the diagram drawn by the indicator, then

$$d : m :: p : \frac{m\,p}{d} =$$

the pressure exerted by the steam in pounds on a circular inch, at the point a of the descent of the piston.   Thus, if by trial two pounds per circular inch causes the piston to descend one inch, then

$$\frac{m\,p}{d} = \frac{m \times 2}{1} = 2\,m\,;$$

or each inch of the indicator would correspond to two pounds per circular inch.

If the distance the tracer moves horizontally be divided into equal parts, and the vertical distance between the lines P Q'and R S be taken at each point of division, and the sum of these distances, less half the distance P S, be taken and divided by the number of divisions, the result will be the mean distance the piston of the indicator moves over ; and calling this mean distance m, it will be

$$\frac{m\,p}{d} =$$

the mean pressure on the steam piston in pounds upon a circular inch.*

562.—*To measure the useful effect of an engine.*   The preceding methods only give the state of parts, but the useful effect depends on the whole being in order, and the most simple and convenient mode of measuring the effect, is by means of friction.†   If the rim of a brake wheel on the engine shaft of a known diameter be pressed with a force producing a known degree of friction, which is exactly equal to the effect of the engine at its working speed ; then it is clear, that if the friction this pressure produces be ascertained, the power of the engine will be equal to the friction multiplied by the velocity of the rubbing surface.

To apply this, let A B be a lever, with a friction strap that may be tightened upon the cylindrical surface of the shaft or wheel C, and let it be tightened by the screw at B, (the lever being stopped by the stop D) till the friction be equal to the power of the engine, when all other work is thrown off, then while the engine is still in motion add such a weight at E as retains the lever in a horizontal position.

FIG. 24.

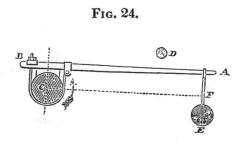

To calculate the power.   Multiply together the length F C of the lever in feet, the

* The indicator appears to have been invented by Watt, (Robison's Mech. Phil. Vol. II. p. 156.) and the tracer applied to it by Field.

† It is for use in those cases where the work itself is not susceptible of accurate measure; and almost all engines for impelling machinery are in this class.   The power of engines for raising water is easily computed.

weight E in pounds, the number of revolutions of C per minute, and the number 6·2832, the result will be the pounds raised one foot per minute; and, divided by 33000, it is the horses' power.*

Thus if a shaft C make twenty-five revolutions per minute, and the length E C of the lever be ten feet, and if it be found that a weight of 240 pounds is sufficient to retain the lever in a horizontal position, then 6·2832 × 10 × 240 × 25 = 376992 pounds raised one foot, or eleven and a half horses' power, very nearly.

The trial is so easily made, and the result so accurate a test of the qualities of an engine, that I strongly recommend it to the notice of those who are desirous of having good engines.

563.—*The counter.* To estimate the saving of fuel by the application of Watt's engines, an apparatus was attached to the beam, to ascertain the number of strokes the engine made in a given time; it is called the counter, and consists of a train of wheel work resembling that of a clock, so arranged that every stroke made by the engine moves one tooth, and the index shews how many strokes have been made between the times of examination. The counter is inclosed in a box and locked, to prevent it being altered during the absence of the observer. If the box be attached to the axis of the beam, the inclination of the beam causes its pendulum to vibrate every time the engine makes a stroke, and thus moves the counter round one tooth for every stroke. The box may also be fixed to the supports of the beam, and then at every stroke a small detent is moved one tooth. The counter is still used in Cornwall, in order that the effect of the engines may be reported on by the inspectors; and it is useful in various instances where a check on the consumption of fuel is desirable.

Of Working Steam Engines.

564.—The first attention should be directed to the qualities of the fuel and the

---

* For let *l* be the leverage the weight *w* acts with, and *r* the radius of the wheel or shaft *c*; the friction *f*, the velocity *v*, and the revolutions of *c* per minute *n*. Then $f v =$ the power, and

$$f = \frac{l\ w}{r},$$

and $v = 6\text{·}2832\ r\ n$, consequently, $v f = 6\text{·}2832\ l\ w\ n$, the power in pounds raised one foot, when *l* is in feet and *w* in pounds.

water. The fuel, of whatever kind, should be dry, in small parts, and free from earth, &c. No coals should be above the size of an egg, and they should contain as little pyritical matter as possible. Wood should be in billets not more than a foot long, nor two or three inches diameter. The water used should be pure and soft when it can be got.

All natural waters contain a quantity of matter, which they derive from the strata through which they flow. The purest springs usually rise in beds of gravel, or in siliceous or argillaceous rocks, and they contain, for the most part, only a minute portion of saline matter, which is principally common salt. The water of limestone or calcareous districts generally contains a much larger quantity of solid matter, most frequently lime in solution, either carbonate or sulphate of lime, which occasions that peculiar quality in waters commonly known by the name of hardness.* The waters of mines are still more impure, they often contain earths, acids, alkalies, and saline compounds. Hence, water in its natural state is often unfit for a steam engine.

Almost the only practicable method of improving water, is that of exposing it long to the air in ponds; a more effective one is to use the same water over and over again, with only such addition as compensates for loss, but even this requires a larger reservoir to allow the water time to cool. Foul river water may be cleared by filtering through sand or gravel.

To prevent the sediment from the water adhering to boilers, it is common to put in crushed potato, the refuse of malt, and the like, and change it frequently.

565.—For sea-boat engines sea water must be used, and it deposits salt after the water is saturated. This may be prevented by letting a small quantity of hot water escape constantly from the boiler. One hundred parts of sea water contains three parts of its weight of saline matter; and is saturated when it contains thirty-six parts,† conse-

---

* Hard waters do not readily dissolve soap, nor form a good lather with it, on the contrary they partially decompose it, and a light flocculent substance is produced which is insoluble in water.

† According to Mr. Faraday's experiments, the deposits take place at the following degrees of saturation and temperature, when 1000 parts of sea water were reduced by evaporation.

Quantity of sea water.	Boiling temperature	Salt in 100 parts.	Nature of deposit.
1000	214°	3	None.
299	217°	10	Sulphate of lime.
102	228°	29·5	Common salt.

quently, if the boiler contain one hundred parts of water, and s parts be used for steam, and n parts let out, then fix on the degree of saturation, for example, to contain a parts of saline matter; then if $3 (s + n) = a n$, the quantity of salt entering, and the quantity quitting in the same time will be equal; hence,

$$\frac{3 \text{ S}}{a - 3} = n.$$

If $a = 30$ the water in the boiler will not reach to a higher degree of saturation when a ninth part of the quantity used for steam is allowed to escape. Now as it requires only about one-sixth of the quantity of fuel to boil water, that is required to convert it into steam, hence, the loss of fuel will be $9 \times 6 =$ one fifty-fourth part.*

566.—*Of working a condensing engine,* The engine is supposed to be at rest, the cylinder quite cold, and the condenser partially filled with water, and the piston at the top. When the water in the boiler begins to boil let steam enter by the valves, or the slide, and the communication which is added when a slide is used; it will then fill the cylinder and pipes, and the injection cock being shut, it will gradually displace the water in the condenser,† by forcing it out at the blow valve, and afterwards the air. When all the air except that mixed with the steam is driven out, which will be known by the sharp cracking noise at the blow valve, this noise must be allowed to continue till it be supposed that there has been time for at least as much steam as fills the engine to run through. Then shut off the steam except to the upper side of the piston, and open the injection cock; if motion does not commence, the injection cock must be shut, and the blowing through of steam repeated. If the engine have a steam case or jacket, that case must be cleared of air and water, and filled with steam before the operation of blowing through be commenced.

---

* An ingenious combination was formed by Messrs. Maudslay and Field, to avoid this loss by causing the water enter_ing the boiler to traverse the central parts of the pipe which removed the saturated hot water; the loss of effect then becomes very small, even when the degree of saturation is much less. Now to prevent the deposition of sulphate of lime we must make $a = 10$, and then three-sevenths of the quantity required for steam must be let out, and one-fourteenth of the effect of the fuel will be lost. It may be necessary to explain that sulphate of lime does not appear to exist in sea water till a change of combination among its constituents takes place through evaporation; also, it requires thirty-six parts of salt to saturate one hundred of water at 226°, but it appears that deposit takes place at 228° with only thirty parts.

† The best method would be to use a cock to let out the water from the condenser, placed so low as to completely drain it, and the water should be let out in the first instance; then the steam should be let on, and as soon as steam issued at the cock it should be closed, till it be evident that the cylinder is as hot as the steam will render it; then open the cock again till steam has passed out a few seconds, and close it, confining the steam to the top of the piston, and open the injection, when motion will commence, if the operation has been properly performed.

The noncondensing species of engines require nothing more than to be heated and freed from water, and atmospheric engines condensing in the cylinder to be freed of air.

567.—*Of the management of the fire.* The chief thing is to obtain as equable a supply of steam as possible; and the object of the attendant must be to render it so with the least occasion for opening the fire door. He must endeavour to preserve a clear free burning fire, which cannot be done if it be allowed to become foul by clinkers accumulating. Every coal which will not pass a ring of about two inches and a half diameter should be broken; and either feed frequently, thinly, and equally over the surface of the fire, or adopt the method pointed out in (art. 249,) but in important works Brunton's method should be applied, (art. 250.)

568.—Great care should be taken to keeping the engine and boiler clean and in good order, and for this purpose frequent and steady attention is more effectual than twice the quantity of irregular labour. In work of this kind that which is done well is twice done; and the most furious zeal is vastly inferior to steady attention, for the one destroys the objects of its care, the other preserves them.

The best kinds of oil, and tallow should be used; for piston grease, tallow is most esteemed, and when cylinders are new, a small addition of very soft black lead in fine powder improves the effect of the tallow. Oil appears to be improved by the addition of a small quantity of wax.

SECTION IX.

OF THE APPLICATION OF STEAM ENGINES TO DIFFERENT PURPOSES.

569.—The immense variety of objects to which steam power is or may be applied, renders it necessary to confine our attention to the most prominent ones for illustration. These are to raising water; to impelling machinery for mining, manufacturing, and agricultural purposes; and to land carriage: the application to navigation being so distinct and important, as to require a separate section. (See Sect. X.)

Of raising Water.

570.—Water is generally raised by means of pumps of the lifting or forcing species. The stroke of a pump should not exceed about eight feet, otherwise the air disengaged from the water, the escape by the bucket or piston, and the defect of pressure on the fluid which is rising after the piston, becomes greater than the escape by the valves. The velocity of the piston should not exceed ninety-eight times the square root of the length of the stroke, (art. 342.)

571.—Owing to the escape at the valves and the disengagement of air, the quantity of water a pump in the best order delivers at one stroke is

$$\frac{\cdot 95 \; l \; d^2 \times \cdot 7854}{144} =$$

$\cdot 00518 \, l \, d^2 =$ the quantity in cubic feet; when l is the length of the stroke in feet, and d

the diameter of the pump in inches; or substituting half the velocity for l, it gives the cubic feet per minute.

572.—The power required to raise water a given height, is found by taking the exact height in feet, from the surface of the water to the point of discharge, and adding one foot and a half for each lift, for the force required to give the water the velocity; and also add one-twentieth of the height for the friction of the piston, and call this quantity in feet h; then $\cdot 341 \, h \, d^2 =$ the load in pounds.

Whence, if $P =$ the mean effective force on the steam piston in pounds per circular inch, we have

$$d \left(\frac{\cdot 341 \; h}{P} \right)^{\frac{1}{2}} = D,$$

the diameter of the steam piston in inches.*

And as 180 feet per minute is a very good velocity for raising water, if Q be the quantity in cubic feet,

$$\left(\frac{Q}{\cdot 00518 \times 90} \right)^{\frac{1}{2}} =$$

$(2\cdot 15 \, Q)^{\frac{1}{2}} = d$; hence

$$\left(\frac{\cdot 7332 \; h \; Q}{P} \right)^{\frac{1}{2}} = D.$$

Example.   Suppose it be required to raise eighty cubic feet of water per minute by a single acting engine, the mean effectual pressure of the steam being eleven pounds per circular inch, and the lift 149 fathoms in six lifts. In this case 149 fathoms = 894 feet;

---

* When a given quantity of water is to be raised ; if Q be that quantity, we have

$$\frac{2 \, Q}{\cdot 00518 \, v} = d^2,$$

and

$$\frac{D^2 \, P}{\cdot 341 \, h} = d^2, \text{ or } \frac{2 \, Q}{\cdot 00518 \, v} = \frac{D^2 \, P}{\cdot 341 \, h}.$$

Also,

$$l = \frac{2 \, D}{12};$$

and $v = 98 \, \sqrt{l}$; therefore,

$$\left(\frac{3\cdot 5 \, Q \, h}{P} \right)^{\frac{2}{5}} = D.$$

hence, $9 + 894 + 44.7 = 948 = h$; and

$$\left(\frac{.7332 \times 948 \times 80}{11} \right)^{\frac{1}{2}} =$$

$(5054.171)^{\frac{1}{2}} = 72$ inches nearly, for the diameter of the cylinder; and $(2.15 \times 80)^{\frac{1}{2}} = 13.115$ inches $=$ the diameter of the pump; the velocity being 180 feet per minute.

Both these diameters ought to be increased five per cent for contingencies.

Drainage of Mines.

573.—In this country the drainage of mines is a subject of vast importance. It is mines which supply the means of employing steam power, and also a large proportion of the materials on which that power is expended. To persons accustomed to mines, it is seldom necessary to state those principles which should direct them in the choice of engines. The absolute necessity of an economical system of drainage is felt and acted upon, and it is by comparison of annual expense, and not by a comparison of the effect, from a given quantity of fuel, that this economy should be estimated.

A mine engine should be simple in construction, durable in use, and made with a view to easy repair. When coals are not expensive, the most simple methods are the most economical; for instance, at the mouth of a coal pit the extra consumption of coals is of less value than the extra wear and tear of a complex engine.

574.—The modes of draining mines are dependent on the nature of the district where they are situated. If it be mountainous, a subterraneous channel or day level drift,* may be made, from the lowest part of the mine to terminate in the nearest valley, to carry off the water, and it is only when this method is impracticable that water is raised by power, and even then the water is raised no higher than to where a day level drift can be obtained. But it frequently happens that the flatness of the country renders any other method impracticable, than that of raising the water to the surface. For example in the coal-field of Northumberland and Durham, many of the large double pits exceed one hundred fathoms in depth, and some are nearly 150 fathoms deep, with no means of drainage by levels. These pits therefore require very powerful engines, and lately they have chiefly erected double engines, some of which are above one hundred horses' power; the largest I saw there was one on the south side of the Tyne which was working with 160 horses' power,

---

* In Cornwall and Devon it is called an *adit*, in some other places a sough.

and was capable of exerting the power of 200 horses in action at once. In Cornwall they have some larger engines, but two engines should always be preferred, when the cylinder of one engine would exceed about sixty inches in diameter, for two engines give many advantages.

575.—When double engines are used for lifting water, they generally work one set of pumps by the outward end of the beam, and another set by a diagonal spear from the piston rod end. And in cases where it has not been convenient to divide the pumps into two sets, the ascending motion of the piston has been employed to raise a weight equal to the pressure of half the column of water in the pumps, but for such cases a single engine should be preferred.

576.—The following table will give some idea of the work done by a given quantity of fuel, and of the nature of the engines most approved of in Cornwall; the results however can be correct only through the different errors of the mode of estimation balancing one another, for the weight of the column of water is less than the resistance, and the counter only registers the strokes and not the actual quantity of water raised.*

---

* In the year 1811, a number of the respectable proprietors of the valuable tin and copper mines in Cornwall, resolved that the work which their respective steam engines were performing, should be ascertained, as it was suspected that some of them might not be doing duty adequate to the consumption of fuel; and for the greater certainty of attaining their object, it was agreed that a counter should be attached to each engine, (art. 563,) and all the engines be put under the superintendance of some respectable engineer, who should report monthly, the following particulars in columns: viz.

The name of the mine; the size of the working cylinder; whether working single or double; the load per square inch in the cylinder; length of stroke in the cylinder; the number of pump lifts; the depth in fathoms of each lift; diameter of pumps in inches; time worked; consumption of coals in bushels; number of strokes during the time; length of stroke in pump; load in pounds; pounds lifted one foot high by a bushel of coals; number of strokes per minute; and lastly, a column for names of engineer and remarks. Messrs. Thomas and John Lean were appointed to the general superintendance; and the different proprietors, as well as the regular engineers of the respective mines, engaged to give them every facility and assistance in their power. The first monthly report was for August 1811. See Philo. Mag. Vol. XLVI. p. 116.

A part of a monthly Report containing six of the most effective Engines, in December 1826, the whole number reported on forty-six.

Mines.	Diameter of cylinder.	Load per square inch on the piston. (lbs.)	Length of the stroke in the cylinder. (ft. in.)	Number of strokes per minute.	Number of lifts.	Depth of lift. (fath. ft.)	Diameter of the pump. (inches.)	Consumption of coals in bushels.	Number of strokes.	Length of stroke in the pump. (feet.)	Load in pounds.	Pounds lifted one foot high by consuming one bushel of coals.	Remarks and engineers' names.
Wheal Ilope.	60 inch single	8·37	9 0	5·5	1 1 1	46 5 11 2 11 2	15 12⅜ 11	1242	261,890	8·0	27,766	46,838,246	Drawing all the load perpendicularly. Main beam over the cylinder. One balance-bob at surface...........*Grose.*
Wheal Vor.	80 inch single	13·37	10 0	5·56	5 4 1 1	135 2 44 0 12 0 11 5	15 16 9¼ 9¾	3274	199 960	7·5	89,607	41,045,698	Drawing perpendicularly 135 fathoms, and on the underlay twenty-seven fathoms. Main beam over the cylinder. Two balance-bobs under ground..*Sims & Richards.*
Consolidated Mines.	90 inch single	9·42	9 11	8·12	1 1 6	6 0 15 0 144 1	12 12 16	4680	304,500	7·5	81,673	39,854,853	Drawing perpendicularly, with main beam over the cylinder. One balance-bob at surface........*Woolf.*
Dolcoath.	70 inch single	10·05	8 9	6·3	1 5 1 3 1	2 0 93 1 22 0 65 1 15 3	8⅛ 11½ 12 11¾ 13	2660	264,970	7·25	35,021	39,375,762	Drawing perpendicularly 179 fathoms, and on the underlay thirty-three fathoms. Main beam over the cylinder. Four balance-bobs under ground, and one at the surface. Sixty fathoms of dry rods in the shaft.........*Jeffree.*
Ting-Tang	63 inch single	13·4	7 9	6 1	2 4 1 1 1	39 0 81 3 20 3 11 3 12 0	9 14 12 9 8	1980	229,520	6·75	48,646	38,063,288	Drawing perpendicularly, with main beam under the cylinder. Fifteen fathoms of horizontal rods under ground....*Sims & Sons.*
Binner Down.	70 inch single	6·12	10 0	8·6	1 1 1	2 5 23 2 40 4	10 9 18	2628	420,550	7·5	31,395	37,680,271	Drawing all the load perpendicularly. Main beam over the cylinder....*Thomas.*

The engineers' names are given who plan the construction and superintend the execution and erection of the engines, for which they are paid in proportion to the power; they also attend to them afterwards, and direct such renewals or repairs as may be necessary, at fixed salaries. The principal manufacturers of engines for the Cornish mines are Messrs. Trevenan, Carne and Wood; Messrs. Harvey and Co; Messrs. Fox and Co; and Messrs. Price and Co.

577.—The depth of the pump shaft of a mine is divided into lifts of not more than twenty-five or thirty fathoms, if it can be avoided, with a cistern at each lift, consequently the water is raised from cistern to cistern. The size of the pumps is seldom greater than sixteen inches in diameter, and it will always be found better to make an additional set than to exceed this size.

578.—The engines most adapted for economy of fuel, are described in (art. 411, and 419,) those which are most simple in (art. 393 and 400,) and as it frequently happens that engines have to be removed from place to place, an engine supported by frames of cast iron is shewn in Plate XI.

579.—For drawing ores and coals, a double engine of from twenty to thirty horses' power is used; the size of the cylinder should be such, that the power shall be equal to the resistance, when the stress is the greatest; hence, engines for this purpose require more fuel to raise the same quantity of matter a given height, and there is also much loss of effect through stoppages, changes of motion, &c. When one pound of coal raises 70,000 pounds of ore, it is about the maximum quantity in irregular work of this kind. The weight of matter drawn at once is from three to seven cwt. The weight of a rope is about $\cdot 27 \, c^2$ pounds per fathom; when c is the circumference in inches: the greatest stress on a rope should not be more than 700 times the weight of a fathom of the rope; and the stress on the engine should be equalized by the rope winding on to a spiral drum,* like the fusee of a watch, by which the expense of the engine, and the expenditure of fuel would be reduced. The engine should work expansively, (art. 419,) and be equalized by a fly wheel, (art. 540,) and regulated by a governor (art. 550.)

When an inclined plane is necessary under ground, a small high pressure engine is sometimes used to draw the coals to the principal shaft, of the kind of engines described in (art. 371.)

580.—Engines are also employed to break ores by means of stampers, a process which seems capable of much improvement. Double engines are employed to raise the stampers by means of cams; and as the power of the engine is nearly uniform, the space

---

* See Ency. Metho. Dict. de Chimie et Métallurgie, Seconde Partíe, Planche 20; or Gilpin's Method, Transactions of the Society of Arts, Vol. XXV. p. 76.

through which the stamper is raised should increase at first in the proportion due to an uniform force; otherwise the motion will be irregular, and the loss of power considerable. The weight of a stamper is usually made about 190 pounds, and the height it is raised about two feet; and not less than two-thirds of the stampers should be rising at any instant of time.

Water Works.

581.—The same formulæ apply to water works as to other modes of raising water, when it is raised perpendicularly; but as this is seldom the case, instead of adding 1·5 feet for each lift, as in (art. 572,) add

$$\frac{v^2\, L}{140\, d}$$

feet to the vertical height; where v is the velocity in feet per second, L the length of the main in feet, and d its diameter in inches: add also one-tenth of the height for the friction of the piston, and proceed in other respects as in the article referred to.

582.—The supply of a town should be ten cubic feet per day for each house, and for the averaged sized houses this is not more than comfort and cleanliness requires; or two cubic feet per day for each individual, besides what is required for watering streets, for breweries, engines, and various purposes; and for these purposes two cubic feet more ought to be delivered in summer, making a total of four cubic feet per day for each person, for the greatest quantity : in small and open towns a less quantity of water is required, but even in these, two cubic feet and a half ought to be calculated upon.*   In raising water by forcing, the air vessel should always be in the direction of the motion of the fluid, and not to one side of it; want of attention to this causes those concussions to take place which tear the joints asunder, break the cranks, and spoil the machinery.   Double engines with fly wheels are the most economical when fuel is dear, (art. 419,) and single engines where it is cheap, (art. 411 and 400 .)   See Plates XII. and XIII.

---

* The following tablet is compiled chiefly from Leslie's Nat. Phil with some additions, and a more reasonable esti-

Of Impelling Machinery for manufacturing Purposes.

583.—*Iron manufacture.* In this manufacture the steam engine is applied to blowing machines, forge hammers, rolling, flatting, and slitting machines, and various other purposes.

584.—*Blowing machines.* The object of this machine is to supply oxygen to furnaces, either for melting, or reducing ores to the metallic state; hence, in order that the effect may be the same, or nearly so, when the same fuel is used, the supply of oxygen should be the same. But in the same bulk of dry air, there is nearly ten per cent less oxygen at 85° than at 32°; and twelve per cent less when the air at 85° is saturated with vapour; consequently, if 1500 feet per minute be a sufficient supply for a furnace in winter, it may require 1625 feet per minute in summer, to have the same effect: and the difference ought clearly to be gained partly by the aperture being enlarged, and partly by increasing the intensity of the blast.

The blast is usually produced by condensing the air, till it will sustain a column of from four to six inches and a half of mercury, (one and a half to two pounds per circular inch,) according to the quality of the coal; and the mean between these is most generally found to answer: the quantity discharged varies from 3000 to 1200 feet per minute.

If v be the velocity of the piston of a blowing cylinder in feet per minute, p the force of compression in pounds per circular inch, and d the diameter of the blowing cylinder in inches, then allowing that the friction increases the power from 1 to 1·25, we have 1·25

mate of the quantity supplied to ancient Rome.

Towns.	Inhabitants.	Supply of water per day.	Each person per day.
London	1,225,694	3,888,000 cubic feet	3·15 cubic feet.
Edinburgh (old service)	138,235	80,640 ———	0·61 ———
Rome (modern)	136,000	5,305,000 ———	39· ———
Rome (ancient)	1,200,000	10,500,000 ———	9· ———
Paris	713,765	293,600 ———	0·42 ———
Plymouth	21,570	33,400 ———	1·56 ———

$p\, v\, d^2 =$ the power in pounds raised one foot high per minute, when the stroke is effective in both directions, and half that when in one direction only.* The capacity of the air chest should be proportioned by the principle given in (art. 211,) and the passages to it should be about one-twentieth of the area of the cylinder. The quantity of air delivered into the chest will be about one-fifth less than the capacity of the cylinder, when taken at atmospheric density, partly through escape by the valves, and by the air not entering till the space within the cylinder is rarefied so as to produce the velocity.

For this as well as all other parts of iron manufacture, the double acting condensing engine, prepared to work either expansively or at full power, will be found the best. (See art. 421.)

585.—*Cotton mills.* The steam engines best adapted for cotton mills are double acting engines working expansively. The mean pressure on the piston of an engine of this kind, using low pressure steam, when working with the greatest advantage, is about five pounds per circular inch, (art. 420,) and each circular inch of the piston may be estimated to drive three spindles of cotton yarn twist, with the preparatory machinery. And for mule yarn with its preparation, if fifteen be added to the number of the yarn,† and the sum be multiplied by ·26, the result will be the number of spindles for each circular inch of the piston. Thus if it be No. 40, then $40 + 15 = 55$, and $·26 \times 55 = 14$ spindles.

It is somewhat more accurate to estimate the power of the engine in horses' power, and then one horse power will drive one hundred spindles with cotton yarn, and the preparatory machinery. And add the number of the yarn to fifteen, and multiply the sum by eight, and the result will be the spindles that are equivalent to one horse power of mule yarn with preparation.

One horse's power will work twelve power looms with preparation.‡

The day's work, supposing it to be eleven hours, ought to be done with about ninety pounds of the best caking coal for each horse power.

586.—*Paper mills.* Steam engines are also used extensively in making paper, for where the supply of water is regular it has acquired a value equivalent to steam power, while the latter possesses many advantages.

A beating machine requires about seven horses' power to work it; the new machines

---

* The rule is only an approximation, but nearly correct for small degrees of compression; in greater ones the principles of the note to (art. 377,) should be applied.

† The number is the hanks to the pound of yarn, and a hank appears to be 120 yards. A spindle produces two hanks per day at an average, and the waste in spinning is about ten per cent.

‡ Brunton's Compendium, p. 109.

for making paper from two to two and a half horses' power; and three and a half horses' power will prepare one ton of old rope in a week, when the machine works ten hours per day.*

Of Impelling Machinery for Agricultural Purposes.

587.—In farming there are few things that admit of the employment of steam power with economy, but where it is employed at all it is an advantage to apply it to as many purposes as possible.

The species of work to which it is susceptible of application are—thrashing and winnowing grain, chaff cutting, grinding bones for manure, and to grinding corn for fatting cattle and for family uses.

The boiler may be further applied to steam food for cattle. No other objects occur to me except to notice that for drainage in fenny districts, and for irrigation in others, it is worthy of the landowner's consideration whether its application would or not repay the expense.†

588.—*Thrashing.* Thrashing machines to be driven by a steam engine are made from four to six horses' power; and the usual proportions are

The feeding rollers thirty-five to thirty-seven and a half revolutions per minute, diameter three inches and a half.

Straw rakes thirty revolutions per minute, diameter three feet and a half.

Drum 300 revolutions per minute, diameter three feet and a half.

---

* Fenwick's Essays, third Edition, p. 62.

† By a single engine, (art. 411.) 280,000 cubic feet of water may be raised one foot high by one bushel of coals, and for any other height divide 280,000 by the height in feet. It will require an engine of one horse's power to work eleven hours and a half per day, to raise that quantity daily. The expense will be about £ 8 per annum for each horse power, to return the first cost and pay for the renewals and repairs of the engine; the fuel is one bushel of coals per day for each horse power, and one man and boy will attend an engine of ten or twelve horses' power, and also partly to the distribution of the water: the quantity required for an acre would be about 500 cubic feet per day, and therefore an engine of ten horses' power would supply 560 acres, if it had to be raised ten feet; at a cost which in few cases could exceed ten shillings per acre if continued for six months per year. On proper lands for the purpose the return would be ample, and a more perfect mode of applying water to land will very probably be discovered.

The drum has four beaters faced with strong iron plate; a cylindric frame with four sets of teeth five inches long, attached to its circumference, forms the straw rake.*

The breadth of the machine, or length of the rollers to receive the feed, is limited by the width to which one feeder can attend in a proper manner, and about from four to five feet is the range: the thickness of the feed cannot be materially altered from that which gives the best effect; and, therefore, there is only the velocity in which the power can be altered beyond the small change in width. The quantity of wheat thrashed by a machine four feet in breadth, varies, according to its quality, from twelve to twenty-four Winchester bushels per hour, and from sixteen to thirty bushels of oats per hour.

The power required is 100,000 lbs. raised one foot per minute when for thrashing, and 133,000 raised one foot per minute when winnowing machinery is also worked; other sized machines nearly in proportion to their breadth: this supposes the machine to be well made and kept in tolerable order.

The proper species of engine for farm use is the double engine (art. 414 and 419,) with slides, and the whole arranged in the most simple and obvious manner.†

589.—*Corn mills.* The mean quantity of power required to grind and dress a Winchester bushel of wheat per hour is 31,000 lbs. raised one foot per minute, and the best velocity for the circumference of a millstone is twenty-three feet per second; and with this velocity a pair of five feet stones will grind from four to five bushels per hour, according to their condition and the state of the grain: the double expansive engine should be used for this kind of work, and when working to the best advantage with low pressure steam, it should grind fourteen bushels of wheat for each bushel of coals, and the average should be eleven bushels and a half for one bushel of coals.‡ The same species of engine with strong steam will of course do more work with a given quantity of fuel. (See art. 419.)

Of the Application of Steam Power to Carriages.

590.—The application of a power within a carriage to move it, is a subject that at an

---

* The straw advances so that each inch receives three strokes of the beaters, and the stroke should be made with a velocity of about fifty-five feet per second, or the beater should move at the rate of 3300 feet per minute; these conditions being attended to the parts may be, in other respects, arranged as the engineer pleases.

† If an engine be applied for irrigation it may have the thrashing machine attached when the situation is convenient; but in that case a double engine should be used.

‡ For 6·5 pounds per hour is equal to a horse's power, (art. 419,) and a bushel of coals being thirteen times this quantity it is equal to thirteen horses' power per hour; and 31000 : 33000 : : 13 : 14, nearly.

early period engaged the attention of speculative men. Some of their schemes are described by Emerson in his "Mechanics," and he gives an example of calculation there (Ex. 20, p. 194,) which seems to be very little understood. The object of it is to determine the power required to move a waggon; but in fact it simply determines the relation of the forces, the power being the same whether it be in the waggon or out of it, provided it does not add to the weight of the waggon. But power cannot be gained without adding weight; and in steam carriages the whole mass of the engine, with its boiler, and fuel, and water, has to be moved as well as the load; and in order to keep the engine as simple and light as possible, and to avoid the weight of water and complexity of a condensing apparatus, high pressure steam is always employed.

The idea of employing steam as a moving power has been considerably ridiculed, and some of the schemes for applying it not without reason. As far as railways are concerned it has however been proved to be applicable, and with as few accidents as in any other of the varied application of steam power.*

591.—*Of the application of steam power to railways.* The power of steam may be applied by means of a fixed engine, or by a moveable one called a *steam carriage.*

Fixed engines have been applied only in the case of inclined planes, and no peculiarity is required in the construction of the engines more than is wanted in one for impelling a machine. Low pressure engines are generally employed for this purpose; and they are obviously the most safe and economical for the end, unless when there is not a convenient supply of water.

The motion of the engine ought to be equalized by a fly wheel, and it should also be provided with a regulating valve.

To proportion the power of the engine to the effect, the area of the piston in inches, mul-

---

* On common roads there are several circumstances which prevent the application of steam power. The undulations of the road render it necessary to provide a power competent to ascend the greatest inclination, and consequently an immense addition must be made to the weight of the engine, so as to make the engine itself on an ordinary road, consume half the power it generates. The resistance of these roads may be reduced by using larger wheels, as it chiefly arises from the wheels sinking into the road; and larger wheels afford a greater surface without increasing the quantity to be depressed, while broad wheels give very little if any advantage. See my book on "Rail roads," p. 44. It may be proved that no species of feet can be applied that will require less power than plain wheels. An animal is contrived to move among obstructions, and when we attempt to copy from the beautiful arrangements of our Creator, we should never lose sight of their object. It is their perfect adaptation to the end, and their accomplishment by the most simple means that excites our admiration, and the more we study the fine examples of the application of power which nature affords, the more we feel the advantage of knowing the first principles which determine the action of natural forces.

tiplied by the effective pressure on an inch in pounds, should be equal to the resistance of the carriages added to the friction of the rope and the engine.

If A be the ascending and D the descending load, and q the resistance from friction at the axis, and i the angle of inclination.

Then A (sin. $i + q$) — D (sin. $i - q$) $=$ sin. i (A — D) $+ q$ (A $+$ D) $=$ the resistance of the carriages.

The weight of the rope or chain and of the moveable parts of the engine being C, its friction and the stiffness of the rope may be represented by C S, hence, if d be the diameter of the piston, and $p =$ the pressure on a circular inch, we have $d^2 p =$ sin. i (A — D) $+ q$ (A $+$ D) $+$ C S.

From this the diameter of the cylinder is easily found, and in all cases

$$q = \frac{r f}{R},$$

where R is the radius of the wheels of the carriage, r that of the axles, and f the friction, when the pressure is 1.

Also

$$S = \frac{x f}{X},$$

when X is the diameter of the pulleys, and x that of the axes. When the railway is level $d^2 p = q,$ (A $+$ D) $+$ C S, when it is vertical $d^2 p =$ A — D $+$ S (A $+$ D $+$ C.)

In these equations the piston of the engine, and the load, are supposed to move at the same velocity; if the carriages move at n times the velocity of the piston, then it will require the square of the diameter to be increased n times.

592.—*Steam carriages.* The engines of steam carriages are double noncondensing engines, of the kind described in (art. 372.) They have generally two cylinders. If the mean effective pressure on the piston be p, the diameter of the cylinders d, and v the velocity in feet per minute; and if i be the angle of inclination of the rails of the road, q the friction of all the axes, W the weight of the carriages and their loads, V their velocity in feet per minute, and E the weight of the engine; then V (W $+$ E) ($q +$ sin. i) $= 2 d^2 p v$, in ascending the inclination: and V (W $+$ E) ($q -$ sin. i) $= 2 d^2 p v$, in descending the inclination. Also

$$\frac{E (\cdot 08 \cos. i - \sin. i)}{q + \sin. i} = W,$$

so that the engine may not slide in ascending; and

$$\frac{E\ (\ \cdot08\ \cos.\ i\ +\ \sin.\ i\ )}{q\ -\ \sin.\ i} = W,$$

when it will not slide in descending.

In either case

$$q = \frac{r}{8\,R}\ ;$$

when r is the radius of the axle, and R the radius of the wheels.

The engines should work expansively when moving at the ordinary rate, and upon the mean inclination, with the power of working at full pressure on the steeper ascents. See Sect. V. (art. 371—380.)

The carriage is described in Plate XX ; and for further information see my "Treatise on Rail Roads."

SECTION X.

OF STEAM NAVIGATION.

593.—On the value of the application of steam to impel vessels, it has become un-necessary to say more than that its employment is extending rapidly at almost every place on the globe where the trade is considerable; and that its use is limited only by its yet imperfect state. If we had intended to have confined our researches to the mere applica-tion of an engine to a vessel already constructed, our labour would have been short, and easily completed: but the construction of vessels is a subject which is capable of improve-ment; and while we think there is a power in science to indicate the steps by which it may be improved, it is our duty to submit it to the reader.

The forms of vessels for stability, speed, capacity and strength; the kinds of vessels for different purposes, the resistance, and modes of propulsion; the nature of the engines adapted for vessels, the strength of their parts, and the species of fuel, and its management to obtain the best effect; are all objects of importance, and each of these we propose to consider.

These inquiries are equally applicable to mercantile and to government purposes, but there is yet another portion of the subject to which it would be desirable to direct atten-tion.

In the case of war, steam boats will become a means of attack, therefore it ought to be considered how far they may become a means of defence, the power of resisting being the best guard against a mode of attack, which will deprive us of many of the advantages of our insular state. Hence, the construction of gun boats for the defence of rivers, and of river navigation, and harbours, would be a proper subject for inquiry, if our limits did not forbid it.

Of the Forms of Vessels for Stability, Speed, Capacity, and Strength.

594.—In considering the properties of a vessel, the orderly arrangement of our subject requires that we should treat, First, Of stability, or the power a vessel has of resisting any change of position when in the water; Secondly, The forms having stability which have the least resistance, and are therefore best adapted for speed; Thirdly, The different methods of propelling vessels; and, Fourthly, The construction for strength.

Of the Stability of Vessels.

595.—A perfectly spherical ball floating in a fluid has no stability whatever, except that which arises from the friction of the fluid against its sides. On the addition of a small weight to any point of its surface, that point would immediately descend, and become the lowest. Such a form would be useless as a vessel. It is obvious however that when a weight has been added, and become the lowest point, the sphere possesses a degree of stability depending on the quantity of weight compared with the weight of the sphere itself. Hence, stability may be given by disposing the weight of a floating body.

Stability may also be given by the form of the floating body; a spheroid for example remains in stable equilibrium when its longer axis is horizontal, and a triangular prism resists change of position with considerable energy from its peculiar form; so does a thin rectangular prism.

596.—Stability is distinguished by its being *longitudinal* or *lateral;* these should be separately considered, and when each is the greatest possible, their joint effect will be a maximum.

597.—For river navigation, the mode of obtaining stability does not appear to be of much importance, but for a sea vessel it must be obtained, so that the vessel may have the least motion possible in consequence of the action of the disturbing forces; hence, it is necessary to consider that the sea is not a level surface at rest, and that at the time when stability is most important to a vessel, the greatest degree of unevenness occurs.

598.—*Longitudinal stability.* A vessel at rest would be least disturbed by the motion of the sea, if its surfaces at the water line were vertical ones; and the fore and aft parts of the same figure; but in motion it is an advantage that the parts should spread above water, both fore and aft, to prevent the vessel burying its head in the wave, or dropping behind as the wave leaves it. The quantity of motion is not increased by this construction, provided the parts produce similar effects, and the degree of inclination should be proportioned to the velocity the vessel is expected to make. It is also obvious that the vessel will be more easy in its longitudinal motions, the more gradually it terminates at its extremities. If the vessel be inclined by the action of a lateral force, the longitudinal motions will be more easy in proportion as the cross section approaches to that of a solid of revolution.

599.—*Lateral stability.* The inequality of the surface of the sea will alone produce considerable lateral motion, if the sides be not sensibly vertical, hence, in sea vessels lateral stability should not be obtained by form at the surface of the water. The next important point is that the stability should be equal throughout the length.

To render it easier to manage the inquiry, we may consider the vessel to be a homogeneous mass of matter, of the same density as water, with vertical or circular surfaces at the water lines when at rest; and that it is of a parabolic form, having the equation $p\,x = y^n$, taking the two cases when the ordinate y is the half breadth, and when it is the depth: for these cases enable us to contrast very opposite forms.

600.—The ordinates being parallel to the depth, we have $y\,\dot{x}\,(\frac{1}{2}\,b - x + y\,\sin.\,i\,)$ $-\,y\,\dot{x}\,(\frac{1}{2}\,b - x - y\,\sin.\,i\,) =$ the difference of the moments of the parabolic parts, when i is the angle the body makes with its position, when B D, (Fig. 3, Plate XVI.) coincides with the water line $=$

$$\frac{n\,y^2\,x\,\sin.\,i}{2 + n}\,;$$

and the difference between this quantity and the moment of twice the area of the triangle B C b is the stability. But

$$\frac{2\,b^3\,\sin.\,i}{8 \times 3} = \frac{b^3\,\sin.\,i}{12} =$$

the moment of the triangles; hence,

$$\frac{b\,\sin.\,i}{12}\,\left(\,b^2 - \frac{6\,n\,d^2}{2 + n}\,\right) =$$

the stability.

The capacity is

$$\frac{n\,b\,d}{1 + n}.$$

601.—If the negative quantity be less than b^2 the body has no stability, hence, we see that a certain relation must hold between the breadth and depth to render a vessel stable; and substituting for d^2; we have

$$\frac{b^3 \sin. i}{12} \left(1 - \frac{6 n p^{\frac{2}{n}} b^{\frac{2-2n}{n}}}{2 + n} \right)$$

602.—If the form be a triangle, then $n = 1$, and putting $S =$ the stability, and $A =$ the area, we have

$$\frac{b \sin. i}{12} \left(b^2 - 2 d^2 \right) = S, \text{ and } \frac{b d}{2} = A.$$

603.—If the form be a common parabola, then $n = 2$, and

$$\frac{b \sin. i}{12} \left(b^2 - 3 d^2 \right) = S, \text{ and } \frac{2 b d}{3} = A.$$

604.—If the form be a cubic parabola, then $n = 3$, and

$$\frac{b \sin i}{12} \left(b^2 - 3.6 d^2 \right) = S, \text{ and } \frac{3 b d}{4} = A.$$

605.—If the form be a parabola where $p x = y^5$, as Fig. 3, Plate XVI. then

$$\frac{b \sin. i}{12} \left(b^2 - 4.3 d^2 \right) = S, \text{ and } \frac{5 b d}{6} = A.$$

606.—The stability and capacity both increase as the ordinate of the parabola becomes of a higher power, but a greater breadth is necessary in proportion to the depth, to give stability.

607.—When the ordinates are parallel to the breadth,

$$y x \left(y + (d - x) \sin. i \right) - y x \left(y - (d - x) \sin. i \right) = \frac{2 y d^2 n \sin. i}{n^2 + 3 n + 2};$$

and

$$\frac{b^3 \sin. i}{12} - \frac{2 y d^2 n \sin. i}{n^2 + 3 n + 2} = \frac{b \sin. i}{12} \left(b^2 - \frac{12 n d^2}{n^2 + 3 n + 2} \right) =$$

the stability. The capacity is

$$\frac{n b d}{n + 1},$$

as before; and in the case of the triangle we have the same result.

608.—But if the form be a common parabola, or $n = 2$, then

$$\frac{b \sin. i}{12} (b^2 - 2 d^2) = S, \text{ and } \frac{2 b d}{3} = A.$$

609.—If the form of the parabola be $p\, x = y^5$, as in Fig. 4, Plate XVI. then

$$\frac{b \sin. i}{12} (b^2 - 1.43 d^2) = S, \text{ and } \frac{5 b d}{6} = A.$$

610.—This species of figure may be easily traced through all the varieties of form, and it has obviously a decided advantage in point of stability, and it is so easy to compute its capacity and to describe it by ordinates, that it is much to be preferred to the elliptical figures which foreign writers have chosen for calculation.*   The breadth should be every where in the same ratio to the depth, to render the stability equal throughout the length, or so that the vessel will undergo no strain from change of position.†

Of the Resistance of Vessels.

611.—The resistance of a vessel moving in a fluid increases from the commencement of the motion, till it be equal to the moving force, and then the motion becomes uniform. It is the resistance at this uniform motion only which we have to consider.

In order to assist in the first steps of the inquiry, let us confine ourselves to a prismatic vessel with flat ends, moving in the direction of its length.

612.—The resistance of such a prism would be nearly equal to the head of water, which would give the water in a canal of the same length, and one and a half times the section of the immersed part of the prism, the same velocity as the prism.

For let A B be that head, then the resistance to efflux at D, must be equal to the resistance to motion at C, the section being the same; otherwise, the motion would accelerate.   But the fluid will rise at C, and fall at D, till the difference be equal to the head

* For a mode of describing curves of this kind, see my "Principles of Carpentry," Sect. I. art. 58.

† For other methods see Bossut's Hydrodynamique, Tom. 1. chap. XIII. et XIV; or Poisson's Traite de Mecanique, Tom. II. p. 389.

FIG. 25.

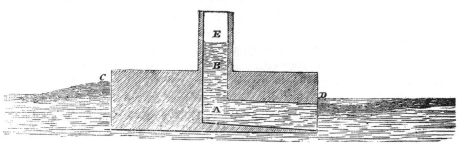

due to the velocity of the boat; and the efflux at D, must both supply a void with fluid equivalent to the velocity of the boat, and supply a resistance equal to the head pressure. This will be the case when two-thirds of A B is the head corresponding to the velocity. Hence, if v = the velocity, A E = h, and B E = x = the head equivalent to the friction,

$$\frac{64\,(h-x)}{1\cdot5} = v^2;$$

where 64, neglecting a small fraction, is the proper coefficient for the motion of a fluid when free from friction or cohesion; consequently,

$$h - \frac{1\cdot5\,v^2}{64} = x.$$

613.—Now if c be the perimeter of the section in contact with the fluid, and a its area, l the length of the vessel, and F the friction when the surface is 1, we have

$$\frac{l\,c\,F\,v^2}{a} = x.$$

The head equal to the friction being as the square of the velocity directly, and the area inversely, and the friction being as the surface of the fluid put in motion, or rubbing the surface of the vessel.

These two values of x must therefore be equal, hence

$$h - \frac{1\cdot5\,v^2}{64} = \frac{l\,c\,F\,v^2}{a},$$

or the whole head,

$$h = \frac{1\cdot5\,v^2}{64} + \frac{l\,c\,F\,v^2}{a}.$$

But the resistance being usually estimated in pounds we have for sea water, $64\,h\,a =$

that resistance, $= v^2 (1\cdot5\, a + l\, c\, \mathrm{F}) = \mathrm{R}$, the resisting force; and the power required is as the force and velocity $= v^3 (1\cdot5\, a + l\, c\, \mathrm{F}) =$ the pounds raised one foot per second, when F is in pounds.

614.—For fresh water put $1\cdot45$ in the place of $1\cdot5$, but the correction is not necessary in practice. The coefficient F is $0\cdot0032$ pounds found from experiment.

615.—If the body have a simple angular prow, and an after body of the same figure, and a be the angle the prow forms with the direction of its motion, and e the angle of the after part, then the pressure on its surface depends on the velocity of the surface, in a direction perpendicular to itself. This velocity before is v sin. a; and behind v sin. e; being to the velocity of the vessel as the sine is to the radius of the angle; and therefore

$$v^2 \left(\frac{2\, a \sin.^2 a + a \sin.^2 e}{2} + l\, c\, \mathrm{F} \right) = \mathrm{R}.$$

The effect of this head in the direction of the motion of the vessel, is as the sine of a is to the radius at the prow, but the quantity of fluid required to fill the void behind is constant for the same angle, hence,

$$v^2 \left(\frac{2\, a \sin.^3 a + a \sin.^2 e}{2} + l\, c\, \mathrm{F} \right) = \mathrm{R},$$

the resistance; and

$$v^3 \left(\frac{2\, a \sin.^3 a + a \sin.^2 e}{2} + l\, c\, \mathrm{F} \right) =$$

the power in pounds raised one foot per second.

This gives the resistance when the vessel is of a wedge shape at both ends, or of any regular pyramidal form, a being the angle the slant side of the pyramid makes with the length; also when the body terminates either in cones or pyramids, the angles being those the slant sides form with the length.

616.—If the section be a triangle, and the ends triangular pyramids, c being the angle the side of the triangle forms with the upper surface, then, if q be the product of the sin. a, by the sin. c, the resistance will be

$$v \left(\frac{a\, (2\, q^3 + q^2)}{2} + 0\cdot0032\, l\, c \right) =$$

the power in pounds raised one foot per second.

The resistance of this figure is less than that of any convex curved solid, but its capacity is also small, and its stability depending on the form at the water lines, it will be subject to roll at sea. Great capacity cannot be obtained with a minimum of resistance.

617 —If the plan of the water lines be composed of circular arcs, the bottom flat, and the radius be m times the half breadth of the boat, and z be the length of the curved part, $r =$ the radius, and d the depth which is uniform; then $3\,z = r\,(\,4\,\sqrt{2\,m} - \sqrt{2\,m - n^2}\,)$, and

$$v^3\,d\,\left(\frac{3\,z - r\,(\,1 - m\,)\,.\,(\,3 - 4\,m\,)\,\sqrt{2\,m}}{4} + \frac{r\,(\,2 - 2\,(\,1 - m\,)\,.\,\overline{(\,1 - m\,)}{}^2 + 3\,m\,)}{3}\right.$$
$$\left. + \frac{\cdot 0032\;l\,c}{d}\,\right) =$$

the power in pounds raised one foot per second, required to keep the vessel in motion at the velocity v.

618.—In canal boats $m = \frac{1}{8} = \cdot 125$, or the radius is four times the breadth, and therefore $v^3\,(\,\cdot 35\,b\,d + \cdot 0032\,l\,c\,) =$ the power in pounds raised one foot per second.

If the radius be equal the breadth, then $m = \cdot 5$, and $v^3\,(\,\cdot 74\,b\,d + \cdot 0032\,l\,c\,) =$ the power in pounds.

619.—Mr. Bevan made some experiments with a canal boat of the form just described, the results of which he has communicated to me for the purpose of comparing theory with practice.

The length of the boat was 69·57 feet, its width 6·83 feet, its floating depth when tried 0·89 feet; the bottom was flat and the sides were parallel to within about 13·75 feet of each end, but the ends were curved, the curves being circles described by a radius of eight times the half breadth of the boat. The whole surface in contact with the water was 540 feet; and the weight was nine tons and three quarters. Putting these numbers in the equation (art. 618,) we have

$$v^2\,(\,\overline{\cdot 89 \times 6\cdot83 \times \cdot 35} + \overline{\cdot 0032 \times 540}\,) = 3\cdot 8\,v^2 =$$

the resistance.

R R

Velocity.		Resistance in pounds.	
Feet per second	Miles per hour	By experiment.	By calculation.
feet.	miles.	lbs	lbs.
1·			3·8
1·31	0·89	·6·1	6·50
1·98	1·35	14·	14·8
2·93	2·00	28·	32·5
3 666	2·5		51·0
4·30	2·92	56·	70·0

The agreement is sufficiently near for practical purposes.

620.—The area of the bottom being 417 feet, a ton will sink it an inch. The increase of section by adding a ton to the load is therefore

$$\frac{6\cdot83 \cdot}{12} =$$

·57 feet; and the increase of surface twelve feet. Adding each ton must therefore increase the resistance about 3·2 pounds, at the velocity of two miles and a half per hour; therefore the load in tons multiplied by 3·2, added to 51 pounds for the boat, will give the force required to draw it. Thus if the load be twenty tons, the force of traction will be $\overline{20 \times 3\cdot2} + 51 = 115$ pounds.

621.—The forms used for vessels are generally curved surfaces of double curvature. To investigate these we may consider them divided into gores, having their bases at the section, and meeting in a point at the water line. A solution on this supposition is fully sufficient for practical objects. Let r be the radius of curvature of the gore, and c its breadth at the base, a being its distance from the axis. Then the fluxion of the area of the section occupied by the gore will be

$$c\,\dot{x} - \frac{c\,x\,\dot{x}}{a};$$

and

$$v^2 c \left(1 - \frac{x}{a} \right) \dot{x} \left(\frac{2 \sin.^3 a + \sin.^2 a}{2} \right) =$$

the resistance from pressure. By using the approximate equation, $2\,r\,x = y$, we have

$$\dot{x} = \frac{y\,\dot{y}}{r}, \text{ and } \frac{y}{r} = \sin. a,$$

hence,

$$v^2 c \left(\frac{y^4\,\dot{y}}{r^4} + \frac{y^3\,\dot{y}}{2\,r^3} - \frac{y^6\,y}{2\,r^5\,a} - \frac{y^5\,y}{4\,r^4\,a} \right) =$$

the fluxion of the resistance. Its fluent is

$$\frac{v^2 c\,y^4}{r^3} \left(\frac{y}{5\,r} + \tfrac{1}{3} - \frac{y^3}{14\,r^2\,a} - \frac{y^2}{24\,r\,a} \right) =$$

the direct resistance. Putting $y = \sqrt{2\,r\,x}$, and $b = 2\,x;\ x = a$

$$= \frac{r}{n},$$

and correcting by comparison with particular cases, we have

$$v^2 c \left(\frac{b}{n} \left(\frac{\cdot 1617}{\sqrt{n}} + 0 \cdot 0833 \right) + \cdot 0032 \left(l + \cdot 29\,b \sqrt{1 + 2\,n} \right) \right) =$$

the resistance in pounds.

622.—Now by taking a radius that will describe an arc nearly agreeing with the form, the resistance will be found with tolerable accuracy, even in the most complicated forms; and in cases where the curve is a circle, it will be very near the truth. To render it more easy to make such calculations, the following table is added with examples to explain its application.

Radius of cur-vature in half breadths.	Equations for different radii of curvature.
1	$v^2 c \left(\cdot 245\, b\ +\ \cdot 0032\ (\ l\ +\ \cdot 5\, b\ )\ \right)\ =$ resistance
1¼	—— ·188 - + ———— + ·545 —— = resistance
1½	—— . ·146 - + ———— + ·58 —— = resistance
1¾	—— ·120 - + ———— + ·616 —— = resistance
2	—— ·101 - + ———— + ·65 —— = resistance
2¼	—— ·086 - + ———— + ·68 —— = resistance
2½	—— ·075 - + ———— + ·71 —— = resistance
2¾	—— ·067 - + ———— + ·74 ——- = resistance
3	—— ·060 - + ———— + ·77 —— = resistance
4	—— ·041 - + ———— + ·87 —— = resistance
5	—— ·032 - + ———— + ·955 —— = resistance
6	—— ·025 - + ——— + 1·05 —— = resistance
7	—— ·021 - + ——— + 1·13 —— = resistance
8	$v^2 c \left(\cdot 018\, b\ +\ \cdot 0032\ (\ l\ +\ 1\cdot 2\, b\ )\ \right)\ =$ resistance

In this table b is the breadth of the vessel at the surface of the water, l the length of the part which is parallel, c the girt from water edge to water edge, of the lower part of the midship section, v the velocity in feet per second, and the result is the resistance in pounds. To find the power multiply the resistance again by the velocity; or, use the cube instead of the square of the velocity in the above table, the result is the pounds raised one foot per second.

623.—Required the resistance of a vessel of which the breadth is 22 feet, the length of the parallel part 80 feet, and the girt of the midship section 31 feet, when the velocity is 10 feet per second, and the radius of curvature equal four half breadths.

In this case by the table we have

$$v^2 c\ (\ \cdot 041\, b + \cdot 0032\ (\ l + \cdot 87\, b\ )\ ) = 10^2 \times 31\ (\ \overline{\cdot 041 \times 22} + \cdot 0032\ (\ 80 + \overline{\cdot 87 \times 22}\ )\ ) =$$

3778·9 pounds for the resisting force, consequently the power is 10 × 3778·9 = 37789 pounds raised one foot per second; and as 550 pounds raised one foot per second is a horse's power, the resistance is equivalent to

$$\frac{37789}{550} =$$

68·7 horses' power.

As ten feet per second is six nautical miles per hour, or nearly seven common miles, and the power required is as the cube of the velocity, it is easily ascertained.

Of the Methods of propelling Steam Vessels.

624.—Much of the advantage of steam power depends on its being commodiously and effectively applied to propel vessels. A slight review of these methods will therefore enable us to judge whether or not the most effective and commodious have been resorted to.

The first and most simple and ancient method of applying a power within a vessel to move it, is by means of oars, and the mode of combining them appears to have been carried to a considerable degree of perfection. Oars, however, are not at all adapted to move a large vessel, they occupy too much space, and would require too complicated a system of machinery to move them. Second, Next in simplicity, and perhaps also next in time, is the method of putting a wheel like a water wheel, with paddle boards on each side of the vessel. This mode is now almost universally followed. Third, An ingenious combination of parts has been proposed to be constantly under water, and to fold up into a small space when they are moved forward, and spread when striking backward. Fourth, Inclined planes placed behind the vessel, and moved with an alternating motion. Fifth, Daniel Bernoulli's method proposed in 1752, consisting of planes immersed in the water, parallel to the sides of the vessel, which, turning in a collar, were to be moved in a plane, perpendicular to the keel. Sixth, A screw, resembling the water screw, working in a cylinder entirely immersed in the water.* Seventh, Or two spirals or screws to work in opposite directions without a cylinder.† And, lastly, A pump to raise and propel water out

---

* It was proposed by Mr. Scott of Ormiston. Dr. Thomson's Annals of Philosophy, Vol. XI. p. 438.

† This method was partially tried by Mr. Whytock, (Brewster's Philosophical Journal, Vol. II. p. 39,) and is alluded to by Col. Beaufoy, who states it to have been brought from China; and he attended to see an experiment on a considerable scale, made in Greenland Dock by Mr. Lyttleton. This gentleman had fixed to the stern-post of a Virginia pilot boat, a frame containing a large copper spiral, which, by a winch turned by two or more men, gave it a rotary motion; the effect was much less than expected, for, notwithstanding the boat was completely empty and considerable exertions used, the progressive velocity did not exceed the rate of two miles per hour.

behind the vessel; this mode was proposed by Bernoulli, and afterwards by Mr. Linaker.* These with numberless variations, the greater part of which are obviously inferior to the methods in their simple form, have been proposed. Some of the best we propose to notice. Our selection must however be limited, because it must be confined to those which afford sufficient power in a convenient manner, and without being liable to injury by the violence of the waves, or to get out of order.

625.—The species may be divided into two classes, viz. 1, Those in which the action is continuous or nearly so; 2. Those which act at intervals. To the first class belong the second, sixth, and seventh methods. To the second the first, third, fourth, fifth, and eighth.

Since when the action is continuous, the area of the surfaces in action, multiplied by their resistance, must be equal to the area of the vessel, by the vessel's resistance when reduced to the same direction, it is obvious that all those which act at intervals only, must require to be of greater area than those which act continually. Hence, unless there be some other manifest advantage, this circumstance alone must determine us to reject all except the first class, and of this to take only the second, sixth, and seventh methods. Most of the others would require complicated action, and be inconvenient in practice. The first class also reduces to two, for the two opposite water screws without a cylinder give about the same effect as one with a cylinder, and this method though it has not been used, deserves attention from the circumstances of its being capable of acting wholly below the surface of the water, and in a direction parallel to the motion of the vessel, and only so far above the centre of resistance as is deemed necessary to stability. I can easily conceive that the trial of an experiment may be the means of condemning a very useful principle, merely through inattention to the proportions and mode of action.

Of the spiral Propeller or Water Screw.

626.—The acting portion is a spiral surface projecting from a cylindrical axis; and in order that it may be at all effective, each point in the surface must revolve so rapidly that the motion of that point in the direction of the axis must be greater than that of the vessel. Also if the angle of the spiral to the axis be constant, it is obvious that by

* Buchanan on Propelling Vessels by Steam, p. 40.

having more than one revolution, the rest add little to the effect, perhaps not equivalent to the additional friction.

Let B A C = a, be the angle which the screw forms with a line A B perpendicular to its axis; then during the time the boat would move from C to B, a point in the surface

FIG. 26.

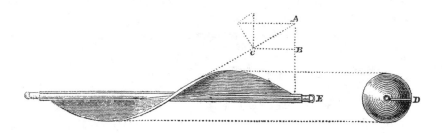

must move from B to A, otherwise it would retard the boat: and in order that it may be effective, it must move at some greater velocity. But the velocity of the boat, v, is to that of a point in the surface when no effect is produced, as B C : A B : : v :

$$\frac{A B . v}{B C} = \frac{v}{\tan . a}.$$

Hence the actual effective velocity must be

$$V - \frac{v}{\tan . a} = \frac{V \tan . a - v}{\tan . a}.$$

Let x be the variable radius of the cylinder, then

$$\frac{2 p x}{\cos . a} =$$

the length of the spiral, and

$$\frac{2 p x \dot{x}}{\cos . a} =$$

the fluxion of its area. Its resistance is

$$\frac{p (V \tan . a - v)^2 (2 \sin . a^3 + \sin . a^2) x \dot{x}}{\cos . a \tan .^2 a}$$

when the vessel is at rest; and when it is in motion it increases in the ratio

$$\frac{V \tan. a - v}{\tan. a} : v;$$

hence, $p v (V \tan. a - v) (2 \sin.^2 a + \sin. a) x \, x = $ the fluxion. The fluent is $\frac{1}{2} p v x^2 (V \tan. a - v) (2 \sin.^2 a + \sin. a) = $ the resistance.

This resistance is to the effect to impel the boat, as the radius is to the tan a; hence, $\frac{1}{2} p x^2 v (V \tan. a - v)^2 (2 \sin. a^2 + \sin. a) \tan. a = $ the force, and $\frac{1}{4} p x^2 v^2 (V \tan. a - v) (2 \sin.^2 a + \sin. a) \tan. a = $ the effect, which should be equal to the resistance of the vessel.

It is a maximum when $v^2 (V \tan. a - v) = $ a max: that is when

$$V = \frac{3 v}{2 \tan. a}.$$

Hence, the effect at the maximum is $\frac{1}{4} p x^2 v^3 (2 \sin.^2 a + \sin. a) \tan. a$. But the power to produce it must be

$$\frac{3 p x^2 v^3 (2 \sin.^2 a + \sin. a)}{8},$$

its velocity being V. Consequently, when $\tan. a = 1$, the power is to the effect as 3 is to 2, as in the ordinary paddle wheel, but if $\tan. a = 1{\cdot}5$ the power and effect are equal; on the contrary if $\tan. a = {\cdot}5$, or the angle C A B about 26° and a half, the power is to the effect as 3 : 1.

Now a little more than one revolution of the spiral would produce this effect, and a second revolution at the same angle could have very little action, because the water would have acquired all the velocity the spiral could communicate. If it be continued it should therefore be made with a decreasing angle.

627.—In practice the size corresponding to these effects is every thing; our next object must therefore be to ascertain it. Taking an angle of 60° for the angle C A B, then $\tan. a = 1{\cdot}732$, and $\sin. a = {\cdot}866$. The effect is in this case

$$\frac{1{\cdot}732}{4} p x^2 (2 \times {\cdot}866^2 + {\cdot}866) = 0{\cdot}73 \, p \, x^2;$$

but $p x^2$ is the area of the end of the cylinder, therefore each foot of surface of the end of the cylinder will act with a force of 0·73 pounds for one foot per second. The length of the cylinder would be $2 p x \tan. a = 10{\cdot}8$ times its radius, or 5·4 times its diameter. The power required for this effect is

$$\frac{3 p x^2 (2 \times {\cdot}866^2 + {\cdot}866)}{8} =$$

·887 p x^2, or 0·887 pounds for each foot of area of the end of the cylinder, for one foot per second.

When $a = 40°$, the effective force is only 0·4368 pounds per foot, and the power to each foot must be ·64 pounds. The power therefore decreases nearly in the same ratio as the length.

These calculations are sufficient to shew that this method may be used with considerable advantage, the action being under water, and the projection from the side not so great as paddle wheels; while the smoothness, and the uniformity of the motion are circumstances much in its favour. On the other hand, the mode of communicating motion and the resistance the parts will offer that are applied for that purpose, are objections; for the present, I shall therefore content myself with recommending it to the notice of my readers.

Paddle Wheels.

628.—The next inquiry is to ascertain the effect of paddle wheels. Of these the commonest species are plain boards, called paddle boards, fixed to the arms of a wheel; these arms are as thin as is consistent with strength, and are connected by one or more thin iron rings to act as braces in giving them firmness; they are sometimes made to slide on the arms, so as to reduce or increase the depth of immersion in the water, according as the vessel is more or less laden.

629.—To determine their power, let V be the velocity of the exterior portion of the wheel, and r its radius; then the velocity at any distance $r — x$ from the centre is

$$\frac{V (r-x)}{r}$$

hence, if v be the velocity of the boat,

$$1·5 \left(\frac{V (r-x)}{r} - v \right)^2 =$$

the resistance to one square foot of the paddle, and which is equivalent to the force to move the boat, or

$$\frac{1·5 (V r - x) - r v)^2}{2} =$$

the force.

But while the paddle acts on the water the vessel moves forward, and the water recedes only at the rate of the difference between the velocity of the wheel and that of the vessel; hence, the quantity of water put in motion is greater as the velocity of the vessel is greater than the excess of the velocity of the paddle.   Therefore,

$$\frac{V(r-x)-rv}{r} : v :: \frac{1\cdot5\,(V(r-x)-rv)^2}{r^2} : 1\cdot5\,v\frac{(V(r-x)-rv)}{r};$$

and making b = the breadth of the paddle,

$$\frac{1\cdot5\,v^2\,(V\,r - V\,x - r\,v)\,b\,\dot{x}}{r} =$$

the fluxion of the effective power.
The fluent is

$$\frac{1\cdot5\,v^2\,b\,(V\,r\,x - \frac{1}{2}V\,x^2 - r\,v\,x)}{r};$$

and when the depth of the paddle is d, it is

$$\frac{1\cdot5\,v^2\,b\,d\,(V\,r - \frac{1}{2}V\,d - r\,v)}{r} =$$

the direct power, which must be equal to the resistance of the vessel.   The loss by oblique action has to be estimated before the power of the engine can be found, but previously we may proceed to determine the best velocity for the wheels in still water.

630.—Let the equation be freed of all its constant multipliers, except those relating to or connected with its velocities; then it is $v^2\,(V\,r - \frac{1}{2}V\,d - r\,v)$; and making v variable, its fluxion is $2\,V\,r\,v\,\dot{v} - V\,d\,v\,\dot{v} - 3\,r\,v^2\,\dot{v} = o$; from whence we have

$$V = \frac{3\,r\,v}{2\,r - d}, \text{ and } v = \frac{V\,(2\,r - d)}{3\,r}.$$

The excess of the velocity of the outer point of the paddle therefore depends in part on its depth compared with the radius, the greater the depth the less excess is necessary.

631.—If this value of V be inserted in the equation for the area of the paddles, then $\cdot75\,v^3\,b\,d$ = effect of the paddles, which must be equal to the resistance of the vessel; and the comparison is easily made by means of the equations for the resistance of vessels, (art. 622.)   The power required to produce the effect is,

$$\frac{3\ r}{2\ r\ -\ d}\ (\ ·75\ v^3\ b\ d\ ) = \frac{2·25\ r\ v^3\ b\ d}{2\ r\ -\ d}.$$

A somewhat greater power is required, because there is also a loss by oblique action, and this will be expressed with accuracy enough for the object by multiplying the power by

$$\sqrt{\frac{2\ r\ -\ \frac{2}{3}\ d}{2\ r\ -\ d}}\ ;$$

when r is the radius C B, and d the depth of the float boards D B. For the centre of

Fig. 27.

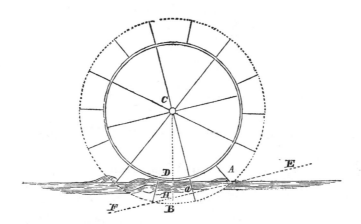

gravity, a of the immersed part A B D, may be considered the actual point where the whole force acts, instead of being distributed over the segment; and its mean direction will be E F, which is perpendicular to the line A C, drawn from the centre of the wheel through the centre of gravity: and the direction of this line will determine the loss by oblique action, for the power is to the effect as A H : A D, and is nearly given by the multiplier above.

An example will give a clearer idea of its effect. Let the radius of the wheel be eight feet, and the depth of the paddles two feet, then

$$\sqrt{\frac{16\ -\ \frac{4}{3}}{16\ -\ 2}} = 1·024,$$

nearly. Hence, as it will not exceed one-fortieth part of the whole, it may be neglected. The mean direction of the action is of greater importance to the motion, than the loss of power.

632.—We have supposed the paddles to be of the same breadth every where, but this may not be the best form, and therefore let

$$d^n : b :: x^n : \frac{b\,x^n}{d^n} =$$

the breadth at any point x, which substituted for b we have

$$\frac{1{\cdot}5\,v^2\,b\,(\,V\,r - V\,x - r\,v\,)\,x^n\,\dot{x}}{r\,d^n} =$$

the fluxion of the power. Its fluent is

$$\frac{1{\cdot}5\,v^2\,b}{r\,d}\left(\frac{V\,r\,x^{n+1}}{n+1} - \frac{V\,x^{n+2}}{n+2} - \frac{r\,v\,x^{n+1}}{n+1}\right);$$

and when $x = d$, it is

$$\frac{1{\cdot}5\,v^2\,b\,d}{r}\left(\frac{r\,(\,V - v\,)}{n+1} - \frac{V\,d}{n+2}\right).$$

If $n = 0$, the form of the paddle is a rectangle with the same result as before.

633.—If it be a triangle, $n = 1$, and the result is less than for the rectangle, the velocity and area being the same.

634.—If $n = \frac{1}{2}$ the form is parabolic and the result is

$$\frac{1{\cdot}5\,v^2\,b\,d\,(\,10\,r\,(\,V - v\,) - 6\,V\,d\,)}{15\,r}.$$

There we obviously gain advantage, by getting an equal resistance with less breadth, and by this form the resistance to the paddle is least when it strikes the water obliquely as at A, and increases as its action becomes more direct. The velocity for the maximum

Fig. 28.

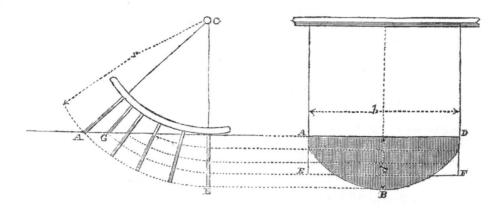

effect is to the velocity of the vessel, as

$$3\,r : 2\,r - 1\!\cdot\!2\,d :: v : V = \frac{3\,r\,v}{2\,r - 1\!\cdot\!2\,d};$$

which is less than for square paddles; if this value of V be inserted in the equations we have

$$\frac{1\!\cdot\!5\,v^3\,b\,d}{3} =$$

the power of the paddles when they are of a parabolic form, with the depth d, and breadth b.

If the form of the exterior edge be more rounded than the vertex of the common parabola, the effect again decreases. I was led to examine this point, by observing the form of those fins of fish which are used for impelling in a similar manner. The lines A D E F, shew the size of a square paddle capable of producing the same effect. It strikes the water at once with its whole breadth, as at G. The parabolic one strikes a little sooner, and gradually acquires its full hold of the water.

635.—The best position for the paddles appears to be in a plane, passing through the axis, as represented in the figure; if they be in a plane which does not coincide with the axis, they must either strike more obliquely on the fluid on entering, or lift a considerable quantity in quitting it.

In the direction of the breadth of the paddle, it is evident the form should be such that the resistance to its motion should be the greatest possible, and the pressure behind it the

least possible. These conditions appear to be fulfilled in a high degree, by making it a
plane in this direction also. A flat curve has been used, the concave surface to strike
the fluid, and perhaps with a very small increase of power. To set the paddles at any
other than a right angle, must obviously be a defect; for the resistance to motion becomes
less when the surface strikes the water obliquely, whereas the greater this resistance, the
greater the effect in impelling the vessel.

636.—It is desirable that the action of the paddles should be as equable and continuous as possible. But in attempting to render the action of the paddles equable, their
number ought not to be increased more than can be avoided, because the construction is
more expensive, and the time for the water to flow between them so as to afford a proper
quantity of reaction is reduced; neither do they clear themselves so well in quitting the
water. If we suppose A E to be the line the water would assume when at rest, the most
favourable arrangement with the smallest number of paddles, appears to be to make

Fig. 29.

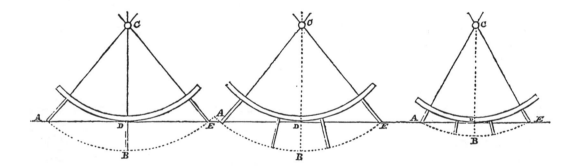

a paddle at A, just entering, when one at B is in a vertical position, and the one E quitting
the water; if a smaller number were employed, there would be a short interval, during
which none of the paddles would be in full action. A still more equable action will be
obtained by dividing the immersed arc into three; beyond this I do not think the advantage will be worth the extra expense, therefore I propose to give general equations, for
any proportion, and particular rules for three to be immersed.

637.—To determine the radius of the wheel or the depth of the paddles, when the
number of the paddles is given, becomes an easy problem, when the preceding conditions
are to be adhered to. For, put B C the radius $= r$, $x =$ the depth B D of the paddles,

n their number, and a the number of parts into which the immersed arc is divided. Then

$$\frac{a\,180^\circ}{n} =$$

the angle A C B, corresponding to half the immersed arc, and

$$r\cos.\frac{a\,180}{n} = \text{C D,}$$

the cosine of the angle being the depth from the centre of the wheel to the surface of the water; and

$$r\cos.\frac{a\,180}{n} = r - x; \text{ or } r\left(1 - \cos.\frac{a\,180}{n}\right) = x = \text{B } d,$$

the depth of the paddles.

And

$$\frac{x}{1 - \cos.\dfrac{a\,180}{n}} = r = \text{B C,}$$

the radius of the wheel.

From these equations we have the following rules for the case when three paddles are immersed.

638.—RULE I.   To find the radius of the wheel, when the number and depth of the paddles are given.   Divide 540 by the number of paddles, which will give the degrees in the angle contained by half the immersed arc.   From unity subtract the natural cosine of this angle, and the depth of the paddles divided by the remainder will give the radius of the wheel.

Or the radius of the wheel multiplied by the remainder will give the depth of the paddles.

639.—RULE II.   To find the number of paddles, when the radius of the wheel and the depth of the paddles are given.   Divide the depth of the paddles in feet, by the radius of the wheel in feet, and subtract the quotient from one.   Find the angle corresponding to the remainder as a cosine, and 540 divided by the degrees in that angle, is the number of paddles required.

If the radius of the wheel be eight feet, and the depth of the paddles two feet, then

$$1 - \frac{2}{8} =$$

·75, which is the cosine of the angle 41° 4′, and

$$\frac{540}{41\cdot4} =$$

13 for the number of paddles.

640.—The size of the wheels depends chiefly on the mode of giving them motion from the engine; they must be so large as to have the proper speed at the circumference, and where large wheels can be admitted they have some advantages: they must necessarily be narrower, and they strike the fluid in a favourable direction, and also quit it better; the paddles having more direct action on the water, they splash it about much less; the weight of the wheel also renders it more effective, as a regulator of the forces acting upon it. On the contrary, there are some strong practical objections to very large wheels for sea vessels; they give the momentum of the waves a greater hold on the machinery, they are cumbersome and unsightly, and they raise the point of action too high above the water line.

641.—When the wheels are on the first motion, the radius is determined by the velocity of the engine. Let that velocity be n strokes per minute, then $3\cdot1416 \times 2\,r =$ the circumference of the wheel; r being its radius, and its velocity per minute is $3\cdot1416 \times 2\,r\,n$: but this is to the velocity of the boat as 3 is to 2;* hence,

$$3\cdot1416 \times 2\,r\,n = \frac{88 \times 3\,v}{2} \text{ or } \frac{21\,v}{r} = n =$$

the number of strokes per minute, when $v =$ the miles per hour.

Also fixing the number of strokes we have

$$\frac{21\,v}{n} = r =$$

the radius of the wheel.

This therefore reduces to this simple rule. The velocity of the vessel per hour multiplied by 21 is equal to the number of strokes per minute, multiplied by the radius of the wheel.

From this exceedingly simple rule it can be known at once whether the wheel becomes too large or not, when the simple crank motion is used; but it ought to be used in preference to a second motion, in all cases where it does not involve some other difficulties than merely the size of the wheel.

* This ratio may be obtained more accurately from art. 630.

Modifications of Paddle Wheels.

642.—Several methods have been tried or projected for getting rid of certain supposed defects of the paddle wheels. The quantity of force lost by oblique action has been greatly overrated, and most of the contrivances are directed to remove it either wholly or in part. The methods proposed are of two kinds. In one a gradual change of position of the paddle is produced by the movement of the wheel; completely forgetting that by loss of the velocity, the decrease of force is as the square, while by variation of direction the loss is only in simple proportion. Mr. Oldham, of the Bank of Ireland, proposed a plan for these revolving paddles, to avoid the defects of the fixed paddles commonly used; and states, that the violent action of the paddles of common wheels, in striking the water in a rough sea, is entirely removed by the use of the revolving paddles, as they enter and rise out of the water, with a peculiarly soft and easy motion. We can only regret that so much ease cannot be obtained without a considerable and constant sacrifice of power.

The other method is to cause the paddles to change at once to a new position at two points in the revolution, by means of proper catches and mechanism. This is a better method for cases where the wheels are to work when deeply immersed in water; but such wheels require to be made so very strong and powerful, that there appears to be small probability of the machinery keeping in order.

The plan of making paddles which seems most plausible, is to have a pair of wheels at each side of the vessel, having two endless chains acting on them, with paddles fixed on these chains. As the chain passes in one direction, the paddle boards are immersed in the water, and return in the opposite direction out of the water; the two wheels around which they pass being partially under water. The whole of the impulse given by these boards from the lower part of one wheel to the lower part of the other, seems as though it would be direct and effectual; and it is stated, that so far as the plan has been tried on a very small scale it has been successful. It is said however by Buchanan, to have been tried on the Duke of Bridgwater's canal, where it did not give satisfaction; and the reason not being assigned, we must endeavour to shew whether or not the arrangement can have greater effect than the common paddle wheel.

If a wheel have a sufficient number of paddles to force the whole of the fluid opposed to the area of the paddle into motion, it is obvious that any continuation of the line of action of the paddles, will be only equivalent to the friction of the stream put in motion

T T

by their first action on it : and this effect is by far too small to be obtained by a complicated arrangement which it would be difficult to render durable; hence, the construction
is imperfect.

Of the Strength of Vessels.

643.—It was not till 1818 that steam vessels were made to perform regular voyages
at sea; and in proportion as they have had experience, the strength of the vessels has
been increased. A vessel should be considered as one slightly flexible frame, and the
strength determined so that the greatest possible stress, acting with the most disadvantage
probable, would not derange the natural elasticity of the parts, nor disturb the connec-
tions. The want of considering the frame as a whole, has often led to weak modes of
construction, and improper modes of bracing. A vessel is also to be considered in the
condition where hydrostatic pressure contributes least to its support. The strains reduce to
those which would take place in a large hollow beam of which we have to find the neutral
axis, and then the resisting forces are easily measured.* When the timbers are filled in
between, it must increase the strength, if it be done in a proper manner; and this increase
might perhaps be obtained with less material, and less addition to the weight of the
vessel, but the advantage of leaving no hollow cavities is of much importance both to
the durability and cleanliness of a vessel.†

644.—In respect to timber, fir has the advantage of lightness, and for straight timbers
it is stronger than a like weight of oak; but for curved timbers the harder woods which
have greater lateral cohesion are better.

Of the Application of Sails.

645.—It is found that sails may be effectively combined with steam power, whenever
the direction is not within four points of that of the wind.

---

* See my " Elementary Principles of Carpentry," Sect. 1 and 2; and " Treatise on the Strength of Iron,
art. 85 a.

† See Philo. Transactions for 1820.

But when the force of the wind becomes considerable, and the sea rough, the wheels often revolve without touching the water in the hollows of the waves, and acquire a great increase of velocity, to be reduced as soon as they meet the wave again to less than the ordinary speed. To lessen the abruptness of these changes it is necessary to diminish the supply of steam, and consequently the power of the engine.

646.—It appears to be impossible to apply so much sail as to give a steam vessel the advantage of being used as an effective sailing vessel, in the event of the engines or coals failing. The proper object of sails in a steam vessel is to save fuel when the wind can be of service,* and to do this with economy, the engines should work expansively, (see art. 419;) hence, the arrangement of the engine should be such as would answer to work at full pressure in a calm. This condition enables us to fix the power of the engine by the rate for still water; and if the vessel has sails sufficient to maintain the speed with about half the power of the engines, when the wind is fair, it will be as much as can be usefully employed. The greatest attention should be given to keep the centre of effort on the sails as low as possible, and to arrange them so that the angle of the vessel's inclination may be inconsiderable, that the wheels may not dip unequally.

647.—The average speed in still water beyond which it does not seem to be desirable to go, is ten feet per second; that is, seven common miles, or six nautical miles per hour: and at this velocity, when the wind is as powerful as it is prudent to carry all the canvas, the direct effect will be only one horse's power for each thirty-two yards superficial.†

648.—A fair wind also contributes to the motion of a vessel, by giving motion to the sea itself; a head wind opposes its motion, and a current has a similar effect. If v be the velocity a vessel is impelled at in still water by the power P, and the velocity of the current be $\pm\,n\,v$, using the upper sign when it is with the vessel, then $v^3 : P :: v$ $(v \pm n\,v)^2 : P\,(1 \pm n)^2$; that is, $P\,(1 \pm n)^2 =$ the power the boat will require.

---

* It is a common notion that the sails should be used in addition to the steam power to gain greater velocity; but this is not desirable, except for post office packets and the like, because an immense extent of canvas affords only a very small power when the vessel moves at a considerable velocity; hence, economy directs to saving fuel, rather than increasing speed.

† To find the effect of the wind in any other direction, and with any other velocity, let V be the velocity of the wind in feet per second, $a =$ the angle it makes with the direction of the vessel's motion, $v =$ the velocity of the vessel in feet per second, $b =$ the angle a perpendicular to the surface of the sail makes with the direction of the motion of the vessel; then it is nearly

$$\frac{3200\ \cos.\ b}{(\,\mathrm{V}\ \cos.\ (\,a + b\,) - v\ \cos.\ b\,)^2} =$$

the yards of canvas equivalent to a horse's power $=$ 550 pounds raised one foot per second.

If the stream be in the direction of the vessel's motion, and half its velocity in still water, then $n = \cdot5$ and $P(1 - \cdot5)^2 = \cdot25 \, P$; or the vessel will require only one-fourth of the power.

If the vessel move against the stream, the stream being half its velocity in still water, then $P(1 + \cdot5)^2 = 2\cdot25 \, P$; or the vessel will require $2\frac{1}{4}$ times the power to preserve its velocity.

649.—But in ascending a stream the difference must be in the velocity, and it is generally so also in the descent; hence, if v be the velocity of the stream, and mv the velocity of the vessel moving in the stream, then

$$v^3 = m \, v \, (m \, v \pm u)^2; \text{ or } v^4 = m \, (m \, v \pm u)^2.$$

The value of m is not easily separated in this equation, but by assuming that the force of the wheels is invariable, we have $v^2 = (m \, v \pm u)^2$; or $v \pm u = m \, v$. The upper sign to be used when the vessel moves with the current.

Hence, if the velocity in still water be eight miles per hour, and that of the stream three miles per hour.

Then down the stream $v + u = 8 + 3 =$ eleven miles; and up the stream $v - u = 8 - 3 =$ five miles.

If the velocity of the paddles alter, the power will not be constant; and if it do not alter this ratio cannot exactly hold.

Rule for the Power of Boat Engines.

650.—The power of boat engines may be ascertained thus. Let p be the mean pressure on the piston in pounds, d its diameter in inches, v its velocity in feet per minute, and $h =$ the number of horses equivalent to its power; then

$$\frac{p \, v \, d^2}{33000} = h.$$

Let the length of the stroke in feet be l, then $v = A \sqrt{l}$, where A is the multiplier found by (art. 336;) hence,

$$\frac{p \, v \, d^2}{33000} = \frac{p \, A \, d^2 \sqrt{}}{33000} = h, \text{ or } \left(\frac{33000 \, h}{p \, A \sqrt{l}} \right)^{\frac{1}{2}} = d.$$

In logarithms, log. $d = \frac{1}{2}$ (log. 33000 + log. h — log. p — log. A — $\frac{1}{2}$ log. l).

For low pressure steam acting at full pressure throughout the stroke we have A = 103, (art. 337,) and $p = 7\cdot1$ (art. 416.) Hence, in this case the rule becomes

$$\text{log. } d = \tfrac{1}{2} (\text{log. } h + 1\cdot667256 - \tfrac{1}{2} \text{ log. } l).$$

Also

$$\text{log. } h = 2 \text{ log. } d + \tfrac{1}{2} \text{ log. } l - 1\cdot667256.$$

651.—If m times the length of the stroke in feet be the diameter in inches, then

$$\text{log. } d = \tfrac{2}{5} (\text{log. } h + 1\cdot667256 + \tfrac{1}{2} \text{ log. } m.)$$

The most common proportion now used is $m = 9$, or the length of the four-thirds of the diameter, hence,

$$\text{log. } d = \tfrac{2}{5} (\text{log. } h + 2\cdot144377).$$

Example. If the resistance be equal to 100 horses' power, with two engines of fifty each;

the log. 50 =	1·698970
	2·144377
	3·843347
	2
	5) 7·686694
log. d = log. 34·5 =	1·537339

Hence, the diameter should be 34·5 inches, and the length of the stroke 34·5 × 4 ÷ 3 = 46 inches.

Of arranging the Proportion of Power for Vessels.

652.—If we now proceed through a calculation of the proportions, and a statement of the conditions to which we ought to attend in arranging the parts of an engine for a vessel, it will form the best illustration of the use of the preceding rules.

653.—*The resistance of the vessel* should be ascertained for the average velocity; now without pretending to fix the best average, I will suppose this to be ten feet per second, or seven miles per hour in still water.* Let the length of the parallel part of the vessel be seventy-two feet, the mean radius of curvature of the ends be six half breadths, the breadth at the midship section 25·7 feet, and the girt of that section in contact with water thirty-eight feet. Then the velocity being ten feet per second, we have by the table, (art. 622,)

$$10^3 \times 38 \, (\, \overline{\cdot025 \times 25\cdot7 + \cdot0032 \, (\, \overline{72 + 1\cdot05 \times 25\cdot7} \,)} \,) =$$

$$10^3 \times 36\cdot5 = 36500 \text{ pounds raised one foot per second, or}$$

$$36500 \div 550 = 67 \text{ horses' power nearly.}†$$

654.—Now if we were to fix on the area and velocity of the paddles for this velocity, it would not be possible to work with advantage against either wind or current; besides there is a disadvantage in large paddles at sea, which is greater than the loss by varying from the maximum. Hence, I would recommend to arrange the paddles for a velocity about one foot per second greater than the average rate. Consequently, (art. 631,) $11^3 \times \cdot75 \times b \, d =$ the power of the paddles; and as the resistance is 36500, we have

$$\frac{36500}{11^3 \times \cdot75} = b \, d = 36\cdot6 \text{ feet,}$$

the area of the paddle boards.

Again, suppose the radius of the wheels to be four times the depth of the paddles, then by the second equation of (art. 631,) we have

$$\frac{2\cdot25 \, r \times 11^3 \times b \, d}{2 \, r - d} = \frac{9 \times 11^3 \times 36\cdot6}{7} = 62633 \text{ pounds}$$

raised one foot per second, or $62633 \div 550 = 114$ horses' power very nearly.

655.—As the vessel requires 114 horses' power, if we have two engines, each will be 57 horses' power; and (art. 651,)

This will be equivalent to an average speed of nine miles per hour, where sailing power is to be used in addition.

† The coefficient ·0032 (see art. 614,) is very likely too high; it is taken from the experiments made by the *Society for the Improvement of Naval Architecture,* and agrees with the more recent experiments of Col. Beaufoy; but I am quite convinced that when water is in motion the friction is less, only the exact measure remains to be determined.

$$\log. 57 = \qquad 1\cdot755875$$
$$2\cdot144377$$

$$\overline{ 3\cdot900252}$$
$$2$$

$$\overline{5)\ 7\cdot800504}$$

$$\log. \text{diameter} = \qquad 1\cdot560101$$

or the diameter equal 36·32 inches; the length of the stroke four-thirds of the diameter, or 48·43 inches; and consequently (art. 336,) the number of strokes per minute will be 25½: hence, by (art. 641,) corrected for the depth of the paddle,* we have, when the velocity is eleven feet per second, or 7·5 miles per hour,

$$\frac{24 \times v}{25\cdot5}, = \frac{24 \times 7\cdot5}{25\cdot5} = 7\cdot1 \text{ feet}$$

for the radius of the wheels; and dividing it by 4 gives the depth of the paddles 1·8 feet. But in order to reduce the breadth of the wheel, it is better that they be made two feet, and the radius of the wheel 7·3 feet; the paddles will then be 2 feet, × 9·2 feet for each wheel, making an area of 36·8 feet for both, or the breadth of the wheel 9·2 feet.†

The other proportions of the engines will be found by the general rule, (art. 415,) except that a somewhat smaller quantity of water produces the steam, owing to its being of a higher temperature, (see art. 90;) but it is only about two per cent less, and the fuel

---

* For when the depth is one-fourth of the radius

$$V = \frac{3\ r\ v}{2\ r - d} = \frac{12\ v}{7},$$

instead of

$$\frac{3\ V}{2}$$

hence,

$$\frac{3}{2} : \frac{12}{7} :: 21 : 24.$$

† The proportions of the vessel in this case are as nearly those of that called the *James Watt*, as I could ascertain them ; and in the tables which follow, the best information I could procure respecting that vessel is given in order to compare the calculated with the reported effect : but the velocity in still water is a very questionable statement ; most likely it was the velocity in a river where that of the current was not deducted.

required is not sensibly altered; there is also a slight advantage by the force of the steam being less in the condenser than when pure water is used. (See table, art. 94.) Hence, the cistern of water to contain the condenser is omitted without loss. The engines should be prepared to work expansively, to be adjusted by hand, (see art. 419 and 481;) and the strength of the parts will be found by (art. 496—527;) the management of the water is treated of in (art. 565,) and the parallel motion (art. 488—495.)

656.—I think it would be desirable to try the effect of giving a considerable degree of elasticity to the arms of the paddles, and to form the boards in the manner shewn in Fig. 28, page 307. The wheels of vessels appear to be kept too forward, so as to keep the fore part of the vessel constantly heaving upwards; and such an action is unfavourable. A vessel should bear firmly in the direction of its motion to move well; and that this re-mark is true in practice as well as theory, may be inferred from the fact that, in the pre-sent construction, they find an advantage in using the sails to steady, and determine the direction of the vessel's motion. In vessels for towing,* this may be adopted with still greater advantage; and in both cases the proper place for the wheels appears to be be-hind the centre of gravity of the vessel. The construction of the boiler is shewn in Plate XVII, and of the engine in Plate XVIII. and XIX.

The following tables are collected chiefly from the evidence printed in the Reports on the Holyhead steam packets, by the committee appointed by the House of Com-mons; and will afford a means of comparing the practice of different manufacturers.

The power required for towing a vessel may be estimated by (art. 622.)

657.—*Table of Steam Vessels.*

Name of engine builder — Maudslay, Son, and Field, London

	Dee.	Enterprise.	Commerce.	Beurs Van Amsterdam.	London Engineer. (Brent.)	Lightning.	Harlequin.	Ivanhoe.	Crusader.
Name of the vessel	Dee.	Enterprise.	Commerce.	Beurs Van, Amsterdam.	London Engineer.	Lightning.	Harlequin.	Ivanhoe.	Crusader.
Name of builder of vessel					Brent.				
Length of deck	166 ft. 7 in.					126 ft. 0 in.			
Breadth (extreme)	30 ft.	26 ft.	22 ft. 4 in.	25 ft. 10 in.	24 ft. 0 in.	22 ft. 4 in.	21 ft. 0 in.	18 ft. 6 in.	16 ft. 2 in.
Draught of water	10 ft.	14 ft.	10 ft. 0 in.	8 ft. 0 in.	5 ft. 0 in.	8 ft. 2 in.	7 ft. 8 in.	7 ft. 0 in.	6 ft. 3 in.
Paddle wheels, diameter	20 ft.	15 ft.	18 ft. 0 in.	16 ft. 0 in.	12 ft. 6 in.	15 ft. 0 in.	13 ft. 0 in.	12 ft. 6 in.	11 ft. 6 in.
Paddle wheels, breadth	10 ft.	7 ft.	7 ft. 0 in.	8 ft. 0 in.	6 ft. 6 in.	9 ft. 0 in.	7 ft. 0 in.	6 ft. 0 in.	5 ft. 6 in.
Paddle wheels, velocity of extremity in miles per hour		12·8 miles							
Paddles, depth									
Tonnage (register)	700 tons	500 tons	400 tons	500 tons	315 tons	296 tons	232 tons	160 tons	95 tons
Total power of engines	200 h. p.	120 h. p.	140 h. p.	120 h p.	70 h. p.	100 h. p.	80 h. p.	60 h. p.	50 h. p.
Velocity in still water									
Coals per hour					630 lbs. Wylem coals	1240 lbs. average			
Engines, number	2 engines	2 engines	2 engines	2 engines	2 engines	2 engines	2 engines	2 engines	2 engines
Engines, diameter of cylinders	53 in.	43 in.	46¾ in.	43 in.	36 in.	40 in.	36 in.	32 in.	29¼ in.
Engines, length of stroke	60 in.	48 in.	54 in.	48 in.	30 in.	48 in.	42 in.	36 in.	36 in.
Engines, strokes per minute	20 strokes	24 strokes	22 strokes	25 strokes	28 strokes	25 strokes	28 strokes	30 strokes	32 strokes
Engines, diameter of air pump									
Used for	Navy	East Indies	Liverpool & Dublin	Amsterdam & London	Margate packet	Navy	Post office packet	Post office packet	Post office packet
Date of construction	1827	1825	1826	1826	1818	1824	1824	1826	1827
Calculated power of engines at the best velocity and full pressure	272 h. p.	160 h. p.	197 h. p.	160 h. p.	88 h. p.	137 h. p.	104 h. p.	76 h. p.	68 h. p.

658.—*Table of Steam Vessels.*

	Soho.	James Watt.	City of Edinburgh.	Shannon.	Sovereign George IV.	Caledonia.	Meteor.	Hero.
Name of engine builder	Boulton and Watt, Soho, Birmingham.							Fenton and Co. Leeds.
Name of builder of vessel		Wood & Co.	Wigram.	Fletcher & Son.	Evans.	Wood and Co.	Evans.	Bancham.
Length of deck	163 feet	146 feet	143 feet	180 feet	126 feet	95 ft. 6 in.		
Breadth (extreme)	27 feet	25 ft. 8 in.	25 ft. 6 in.	49 feet	21 ft. 10 in.	15 ft. 0 in.	20 feet	
Draught of water		10 ft. 0 in.			8 ft. 6 in.	4 ft. 6 in.		6 feet 4 inches
Paddle wheels, diameter	15 ft. 8 in.	18 ft. 0 in.	18 feet		16 feet			14 feet 0 inches
Paddle wheels, breadth	8 ft. 0 in.	9 ft. 0 in.	8 feet		8 feet		8 feet	8 feet 0 inches
Paddle wheels, velocity of extremity in miles per hour	14·6 miles	12 miles	12 miles					15 miles
Paddles, depth	2 feet	2 ft. 0 in.	2 feet					1 foot 6 inches
Tonnage (register)	510 tons	448 tons	400 tons	513 tons	210 tons	102 tons	190 tons	233 tons
Nominal power of engines	120 h.p.	100 h.p.	80 h.p.	160 h.p.	80 h.p.	28 h.p.	60 h.p.	90 horse power
Velocity per hour in still water		10 miles			9¾ miles	8¾ miles		11¾ miles
Coals per hour					896 lbs.		560 lbs.	2240 pounds
Engines, number	2 engines	2 engines	2 engines	2 engines	2 engines	2 engines	2 engines	2 engines
Engines, diameter of cylinders	42 inches	39 inches	36 inches					
Engines, length of stroke	48 inches	42 inches	42 inches					
Engines, strokes per minute	26 strokes	27½ strokes	27½ strokes					30 strokes
Engines, diameter of air pump	23 inches	21 inches	19½ inches					
Used for	Passengers	Passengers	Passengers	Passengers & 200 tons goods	Post office packet		Post office packet	Margate packet
Date of construction	1823	1821	1821	1826	1821	1815	1821	1821
Calculated power of engines at the best velocity and full pressure	151 h.p.	122 h.p.	104 h.p.					

659.—*Table of Steam Vessels.*

	United Kingdom.	Majestic.	Superb.	Talbot.	St. Patrick.	Prince Llewellyn.	Albion.	Duke of Lancaster.	Cambria.
Name of engine builder	Napier, Glasgow.				Fawcett and Littledale, Liverpool.				
Name of the vessel	United Kingdom.	Majestic.	Superb.	Talbot.	St. Patrick.	Prince Llewellyn.	Albion.	Duke of Lancaster.	Cambria.
Name of builder of vessel	Scott & Co.	Scott & Co.	Scott & Co.	Wood & Co.	Mottershead and Hayes.				
Length of deck	175 ft.			92 ft.	130 ft.		103 ft. 6 in.	103 ft.	91 ft. 2 in.
Breadth (extreme)	45 ft. 6 in.			17 ft. 11 in.	22 ft. 1 in.		18 ft. 1 in.	17 ft.	17 ft. 6 in.
Draught of water					13 ft. 8 in.		9 ft. 6 in.	9 ft. 6 in.	8 ft. 4 in.
Paddle wheels, diameter									
Paddle wheels, breadth									
Paddle wheels, velocity of extremity									
Paddles, depth									
Tonnage (register)	1000 tons	350 tons	241 tons	140 tons	200 tons	170 tons	103 tons	94 tons	86¾ tons
Power of engines	200 h. p.	100 h. p.	70 h. p.	60 h. p.	100 h. p.	70 h. p.	60 h. p.	50 h. p.	50 h. p.
Velocity per hour in still water		10 miles	9 miles						
Coals per hour	2240 lbs.	2240 lbs. Scotch coal	1670 lbs. Scotch coal	784 lbs. Scotch coal					
Engines, number	2 engines	2 engines	2 engines		2 engines	2 engines	2 engines	2 engines	2 engines
Engines, diameter of cylinders					42 in.		32 in.		30 in.
Engines, length of stroke					42 in.		33 in.		30 in.
Engines, strokes per minute		28							
Engines, diameter of air pump									
Used for	Edinburgh packet. 175 passengers			Post Office packet.					
Date of construction	1826	1816	1820	1819	1822	1822	1822	1822	1822
Calculated power of engines at the best velocity and full pressure					142 h. p.		73 h. p.		67 h. p.

The average consumption of coals is that required when the engine is in full action, and including all delays, waste, &c.; and is to be understood as that which multiplied by the hours it requires for the average passage, would give the quantity consumed for each passage; and the store ought evidently to be for the longest passage.   In the best engines (of this time) it will be found to vary from twelve to sixteen pounds of Newcastle coals per hour for each nominal horse power, and in inferior engines it may extend to twenty pounds.

When the consumption is stated at less than it amounts to at twelve pounds per hour for each nominal horse power, it may be fairly esteemed an experimental trial; and of course the fires are more carefully attended, with every precaution to prevent waste and give effect.   The last column of Table III will nearly give the fuel required per hour if the nominal power be taken in the first column, (art. 664,) when applied to steam boats.

The velocity of sea vessels appears to average about ten miles per hour; their power to face a wind is inconsiderable, because the wind gives the surface of the water so much velocity that the paddles act with less force in proportion as the velocity of the water approaches to the difference between the velocity of the paddles and that of the vessel; and when these are equal the boat will commence moving backward; and it is also with much reason supposed that the action of the wind itself tends greatly to retard a vessel's motion when it is directly opposed to it: for if a vessel of one hnndred horses' power has a surface of sixty yards above water,* and the velocity of the wiud be fifty feet per second (in which vessels are under their courses) then by the equation (art. 647, note)

$$\frac{2500 \times 60}{3200} =$$

47 horses' power for the resistance offered to motion when the vessel is at rest; and as with whatever velocity the vessel moves against the wind, this velocity should be added to that of the wind, the plus sign being the proper one, in the equation for this case it will appear that the power to move forward is extremely limited with so much surface above water.

The only vessels in the table (art. 657,) which have dimensions to enable us to approximate to their speed are the Lightning, and the Dee; and, notwithstanding the great quantity of power to be placed in the Dee, I expect that its velocity in still water will be one-eighteenth less than that of the Lightning in similar circumstances; and it would require the engines of the Dee to be 230 horses' nominal power to render them of equal speed. The Dee spreads above water at an angle of about fifty degrees with the water line, to a width of about five feet on each side, in the same manner as in boats for the accommodation of passengers.

* This is estimated from a vessel in use, and esteemed.

660.—Much has been said respecting American steam vessels, and these vessels as far as excellent workmanship, neatness of fitting up, and convenience is concerned, do appear to be superior to our own. Their best engines however seem to be not more than equal to the British ones, if so good, as many of the reports respecting them carry internal evidence of their inaccuracy. The best I have met with is that of the steam vessel called the Chancellor Livingston, constructed by Mr. Fulton for the Hudson River, from New York to Albany. It is one of their largest vessels: the keel is 154 feet long; deck 165; breadth thirty-two feet; draught of water about seven feet three inches, and burden 520 tons; the principal cabin fifty-four feet long, seven high; ladies' cabin, above the other, thirty-six feet long, with closets; the forward cabin thirty feet long, and seven high. The number of sleeping-berths, in the principal cabin, is thirty-eight: in the ladies' cabin twenty-four; in the fore cabin, fifty-six; in the captain's cabin on deck two; in the engineers', and pilot's three; in the forecastle six; and for fire-men, cooks, &c. six; being a total of 135. The engine is of seventy-five horses' power; the diameter of cylinder forty inches, length of stroke five feet; the boiler is twenty-eight feet long, twelve broad, with two funnels; the paddle wheels seventeen feet diameter; paddle boards five feet ten inches long; they have two fly-wheels, each fourteen feet diameter, connected by pinions to the crank shaft. The machinery rises four feet and a half above the deck. Average rate of sailing is said to be eight and a half to eight and three quarter miles an hour. With a strong wind and tide in her favour she has made twelve, but with wind and tide against her not more than six miles per hour. As for low pressure steam the engine is estimated at the greatest power of the cylinder, it has been imagined that the vessel is moved by less power than the British vessels of equal magnitude.

The following arises out of erroneous methods of measuring vessels to register their tonnage.

To ascertain the Register Tonnage of a Steam Vessel.

661.—The breadth is to be taken at the broadest part of the vessel, whether it be above or below the main wales, and is to be from the outside to outside of the plank the length is to be the horizontal distance between the back of the main stern post and the fore part of the main stem, under the bowsprit; and calling this length l, the breadth b, and $r =$ the length of the engine room, the rule is

$$\left(\frac{l - r - \frac{3}{5} b}{188} \right) b^2 = \text{the tonnage.*}$$

* Sect. 59th Geo. III. cap. 5.

Example.   If the breadth be thirty-two feet, the length 162 feet, and the length of the engine room forty-seven feet,

$$\left(162 - 47 - 19 \cdot 2 \right)\frac{32^2}{188} = 520 \text{ tons.}$$

The register tonnage is the same whether the draught of water be seven feet or fourteen feet; and it is the same whatever may be the form of the vessel: now it is an unpleasant task for an Englishman to mark as fallacious the modes of measuring the capacity of vessels adopted by his Government, but it is necessary for the purposes of science not only that it should be pointed out, but that the error should be corrected.

662.—Table I. Of the Properties of the Steam of Water of different Degrees of Elastic Force.

Total force of steam.			Excess of force above the atmosphere.							
In atmospheres.	In inches of mercury.	In pounds per circular inch.	In pounds per circular inch.	In pounds per square inch.	Temperature, Fahrenheit.	Volume in cubic feet, the water being 1.	Weight of a cubic foot in grains.	Specific gravity, air being 1.	Velocity into a vacuum in feet per second.	Heat of conversion from water of 52°, to steam in degrees.
·0183	·55	·21	—11·33	—14·4	60°	72190	6·1	·0115	1377	1008°
·0333	1	·385	—11·155	—14·2	77	41010	10·7	·0202	1400	1025
·0667	2	·77	—10·77	—13·7	98·7	21400	20·5	·0388	1427	1047
·1	3	1·15	—10·39	—13·2	112·5	14570	30	·0568	1445	1061
·133	4	1·54	—10·0	—12·7	123	11130	39	·0744	1458	1071
·25	7·5	2·88	—8·66	—10·99	147·6	6187	71	·134	1499	1096
·5	15	5·77	—5·77	—7·33	178	3249	135	·255	1526	1126
·75	22·5	8·65	—2·89	—3·66	197·4	2232	196	·371	1549	1146
1·00	30	11·54	0·	0·	212	1711	254·7	·484	1566	1160
*1·17	35	13·46	1·92	2·44	220	1497	292	·553	1575	1168
1·5	45	17·31	5·77	7·33	233·8	1178	363	·687	1591	1182
1·75	52·5	20·19	·8·65	10·99	242·5	1022	427	·81	1601	1191
2·0	60	23·08	11·54	14·65	250·2	905	483	·915	1610	1199
2·5	75	28·85	17·31	21·98	263·5	737	593	1·123	1625	1212
3·0	90	34·62	23·08	29·3	274·7	623	700	1·33	1638	1223
3·5	105	40·39	28·85	36·63	284·5	542	810	1·53	1649	1233
4	120	46·16	34·62	43·95	293·1	479	910	1·728	1658	1241
5	150	57·7	46·15	58·60	308	391	1110	2·12	1674	1256
6	180	69·24	57·7	73·25	320·6	331	1317	2·5	1688	1269
7	210	80·78	69·24	87·90	331·5	288	1520	2·88	1700	1280
8	240	92·32	80 78	102 55	341·2	255	1660	3·25	1710	1289
9	270	103 86	92·32	117 20	350	229	1910	3·61	1720	1298
10	300	115·4	103·86	131·85	358	209	2100	3·97	1729	1306
20	600	230 8	219·26	278·35	414	111	3940	7·44	1786	1362
30	900	346 2	334·66	424·85	450	77	5670	10·75	1823	1398
40	1200	461·6	450 06	571·35	477	60	7350	13·88	1850	1425
1	2	3	4	5	5	7	8	9	10	11

The mode of obtaining the first five columns is obvious; and in the fourth and fifth the negative sign indicates that the force of the steam is less than atmospheric pressure, and the numbers shew how much less: the sixth column is calculated by (art. 89,) the seventh by (art. 121,) and from it the eighth and ninth: the tenth column is calculated by the equation in the note to (art. 136,) with the allowance for contraction of the aperture; and the last column is obtained in the manner shewn in the note to (art. 190.)

* The usual force of low pressure steam.

663.—Table II. Of the Proportions of single acting steam engines equivalent to different numbers of horses; the horse power being 33000 pounds raised one foot high per minute, and the elastic force of the steam in the boiler = 35 inches of mercury.

Steam acting expansively.								Steam at full pressure throughout the stroke in the same engine.	
Number of horses' power.	Diameter of the steam piston in inches.	Mean pressure on the piston in pounds, at 5½ lbs. per circular in.	Velocity of the steam piston in feet per minute.	Length of the stroke in feet.	Number of strokes per minute.	Water required per hour to supply the boiler.	Coals consumed per hour in lbs.	Number of horses' power.	Coals consumed per hour in pounds.
	inches		feet.	feet.		cubic feet.	lbs.		lbs.
10	26·4	3850	174	4·4	19¾	11·1	114	11·2	152
15	31·1	5324	187	5·2	18	16·7	164	16·8	220
20	34·9	6702	197	5·8	17	22·3	213	22·5	285
25	38·1	8012	203	6·3	16	27·7	257	28·	343
30	41·1	9270	214	6·8	15¾	33·3	307	33·5	410
35	43·7	10490	221	7·3	15¼	39·	356	39·2	475
40	46·1	11670	227	7·7	14¾	44·5	401	45·	536
45	48·3	12820	232	8·0	14½	50·	450	50·5	600
50	50·4	13950	237	8·4	14¼	55·5	500	56	670
55	52·3	15050	242	8·7	14	61·2	551	62	735
60	54·2	16140	246	9·0	13¾	66·7	600	67	800
65	56·0	17210	250	9·3	13½	72·1	649	73	865
70	57·6	18260	254	9·6	13¼	78·	702	78	940
75	59·2	19290	257	9·8	13	83·3	750	84	1000
80	60·8	20310	260	10·1	13	89·	801	89	1070
85	62·3	21330	264	10·4	12¾	94·5	851	95	1140
90	63·7	22320	267	10·6	12½	100	900	101	1200
100	66·5	24290	272	11·0	12¼	111	999	112	1330
120	71·5	28100	283	11·9	12	133	1197	134	1600
140	76·0	31790	291	12·6	11½	156	1404	157	1860
160	80·2	35380	299	13·3	11¼	178	1602	179	2140
180	84·1	38870	307	14·0	11	200	1800	201	2400
200	87·7	42300	313	14·6	10¾	222	1998	224	2650
213½	90	44550	318	15·0	10½	237	2133	265	2860

664.—Table III. Of the Proportions of double acting Steam Engines equivalent to different numbers of horses; the horse power being 33000 pounds raised one foot high per minute, and the elastic force of the steam in the boiler = 35 inches of mercury.

Steam acting expansively.								Steam acting at full pressure throughout the stroke in the same engine.	
Number of horses' power.	Diameter of the steam piston in inches.	Mean pressure on the piston in pounds, at 4·8 lbs. per circular in.	Velocity of the steam piston in feet per minute.	Length of the stroke in feet.	Number of strokes per minute.	Water required per hour to supply the boiler.	Coals consumed per hour in lbs.	Number of horses' power.	Coals consumed per hour in lbs.
	inches.	lbs.	feet.	feet.		cubic feet.	lbs.		lbs.
1	7·8	289	114	1·3	44	·8	15	1·46	31·5
2	10·25	516	131	1·75	37½	1·57	23	2·95	48
3	12·05	697	141	2·	35	2·36	30½	4·4	64
4	13·52	877	149	2·25	33	3·13	38	5·9	80
5	14·9	1049	157	2·5	31½	3·92	45	7·4	94
6	15·9	1214	162	2 65	30½	4·7	53	8·85	111
7	16·9	1373	167	2·8	29¾	5·5	60	10·3	126
8	17 85	1527	171	2·97	29	6·3	67	11·8	140
9	18·7	1678	175	3·1	28¼	7·05	73	13·3	153
10	19·5	1826	180	3·25	26¾	7·82	80	14·6	168
12	20·9	2113	186	3·5	26½	9·4	95	17·7	199
14	22·3	2390	191	3·7	25¾	11·0	109	20·7	230
16	23·6	2659	196	3·9	25	12·6	122	23·6	256
18	24·7	2922	201	4·1	24½	14·1	135	26·5	283
20	25·75	3179	206	4·3	24	15·7	149	29·5	312
22	26·75	3431	211	4·5	23½	17·3	163	32·5	341
24	27·7	3678	213	4·6	23¼	18·8	176	35·5	370
26	28·6	3922	216	4·75	22¾	20·4	189	38·4	395
28	29·45	4161	220	4 9	22½	22·	203	41·3	425
30	30·27	4397	222	5·04	22¼	23·5	216	44·2	451
32	31·1	4630	226	5·2	21¾	25·1	230	47·3	480
34	31·82	4860	229	5·3	21½	26·7	243	50·	510
36	32·56	5088	232	5·43	21¼	28·3	256	53·	535
38	33·3	5313	234	5·55	21	29·7	269	56·	561
40	34·	5535	237	5·67	21	31·4	283	59·	596
42	34·63	5756	239	5·77	20¾	33·0	297	62·	624
44	35·13	5919	241	5·85	20½	34·5	311	65·	652
46	35·9	6190	244	6·0	20½	36·2	324	67·5	680
48	36·5	6404	246	6·1	20¼	37·7	338	70·5	709
50	37·13	6617	248	6·2	20	39·3	353	73·5	739
52	37·7	6828	250	6.3	20	40·7	367	76·4	768
54	38·3	7036	252	6·4	19¾	42·4	381	79·3	798
56	38·85	7245	254	6·49	19½	44·0	396	82·2	827

Table III. continued.

Steam acting expansively.								Steam acting at full pressure throughout the stroke in the same engine.	
Number of horses' power.	Diameter of the steam piston in inches.	Mean pressure on the piston in pounds, at 4·8 lbs. per circular in.	Velocity of the steam piston in feet per minute.	Length of the stroke in feet.	Number of strokes per minute.	Water required per hour to supply the boiler.	Coals consumed per hour in lbs.	Number of horses' power.	Coals consumed per hour in pounds.
	inches	lbs.	feet.	feet.		cubic feet.	lbs.		lbs.
58	39·4	7453	255	6·57	19½	45·4	409	85·1	850
60	39·9	7656	257	6·65	19½	47·0	423	88·1	887
62	40·5	7860	259	6·75	19¼	48·6	437	91·0	916
64	41·0	8062	260	6·83	19	50·2	452	93·9	946
66	41·5	8263	261	6·9	19	51·8	466	96·8	975
68	42·0	8462	263	7·0	18¾	53·4	481	99·7	1005
70	42·5	8662	265	7·1	18¾	55·0	495	102·7	1035
72	43·0	8858	266	7·17	18½	56·6	509	105·6	1064
74	43·4	9043	268	7·23	18½	58·1	514	108·5	1094
76	43·9	9250	269	7·3	18½	59·8	538	111·4	1123
78	44·4	9444	270	7·4	18¼	61·5	554	114·3	1153
80	44·8	9637	272	7·47	18¼	62·5	563	117·3	1182
85	45·9	10120	275	7·65	18	66 5	599	124·6	1256
90	46·97	10590	279	7·83	17¾	70·5	635	131·9	1330
95	48 0	11060	282	8·0	17¾	74·4	670	139·2	1404
100	49·	11520	284	8·16	17½	78·2	704	146·	1478
105	49·95	11980	287	8·32	17¼	82·1	739	153·3	1552
110	50·9	12430	290	8·5	17	86·0	774	161·6	1626
115	51·6	12760	292	8·6	17	89·9	809	167·9	1700
120	52·7	13330	294	8 8	16¾	93·8	844	175·2	1774
125	53·6	13760	297	8·9	16¾	97·7	879	182·5	1848
130	54·4	14210	299	9·0	16½	101·7	915	189·8	1921
135	55·3	14740	300	9·2	16¼	105·6	950	197·1	1995
140	56·1	15080	302	9·35	16¼	109·5	986	204·4	2069
145	56·84	15510	306	9·47	16¼	113·4	1021	211·7	2143
150	57·6	15930	308	9·6	16	117·3	1055	219 0	2217
155	58·4	16360	310	9·7	16	121·2	1091	226·3	2291
160	59·1	16780	312	9·83	15¾	125·2	1127	233·6	2364
175	61·3	18030	318	10·2	15½	129·1	1162	240·9	2438
180	62 0	18440	320	10·3	15½	133·0	1197	248·4	2512
200	67·7	22000	334	11·3	14¾	156·4	1408	292·	2956

EXPLANATION

OF THE

PLATES;

WITH

REFERENCES TO THE ARTICLES WHERE THE CONSTRUCTION IS
INVESTIGATED.

PLATE I.

Fig. 1. is an isometrical projection of a rectangular steam boiler, (art. 225, 226,) with part of the top plates of the boiler, and part of the brickwork removed to shew the internal parts. A is the boiler; the half of the doorway to the fire is at B, and the fuel rests on the fire bars G, and against the back F: the flame passes over F, and along under the bottom; it rises at H, and returns in a flue by the nearest side, passes round the end by the flue I, and along a flue by the further side to the chimney at L; a horizontal damper K regulates the aperture to the chimney, (see art. 257.) The door to the ash pit C should shut perfectly close, and the supply of air for the fire should enter by a passage E, the aperture of which is regulated by the force of the steam acting by the chain n, n, (art. 258.) In the figure the air is supposed to enter by the grating at D. The supply of water enters by the pipe M N, the end N being turned along the bottom of the boiler that the water may acquire heat before it mixes with the rest; the admission is regulated by the stone float c, (art. 251—253.) The steam passes to the engine by the pipe S, and when it is not required, it goes off by the safety valve V, (art. 260,) and through the pipes T W. The internal valve is on the man-hole plate at a b, (art. 259.) The steam gauge is at h, (art. 558,) the gauge cocks at k, i, and a cock to clear the boiler of water is at R. Opposite each flue, as at Q, there must be an aperture to clear it out at.

Fig. 2. shews a method of admitting water to a boiler where high pressure steam is used. The pump forces the water by the pipe D, through the valve aperture A, into the boiler, but when the quantity is in excess, the copper float F closes the valve, and opens the valve B to the waste pipe C, by which the surplus passes off. The parts must be balanced on the axis by the weight G. See (art. 253.)

Fig. 3. 4. 5. and 6. are tops for engine chimneys. Fig. 3 is a plain obelisk; the proportions of an Egyptian obelisk are very well adapted for a chimney, and if the faces were stuccoed and covered with sunk figures, it would render them novel and ornamental. Fig. 4 is an octagon top for a square shaft; Fig. 5, an octagon chimney, and Fig. 6, a chimney to represent a column. (See art. 274—278.)

PLATE I.

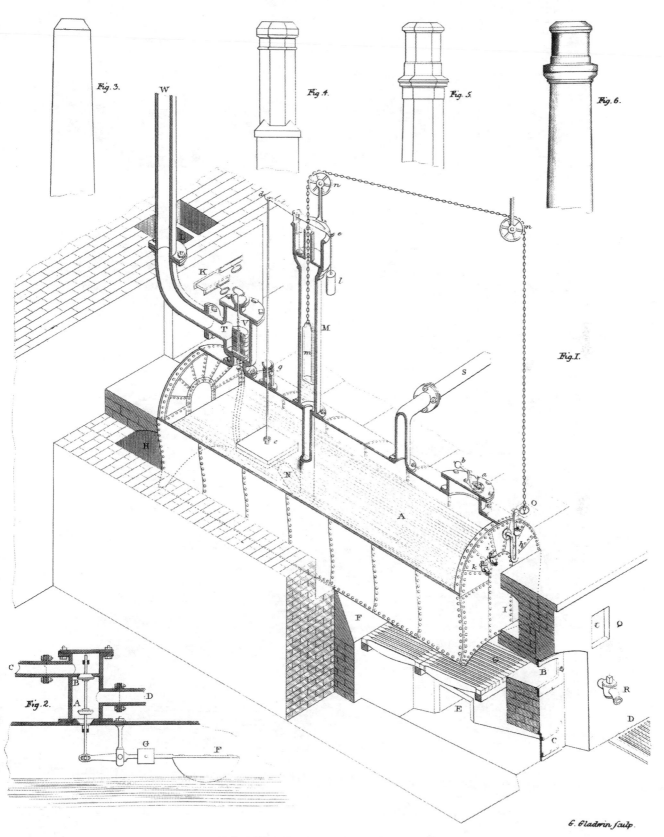

Fig. 3.

Fig. 4.

Fig. 5.

Fig. 6.

W

Fig. I.

Fig. 2.

G. Gladwin sculp.

London, Published 1827, by I. Taylor, Architectural Library, High Holborn.

PLATE II.

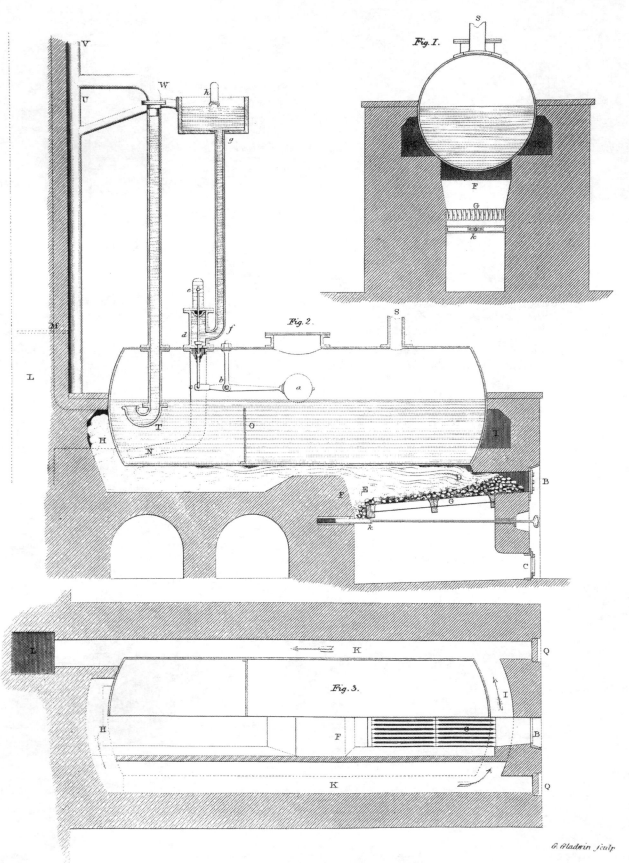

Fig. I.

Fig. 2.

Fig. 3.

G. Gladwin Sculp.

London. Published 1827. by J. Taylor, Architectural Library. High Holborn.

PLATE II.

This plate represents a plan and two sections of a cylindrical steam boiler, (art. 227—230.) Fig. 1 is a cross section, Fig. 2, a longitudinal section, and Fig. 3, the plan; in the latter, half of it is the plan above the level of the boiler bottom, and the other half below it. The fuel is put in at the fire door B, it inflames at D, and the smoke in passing over the hot current of air rising through the red fuel at E is consumed. The ash pit door is supposed to be provided with a register to regulate the admission of air, but it would be better to make it regulate itself as in the preceding plate. By pushing back the plate *k* by the handle *i* the clinkers are let out behind; (art. 248.) The supply of water is regulated by a hollow copper ball float, and the supply is continuous except, when by the water rising, the valve is closed, (art. 254.) To prevent sediment depositing over the fire, I would recommend a division to be placed across the boiler as at O; the safety pipe T W is recommended instead of the valve, on account of the certainty of its action. The first effect of strong steam is to displace the water down to the level of the mouth of the pipe at T; this sets the feed pipe into action, and steam and water rise by the pipe T W till the boiler be cooled to its proper temperature, (art. 264;) no internal valve is required: S is the steam pipe leading to the engine. The same letters refer to the same parts in all the figures. For the area of the grate, and ash pit, see (art. 197—199;) size of the boiler, (art. 229;) the area of the chimney, (art. 274—278,) and the strength of the boiler, (art. 525.)

Note. In Section III. I neglected to remark that the boilers formed of small pipes cannot possibly produce more effect than others, and that every boiler must contain a certain quantity of water and steam, otherwise the slightest neglect of the fire would cause the engine to stop. It has been pretty clearly ascertained that not one of the combinations hitherto proposed, has equalled the kind of boiler above described.

PLATE III.

This plate represents Brunton's apparatus for feeding furnace fires by means of machinery. The general principles of the method and its advantages have been stated, (art. 250,) and it only remains to describe the parts of the figure. The apparatus was added to two boilers of Boulton and Watt's construction. To the original boilers A, A, two additional boilers B, B, are attached, which are prepared for the purpose of being over the revolving fire grate; the smoke from which passes over and under the bridges *d, d,* and round the boilers A, A, by the flue C, and D, D are the flue doors. The coals broken to a proper size, are put into the hoppers E, E, and fall through the openings F, F, and through the top of the boiler to the grate. The door H to examine or repair the fire place is attached to the boiler by a cement joint. The additional boilers are connected to the main boilers by the steam pipes G, G.

To clear the dust away that falls over the edge of the revolving grate, there are doors at I, I; they also admit a small quantity of air to the burning fuel. The axis of the grate K is turned by the pinion and wheel at L, turned by the upright shaft N, which receives its motion by the shaft R from the engine.

The pivot of the shaft N, and of the spindle K, are in the foundation plate M. The grate bars are surrounded by fire bricks *h,* and a thin hoop projects below the frame, and moves in sand in a trough *f,* and prevents air entering by any other passage than through between the bars; a scraper attached to the grate, and consequently moving with it, keeps the channel *i* clear of dust.

To regulate the fire the chains S S are connected to the damper chains, and raise or depress the wedge U by the lever T, and thus increase or diminish the supply of coals according to the force of the steam. (See art. 257.)

The feed pipe O, with its stone float *c* and balance *l,* are as in other boilers, (art. 251.) The gauge cocks are at Z, the man-hole at *a,* with an internal valve at *b;* the safety valve is at V with a pipe Q to convey away the steam; P is the pipe for conveying steam to the engine, with a self-acting stop valve W, to prevent the steam passing from one boiler to the other, when both are in action; and X Y is a lever handle for closing the aperture when only a small supply of steam is required.

The construction will admit of considerable variation; and its advantages in saving fuel, in regularity of action, and in consuming the sooty matter of smoke, render it a desirable addition to a large boiler.

Plate III.

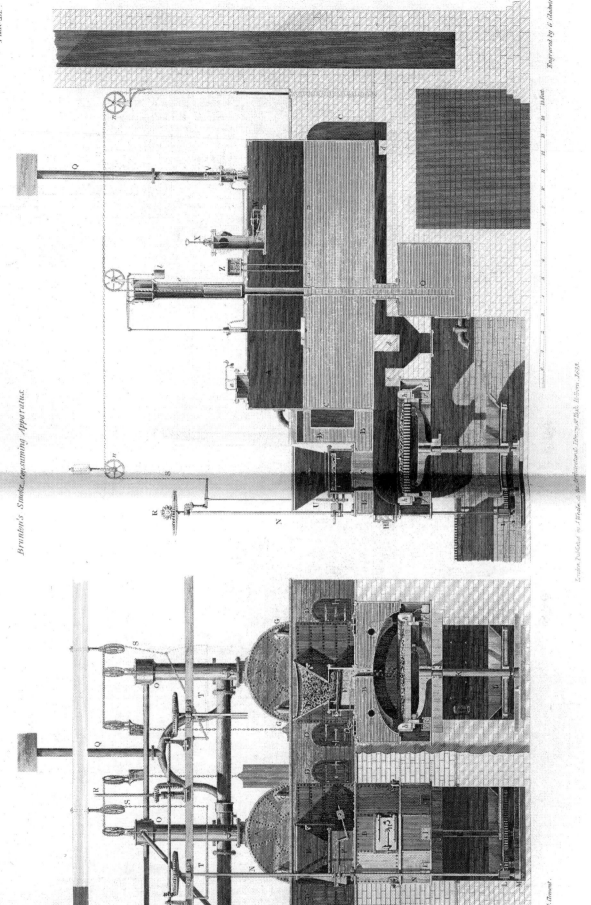

Brunton's Smoke consuming Apparatus.

Engraved by G. Gladwin.

Drawn by J. Brunton.

The material originally positioned here is too large for reproduction in this reissue. A PDF can be downloaded from the web address given on page iv of this book, by clicking on 'Resources Available'.

PLATE IV

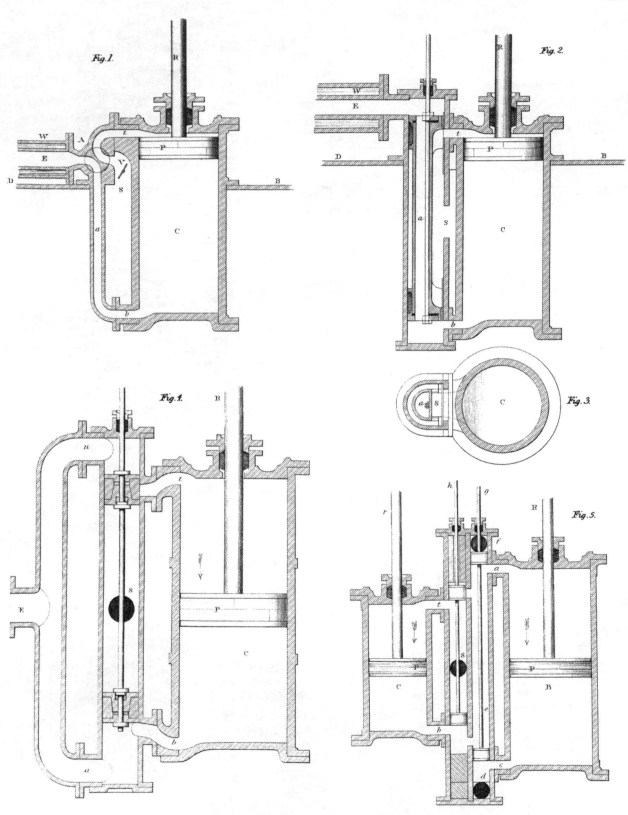

Fig.1.

Fig.2.

Fig.3.

Fig.4.

Fig.5.

Tredgold. del.

London, Published 1827, by I.Taylor, Architectural Library, High Holborn.

Gladwin. sc.

PLATE IV.

Fig. 1. is a section of the parts of a high pressure engine with a four passaged cock. The engine is supposed to be partly within the boiler, of which D B is the top plate. P is the steam piston, and R the piston rod, A is the four passaged cock; the steam enters from the boiler at S, and passes through *t* to the top of the piston, and the steam below escapes through the passage *b*, and pipe *a* and E, to the atmosphere; the pipe E is surrounded by water which the escaping steam warms ready for the boiler. By turning the cock the motions are reversed, but it is obvious we cannot in this engine employ the expanding force of the steam. The motion is regulated by a throttle valve V. (See art. 356—361.)

Fig. 2. and 3. shew a section and plan of a similar engine, with a dee-slide instead of a cock; the steam enters from the boiler at S, and by the passages being shut and opened close to the extremities of the cylinder, there is no loss by the communicating pipes being filled with strong steam. (See art. 364.) This engine will not work expansively unless the construction of the slide be altered. (See art. 371.) Contrary to the usual practice, the packing of the slide is on the sliding part; the advantage of this plan is obvious, but the practical difficulty of boring a semi-cylinder is incurred.

Fig. 4. is a simple arrangement of the high pressure engine by which the expanding power of the steam may be used; it is the invention of Messrs. Taylor and Martineau. The passages are opened and closed by pistons sliding in a pipe: the steam enters this pipe at S, and the steam is supposed to be just shut off by the upper piston, so that by the expansion of that in the cylinder the rest of the stroke is completed, the passage E to the atmosphere being still open, (see art. 371—380.) The slide would be improved by making it of the form of a dee-slide.

The construction of the pistons of the slide is a suggestion which may perhaps answer better than the common ones, (art. 450 and note.)

Fig. 5. is an arrangement to illustrate the action of a high pressure engine to work expansively by means of combined cylinders. (See art. 381—383.)

PLATE V.

Fig. 1. is a section of a double acting condensing engine, with a slide adapted for working by the expand ng force of steam; the slide being, in Fig. 1, in the position for letting the steam on at the top. Fig. 2 shews the' steam shut off and the passage to the condenser still open, and Fig. 3, the position when the steam is let on at the bottom. (See art. 448.) The steam enters at S, and a pipe of communication between the steam pipe and the condenser is necessary, to allow steam to enter the condenser when the engine is about to be set to work, (art. 414.)

Fig. 4. is a section of a single acting condensing engine, with valves to the passages, (see art. 406;) and Fig. 5, a different arrangement of the valves for a single engine.

In all these figures the same letters indicate the same parts. C is the steam cylinder, P the steam piston, R the piston rod, B the condenser, with a jet of water playing into it from I the injection cock; A is the air pump, and p its piston; G is the foot valve between the condenser and air pump; M the air pump, and Q the discharge valve of the air pump, through which the air and hot water are forced into the hot well K, from whence a part is raised by a small force pump to the boiler feed head, and the rest runs off by a waste pipe. H is the blow valve to the condenser, (art. 566.) The condenser and air pump are placed in a cistern which is constantly supplied with cold water by a pipe N.

The jet should be made through a rose on the end of the pipe; for to produce speedy and perfect condensation, the cold fluid should present the greatest possible surface to the steam it is to condense, (see art. 280;) and it should be impelled into the condenser with greater force than the ordinary head in the cistern admits of.

In large engines the connecting eduction pipe E, Fig. 1, may be on the outside of the steam pipe S, and the parts of the slide connected only by a rod, as mentioned in (art. 447.)

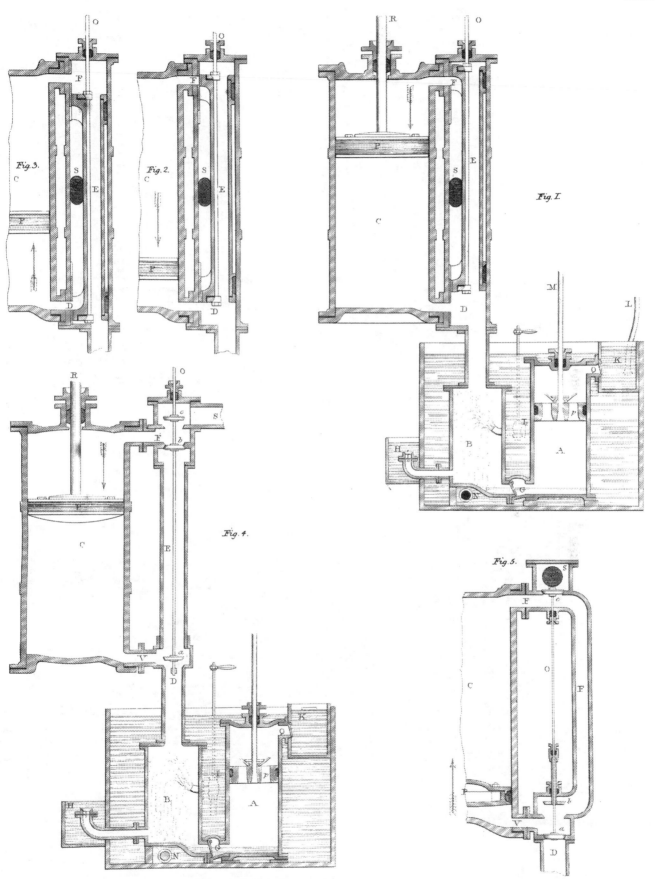

PLATE V.

Fig. 3.

Fig. 2.

Fig. I.

Fig. 4.

Fig. 5.

London, Published 1827, by I.Taylor, Architectural Library, High Holborn.

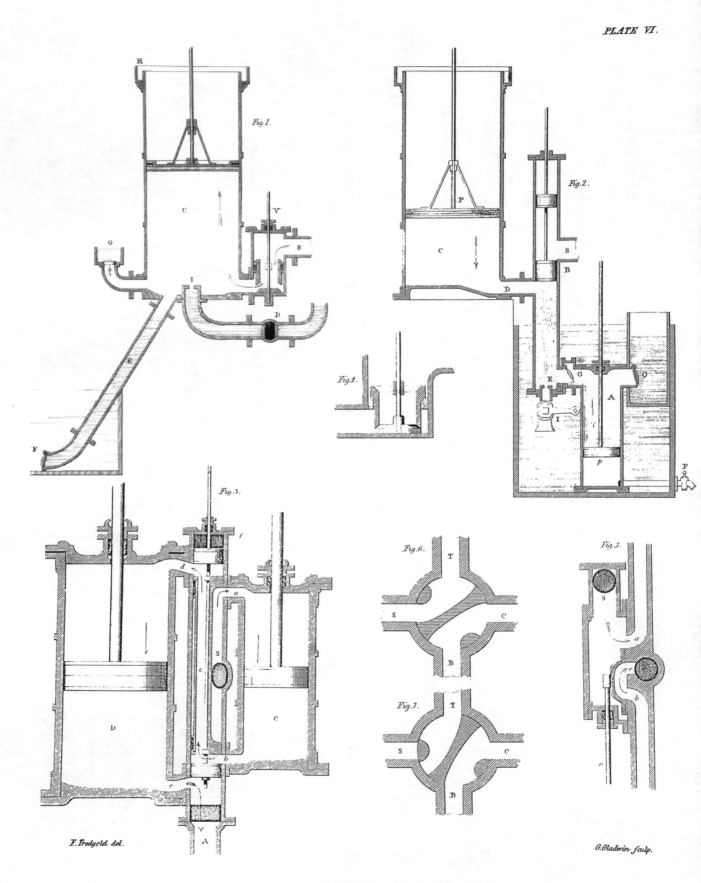

PLATE VI.

Fig. 1.

Fig. 2.

Fig. 4.

Fig. 3.

Fig. 6.

Fig. 7.

Fig. 5.

T. Tredgold del.

G. Gladwin sculp.

London, Published 1827. by J. Taylor, Architectural Library. High Holborn.

PLATE VI.

Fig. 1. is a section of a common atmospheric steam engine; C is the cylinder, and the piston is supposed to have a wooden bottom, according to the practice of Smeaton, (art. 466;) the steam is let on by a modification of Hornblower's valve, (art. 442,) instead of the common regulator, (art. 461.) For the proportions of the engine, see (art. 393—399.)

Fig. 2. is a section of an atmospheric engine with a separate condenser and air pump; see (art. 400—405.) An elevation of this engine is shewn in Plate XI.

Fig. 3. is a section of a combined cylinder engine, on Hornblower's principle, (art. 32,) where the steam passages are opened and closed by a combination of slides in one pipe. See (art. 425—429. Woolf's engines have two cylinders, but the passages are opened by valves.

Fig. 4. is a section of Hornblower's double seated valve. See (art. 441.)

Fig. 5. represents a section of Murray's slide; it is a sliding cover which alternately covers the passages $a\,c$, and $c\,b$: the disadvantage of this construction is that the pressure of the steam is nearly three times as great on the moving surfaces, as it is in Murdoch's arrangement shewn in the last plate. (See art. 446.)

Fig. 6. and 7. shew the mode of forming the apertures of a four passaged cock, so that the steam may be shut off at any period of the stroke without closing the passages to the condenser. See (art. 456.) T is the passage to the top, and B that to the bottom of the cylinder; the steam enters at S, and C is the passage to the condenser. Fig. 6 shews the position of the cock when the steam is entering, and Fig. 7 when it is shut off.

Y Y

PLATE VII.

This plate is to represent the construction of pistons.

Fig. 1. represents a section, and part of the plan, of the *common packed piston*, which is tightened by the screws S when it wears. See (art. 467.)

Fig. 2. shews one of *Woolf's* methods of tightening the whole of the screws at once, and consequently in the most regular manner, without having to take off the cylinder lid. See (art. 468.)

Fig. 3. is the plan and section of *Cartwright's* metallic piston, (art. 43,) as at first executed; it was afterwards made with spiral and other springs, acting only on the interior segments of rings *b b*, but it required that the outer ones should be moved, and this has been done by making the interior segments in short pieces to cover the joints of the outer ones, so as to leave space to insert springs between them, to act directly on the outer segments, by Mr. Lloyd: it has also been done by inserting a short cylindrical piece against each joint of the outer ring, both the cylindrical pieces and the outer segments being pressed outwards by spiral springs; this method is employed by Messrs. Hall. See (art. 469.)

Fig. 4. represents *Barton's* construction of metallic pistons in plan and section. The points of the wedges G expand faster than the segments E in the ratio of *m n* to *o n*, and hence, wear the cylinder unequally. To prevent this the points are shortened, and two elastic hoops *b, b*, Fig. 5, are put round, neatly fitted with a loose tongue joint, as shewn at Fig. 6. See (art. 470.)

Fig. 7. shews a method of avoiding the defect of Barton's piston, by keeping the points of the wedges as much within the segments of the piston as may be necessary for wear, before a change of parts becomes necessary, and making two series to break joints. See (art. 471, 472.)

Fig. 8. represents a section of *Jessop's* piston, and Fig. 9 the expanding coil of metal which rubs against the cylinder. See (art. 473.) I am of opinion that a more perfect method of pressing the packing against the metal coil, is the only thing wanted to render this decidedly the best kind of piston.

The friction of pistons is investigated in (art. 474.)

PLATE VII.

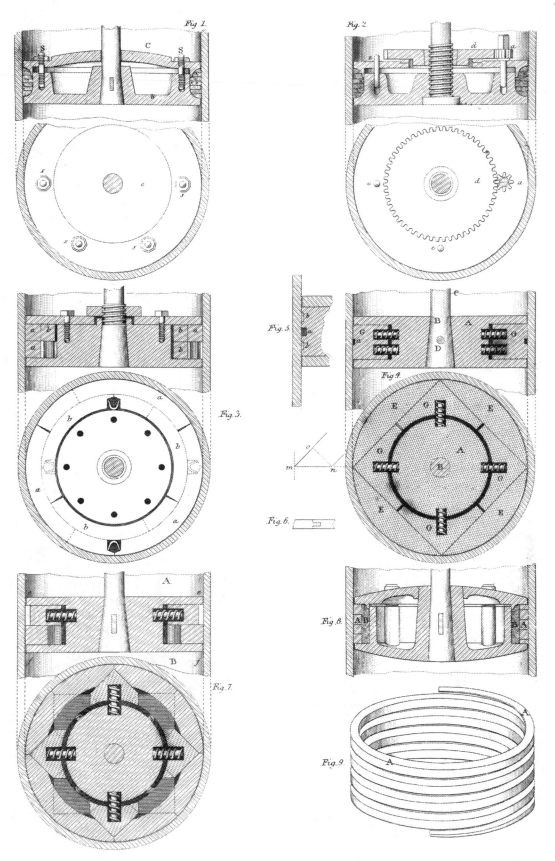

Fig. 1.

Fig. 2.

Fig. 3.

Fig. 4.

Fig. 5.

Fig. 6.

Fig. 7.

Fig. 8.

Fig. 9.

London, Published 1827, by J.Taylor, Architectural Library, High Holborn.

Plate VIII

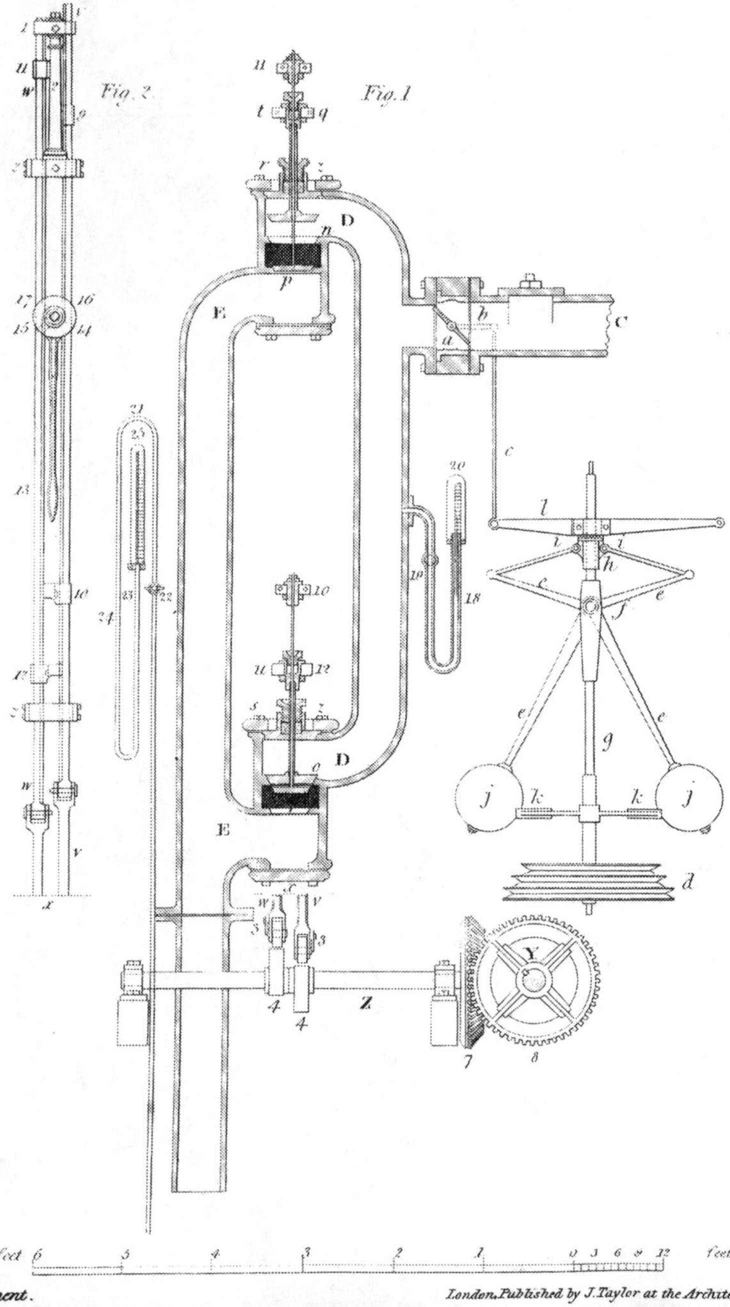

Section of the Cylinder and working Valves to Maudslays Engine.

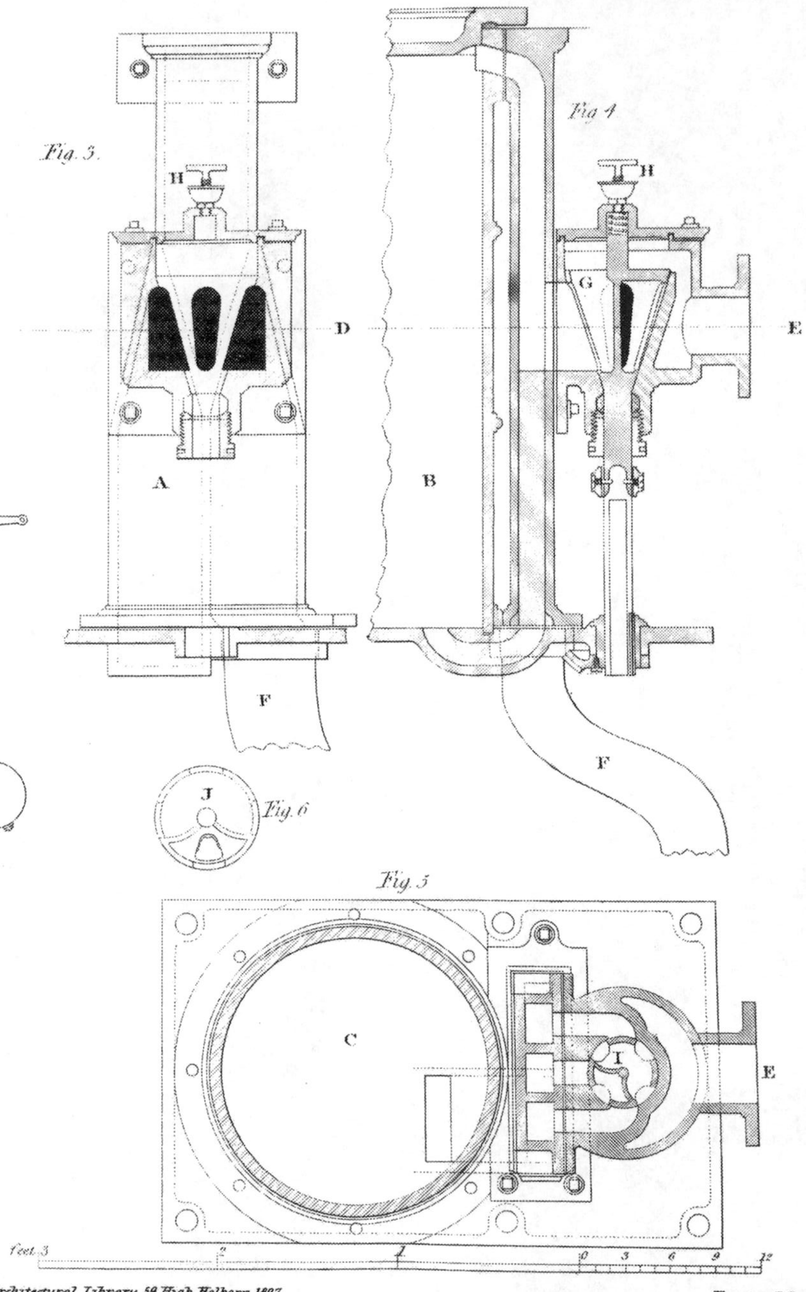

Section of Mr G.R. Fenton & Compy Steam pipes and Valves.

Fig. 1.

Fig. 2.

Fig. 3.

Fig. 4.

Fig. 5.

Fig. 6.

Drawn by J. Clement.

London, Published by J. Taylor at the Architectural Library, 59 High Holborn, 1827.

Engraved by G. Gladwin.

PLATE VIII.

Fig. 1. represents a section of the steam pipes, and valves, of Messrs. Fenton and Murray's double engine, Plate XIV.; and Fig. 2, the communicating rods. The steam enters by the pipe C which has a throttle valve at *a*, to regulate the supply of steam to the engine, (art. 544.) It is regulated by the action of the governor balls on the lever *b*, by the connecting rod *c*; the rotary motion is communicated to the axis of the governor, by means of a band passing from a pulley on the crank shaft to a similar pulley *d* on the axis of the governor. The governor consists of two bent levers *e e*, passing through a slit in the middle of the spindle, and turning upon an axis at *f*. The upper part of the spindle has a slide *h*, which is connected to the levers by the rods *i i*, and ascends when the centrifugal force of the governor increases, so as to cause the balls to rise, and descends when it decreases; and the lever *l* moves with it, and consequently, the valve. When the engine is at rest the balls *j j* rest against the arms *k k;* the upper end of the levers *e e*, are nearer to each other; and the rod *c* is raised so that the throttle valve may be quite in a horizontal direction, and the pipe completely open for the passages of the steam. See (art. 550.)

By the pipe D D, the steam passes either to the top or bottom of the cylinder from the throttle valve *a*, and by the eduction pipe E E the steam from either passes down to the condenser. The valves *n, o*, have each a cylindrical tube or spindle passing through the stuffing boxes *r, s;* the upper end of each of these has a stuffing box, the upper one at *t* the other at *u*, for the rods of the valves *p, q*, which open to the eduction pipe E, so that either the steam or eduction valves may be opened without allowing steam to escape.

Fig. 2. is a front view of the two sliding rods which give motion to the valves *n, o, p*, and *q*. These rods are kept in a perpendicular direction by the pieces *z, z*, and the guide 1; the lower end of the rods have friction rollers 3, 3, which are acted upon by the two eccentric pieces 4, 4, on the horizontal shaft Z, which derives its motion from a shaft Y, placed at right angles to, and communicating by means of beveled wheels with the crank shaft. Four arms, 9, 10, 11, and 12, are fixed to the rods *v, v*, and *w, w*, for the purpose of moving the valves, and a lever or handle (13) turning on a pin screwed on the pipe E, is used to open and shut the steam valves when the engine is first set to work, see (art. 566;) the steam gauge (21) is for measuring the pressure of the steam above that of the atmosphere, (art. 558,) and the condenser gauge (24) is for measuring the force of the vapour in the condenser, (art. 559.)

Fig. 3, 4, 5, and 6, represent the four way cock as executed by Maudslay. Fig. 5 is the plan taken at the horizontal line D E, in Fig. 3 and 4, which are sections through the steam cock. In these figures E is the steam pipe, and F the pipe leading to the condenser G. Fig. 4 is the steam cock or cone ground into its seat, and provided with a grease cup H to afford a regular supply.

Fig. 6. is a plan of the upper side of the steam cone. See (art. 457) and Plate XV.

PLATE IX.

The apparatus for opening and closing the steam passages is of more importance to the perfection of the steam engine, than any other part of its mechanism. In the present state of the engine the action is either very complicated or imperfect; my object in this plate is to shew how the imperfection of the most simple method may be avoided, and also how it may be applied to reciprocating movements.

Fig. 1. shews a section, and Fig. 2. a plan, of an apparatus for opening and closing the steam passages by means of the rotary motion of the crank shaft D; the object of the method is to give such a form to a wheel on the shaft D as will move the pin e twice during the stroke, and in the easiest and quickest manner. For this purpose the shaft passes through a rectangular frame which rests in grooves on the shaft to keep it in its place, with liberty to slide backward and forward; and it is provided with two rollers to be acted on by the curved surfaces of a wheel fixed on the shaft. The curve H G moves the valves at the termination of the stroke, and I K to shut off the steam, that the engine may work expansively the rest of the stroke. In order that there may be the power of varying the time of cutting off the stroke, the curve I K may be on a separate piece M, Fig. 2, capable of being moved from N to O. The apparatus is supposed to move a slide of the kind represented in Plate V. Fig. 1; but is equally applicable to move the four passaged cock, (art. 456 and 458) or valves. For the nature of the curves, (see art. 481.)

Fig. 4. and 5. represent the same principle applied to a reciprocating motion. The plug tree A B is kept in its place by guides on the brackets; these guides slide in the dark grooves. The curved parts H I, C D, K L, &c. successively move, horizontally, the frame C on four rollers supported by the brackets attached to the engine; and by the backward and forward motion turns the axis E, and raises or depresses the lever F, which acts on the rod of the slide. The same movement would obviously open valves or turn a cock. By the lever on the right hand side, the roller frame may be moved by hand; but by not reversing Fig. 5, it has been shewn there on the left. (See art. 483.)

Fig. 3. represents the method of opening valves by weights, the plug tree A B being made a means of raising the weights, and of disengaging them by the tappets f, d. (See art. 478.)

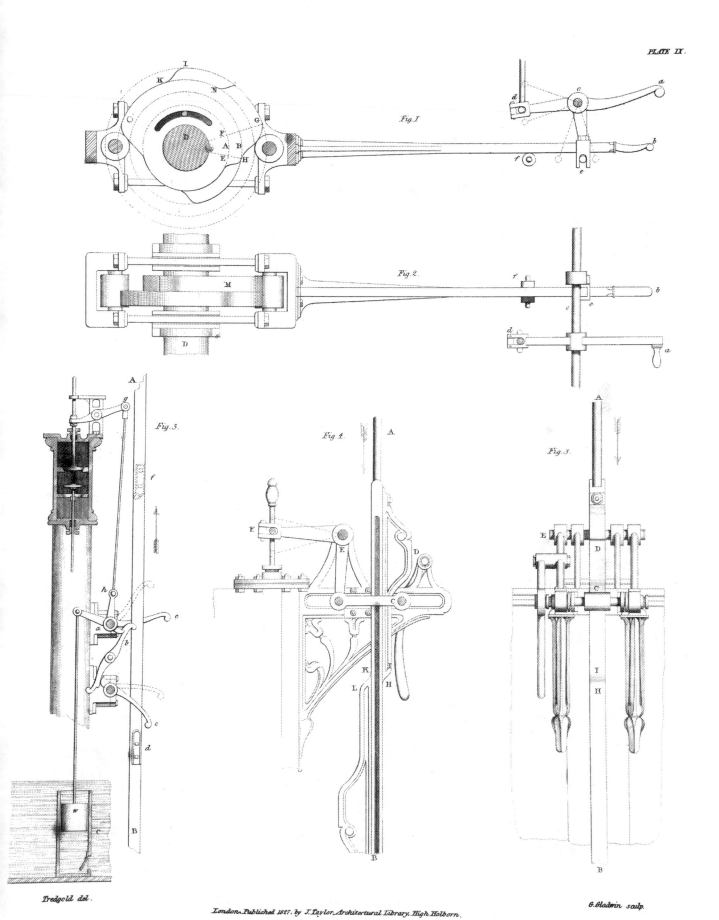

PLATE IX.

Fig. 1.

Fig. 2.

Fig. 3.

Fig. 4.

Fig. 5.

Tredgold del.

London, Published 1827, by J. Taylor, Architectural Library, High Holborn.

G. Gladwin sculp.

PLATE X.

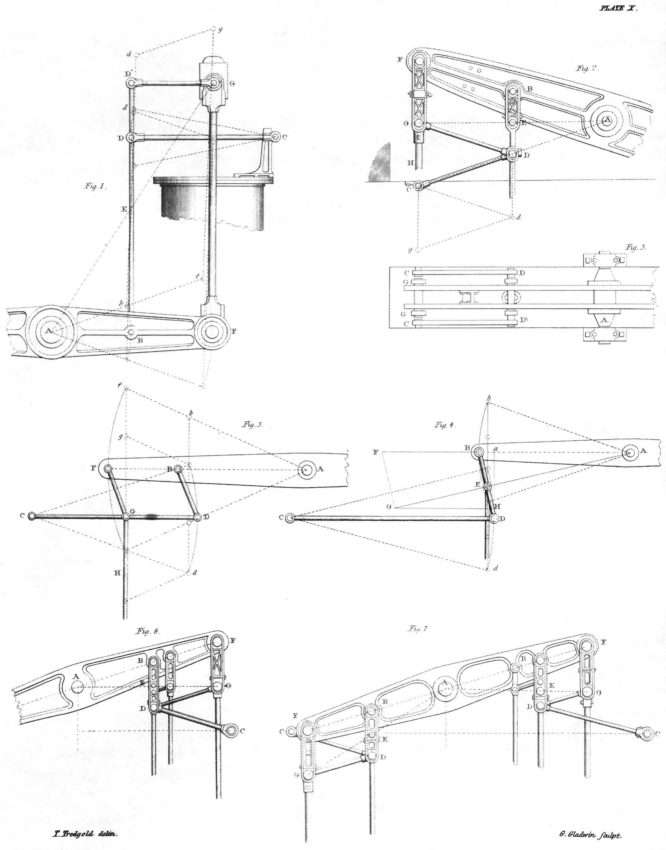

Fig. 1.

Fig. 2.

Fig. 3.

Fig. 5.

Fig. 4.

Fig. 6.

Fig. 7.

T. Tredgold delin.

G. Gladwin sculpt.

London, Published by J. Taylor, Architectural Library, High Holborn, 1827.

PLATE X.

The figures of this plate are to illustrate the combinations used to produce rectilinear motion, from motion in a circular arc.

Fig. 1. is the parallel motion used for steam boat engines. The beam A F is below the cylinder; from G, the end of the cross head, draw a line from G to A the centre of the axis of the beam, and it will cut the rod D B in E, and the length of the radius bar D C may be found by the rule, (art. 491;) when E B is equal to E D the length of the bar C D is equal to A B, and this is the best though not always the most convenient form. The rod D´ G may be at any height, provided it be parallel to A F, and B may be at any point in A F.

Fig. 2. shews the most common construction for engines with the beam above the cylinder. H is the piston rod connected at G, and C D is the radius bar. The line G A cuts the link B D in E, the proper point for the air pump rod. Fig. 3. shews a plan of the upper side of the beam, where C D, C D are the radius bars, and the beam is in two parts, as is usual in large engines. (See art. 492.)

Fig. 4. is a diagram to illustrate the investigation of the properties of the combination in its most simple form. (See art. 489.)

Fig. 5. is a diagram for the apparently more complicated case, when the rod is fixed to one angle of a parallelogram. (See art. 492.)

Fig. 6. shews how to arrange for three piston rods to move parallel, as for Woolf's engine; the points of suspension must be all in the line A G.

Fig. 7. shews another arrangement for three rods at one end, and two at the other end of the beam. (See art. 495.)

In all the cases the corresponding points are marked by the same letters. and therefore by referring to the investigation of Fig. 4, the relations may be traced; the particular forms of different engine makers, will be found by turning the part upside down, altering the place of the parallel bar G D, or altering the proportions of the parts. In every combination where the bar C D is not equal to A B, the variation from rectilinear motion increases with the extent of the angle described.

Other variations of the parallel motion are exhibited in Plate XI. and Plate XIX.

PLATE XI.

This plate represents a plan and elevation of an atmospheric engine, for raising water from a mine. The beam is supported by a frame of cast iron, designed so that it may be taken apart when it is necessary to move the engine to another mine, (art. 578.) The steam comes from the boiler by the steam pipe S, and is admitted to the cylinder C by a sliding piston in B, (see Fig. 2, Plate VI.) and then the piston in the cylinder C rises to the top of the stroke, and the piston rod forces the end f of the beam up with it; as the beam rises it draws the rod F G with it, which near the end of the stroke moves the sliding frame H, (art. 483,) and by the rod O the slide in B so as to shut the steam off, and open the passage to the air pump A; into the passage a jet of water plays at I. The piston then begins to descend by the pressure of the atmosphere and raises the pump rods, and at the end of the stroke the part F of the rod F G, moves the slide and shuts the passage to the air pump, and opens that for the steam. (See Plate VI. Fig. 1.)

The parallel motion is guided by the radius rod $c\,d$, attached to the frame, (art. 491 and 495,) and by connecting the rods h, i; the same rod does for both ends of the beam. The sliding frame H is supported by a cross bar beneath H, and another at K, and the slide may be moved by hand by the lever M. Cold water for injection is supplied by the pump E, and water is raised from the hot well by the pump D, and passes to the boiler by the pipe Q, with a small branch pipe at P, to give water to the top of the piston. (See art. 400—405.)

PLATE. XI.

Fig. 1.

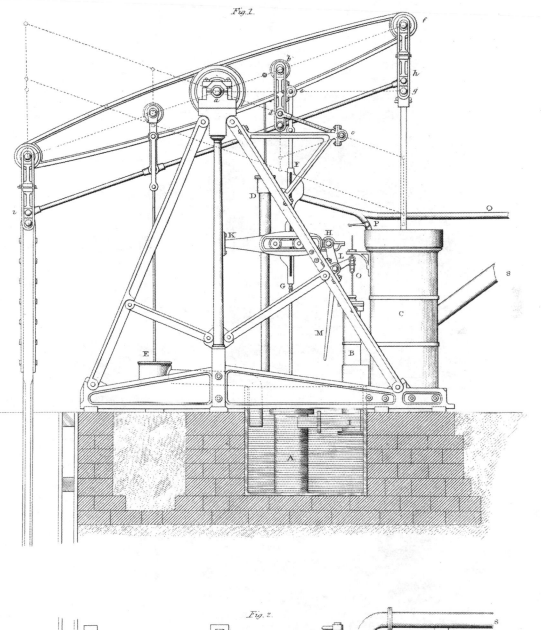

Fig. 2.

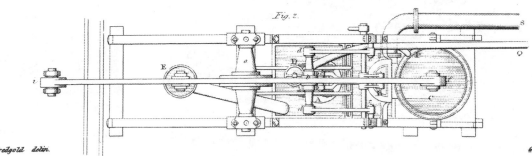

T. Tredgold delin.

G. Gladwin sculpt.

London, Published by J. Taylor, Architectural Library, High Holborn. 1827.

PLATE. III.

Single Acting Engine by Boulton and Watt

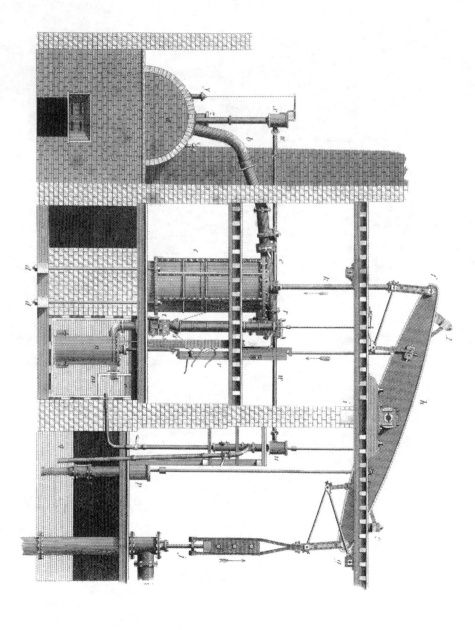

Drawn by J. Clement.

London, Published by J. Taylor, at the Architectural Library, 59 High Holborn, 1827.

Engraved by G. Gladwin.

PLATE XII.

This plate shews an elevation of a single acting engine, as executed by Messrs. Boulton and Watt. The boiler *a* is inclosed in a case of brickwork, and the steam passes by the steam pipe *b*, to the cylinder *c*, which is firmly attached to the floor of the engine room, by the bolts *d*, *d*; its upper end is covered by the lid *e*, through which the piston rod *k* slides in an air-tight box called a stuffing box. The beam *f g* moves on its axis or gudgeon at *h*; and the bearings in which the gudgeons work are sustained by the floor and wall *i*.

The pump rod *j*, carrying a counter weight, is suspended at the end *g* of the beam; and both it and the piston rod *k*, are connected by a parallel motion apparatus to the working beam *f g*. (See art. 492.)

The condensing cistern is at *m*, and contains the air pump *n*, the condenser, and hot well *o*; a continued supply of cold water is procured by the action of the cold water pump *p*, and the excess is carried off by a waste pipe to the well *q*; the whole of the external part of the apparatus being kept by that means at the lowest temperature possible.

The upper steam valve is at *r*, and the lower at *s*, and the exhaustion valve at *t*. (See Fig. 5, Plate V.) These valves are moved by the plug tree *v*, which is furnished with tappets to give motion to the levers acting on the valves *r*, *s*, *t*. (See art. 478.)

The pump to raise water from the hot well *o* to supply the boiler is at *u*, and the water is conveyed by the pipe *w w* to the small cistern *x* on the top of the feed pipe, which is provided with a valve, and acted on by a lever connected by a wire passing down through a stuffing box *y*, to a stone float in the boiler, which by its descent opens the valve, and allows the admission of an additional supply of water when it is required, (art. 251.)

In order to prevent concussion two blocks 1, 2, are fixed across the upper side of the beam, and extend on each side so as to strike on four wooden springs, on the floor which supports the beam. (See art. 549.)

For large engines the beam is in two parts, with a space between them, as Fig. 3, Plate X.

The proportions of the single engines are given in Sect. VI. (art. 406—413;) the application in art. 572, 573, 582, and 587; their effects, (art. 576;) and their power, and consumption of fuel, Table II. (art. 663.)

PLATE XIII.

This plate represents a double acting steam engine for raising water, (art. 570, 582.)

The steam from the boiler enters by the steam pipe S, passes through the top valve *a*, to act on the piston *p*, and forces it down; and just before it arrives at the bottom of the cylinder, the plug on the rod R will come in contact with a lever and shut the valves *a*, *b*, and open *c*, *d*, which were shut; the steam will now continue its course down the pipe S through the valve *d*, act on the bottom of the piston *p*, and again force it up to the top of the cylinder, while the steam which forced it down will make its escape to the condenser B, through the valve *b*, by a pipe which conveys the steam from the valves to the condenser.

Having described one double stroke of the engine, it is only necessary to remark that its continued motion is made of a repetition of the same thing over again. The steam which passes to the condenser B, meets with a jet of cold water from the injection cock I, and the greater part of it is reduced to the state of water, and it is the office of the air pump A to clear it away. The rod R works the air pump which draws out all the injected water, air, and condensed steam, and discharges it through the valve at the top of the air pump into the hot well K, where a certain portion of the water is again forced back to the boiler by the force pump L, and the remainder runs to waste: at the same time, to the other end of the great beam is attached the rod of the cold water pump N, which supplies the cistern containing the condenser and air pump with cold water, and from this source the injection cock has its supply.

The governor Q is put in motion by beveled wheels on the shaft of the fly wheel P, (art. 540,) and regulates the throttle valve in the steam pipe S, so as to close it a little for a less admission of steam when the speed is increasing; and on the contrary, when the engine relaxes in its speed, the balls will again begin to fall, and open it a little so as to admit more steam; by this means the work may vary, and yet a uniform motion be kept up by the engine, (art. 550.)

In the pumping apparatus, the rod M is the pump rod for the purpose of raising water; when the rod descends, the water will be forced through G into the upper air vessel E, from whence it passes in a continued stream to the reservoir, at a distance and height proportional to the power of the engine. The barrel of the pump will be refilled from F, the source which communicates with the lower air vessel H; when the rod rises the opposite valve at the top will open, and the water will be forced through into the air vessel E, and the supply for the descending stroke will rush in by the bottom valve from F, and as before be discharged through G.

For the proportions of this kind of engine, see (art. 414—423;) and I would recommend the adoption of Field's valve instead of the throttle valve. (See art. 547)

PLATE XIII.

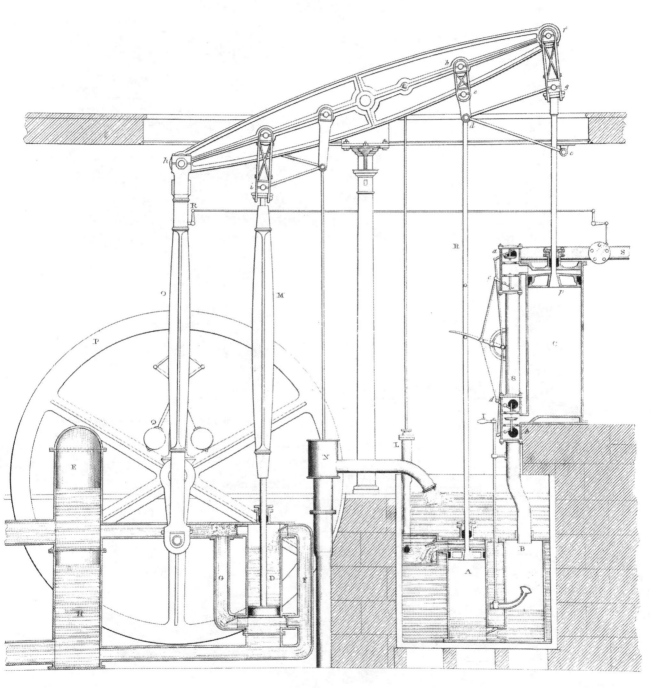

J. White delin.

G. Gladwin sculp.

London Published by J. Taylor. Architectural Library High Holborn. 1827.

Plate IV.

A STEAM ENGINE of 20 Horse Power constructed by FENTON & Cº LEEDS.

Drawn by J. Clement.

Engraved by G. Gladwin.

PLATE XIV.

This plate represents a double acting steam engine, for impelling machinery, as executed by Messrs. Fenton, Murray, and Co. of Leeds.

The engine is supported by the walls A A A A, part of which form the walls of the engine house. The steam cylinder B is secured to the wall below it by bolts; it is enclosed in a jacket or casing of cast iron, a little larger than the cylinder, and the space between them is supplied with steam, to keep the temperature of the cylinder as near to the temperature of the steam as possible. (See art. 155.) The steam comes from the boiler by the steam pipe C, C, to the valve pipe D, D, and passes to the condenser by the eduction pipe E, E, leading to the condenser F, which with the air pump G is immersed in the cold water cistern H, supplied through the pipe J. The cold water pump I, is worked by the rod O attached to the engine beams. The air pump is worked by the rod N, and delivers the water into the hot well, from whence the hot water pump K, worked by the rod P, the upper part of which is connected with the working beam at Q, raises a quantity of hot water to supply the boiler.

The working beam Q is supported by the cast iron column R, and is connected to the piston rod L by the parallel motion M M., see (art. 188;) the other end of the beam gives a rotary motion to the crank shaft by means of the connecting rod S, the lower part of which is attached to the crank T; and a spur wheel U on the crank shaft, working into a pinion on the shaft V, gives motion to it, and to the fly wheel W, (see art. 540.) By means of a train of shafts and beveled wheels X Y Z, moved by the crank shaft, the axis, carrying the eccentric rollers which move the valves, is kept in motion, and the rods a, b, which are connected to the valves, are raised and depressed at the proper times, or by hand by the lever e, in the manner more fully described in Plate VIII. In some of their engines they use slides nearly of the kind mentioned in (art. 447.) The injection of cold water to the condenser is regulated by a cock, moved by the handle c on the spindle d.

The governor g (see art. 550,) is kept in motion by bands from the crank shaft, and opens or closes the throttle valve in the steam pipe C by means of a lever h h. (See Plate VIII. Fig. 1.)

The proportion of the parts for a double engine of this kind may be ascertained by (art. 414—423.) It ought to be made to work expansively, but in this case it has not been done; the saving resulting from causing the engine to work expansively, will be in the ratio of 10 to 7 to do the same work, (art. 422.) See Table III. (art. 664.)

z z

PLATE XV.

As an example of a portable condensing engine for impelling machinery, I have taken that of Mr. Maudslay, where the beam is omitted, and the crank connected directly with the piston rod. Fig. 1, is a front elevation of an engine; Fig. 2, a longitudinal section; and Fig. 3, an end view. The cylinder B is supported by a cast iron frame A, and C is the piston, with the rod D, connected to a cross head at E, and guided by the wheel F which keeps the piston rod in a vertical direction, by moving in the frame G; the side rods H, H, connect the cross head E with the double crank I, I, which turns in the plummer blocks or bearings J, J, one on each side of the frame, and to which the fly wheel shaft K, carrying a fly wheel M, is connected by a coupling box, or clutch L, at the end next the engine.

Two eccentric wheels N, N, on the crank shaft K, give motion to the levers O and T, by means of the connecting rods P, P. The lever O, supported by the bearing Q, works the cold water pump S, by the rod R; while the beam T, working on the centre V, works the air pump X by the slings v, and the hot water pump Y; which by the feed pipe Z supplies the boiler with hot water. On the cross rail (a) the guide is fixed to confine the air pump rod to a vertical motion; the condenser b surrounds the air pump, and is again surrounded by one of the cold water cisterns c; the two cisterns are connected by a pipe d: the steam from the cylinder passes to the condenser through the eduction pipe e, and the cold water for injection is admitted to the condenser by the cock f. The air and condensed steam ascends through the foot valve g in the bottom into the air pump. The mechanism for opening and closing the steam passages consists of an eccentric piece k, fixed on the crank shaft, the action of which communicates a reciprocating motion to the rod i, which by a bent lever moves the connecting rod l, and lever m, that is fixed at one extremity of a spindle having a beveled wheel at the other; which works into another on the spindle n of the steam cock o, by which the steam is admitted from the steam pipe to the cylinder. The engine is worked by hand by means of the handle h.

A pair of beveled wheels on the fly wheel shaft gives motion to the governor balls, and these raise or depress the piece marked p on the axis, which by its peculiar form acts on a lever, and retains the valve in the steam pipe at q open for a longer or shorter time for the admission of steam according to the velocity of the engine, (see art. 547.) There is also an apparatus for working the throttle valve by the same governor.

This engine is in a very compact form: it is however a little too complex, and its frame reminds one of an antiquated style of cabinet work; but the beauty of the workmanship is unequalled, and is faithfully represented by Mr. Clement's drawings.

To proportion the parts of an engine of this kind, (see art. 419—422,) and the articles there referred to; and Table III. (art. 664.) It is adapted for engines of from 2 to 30 horses' power.

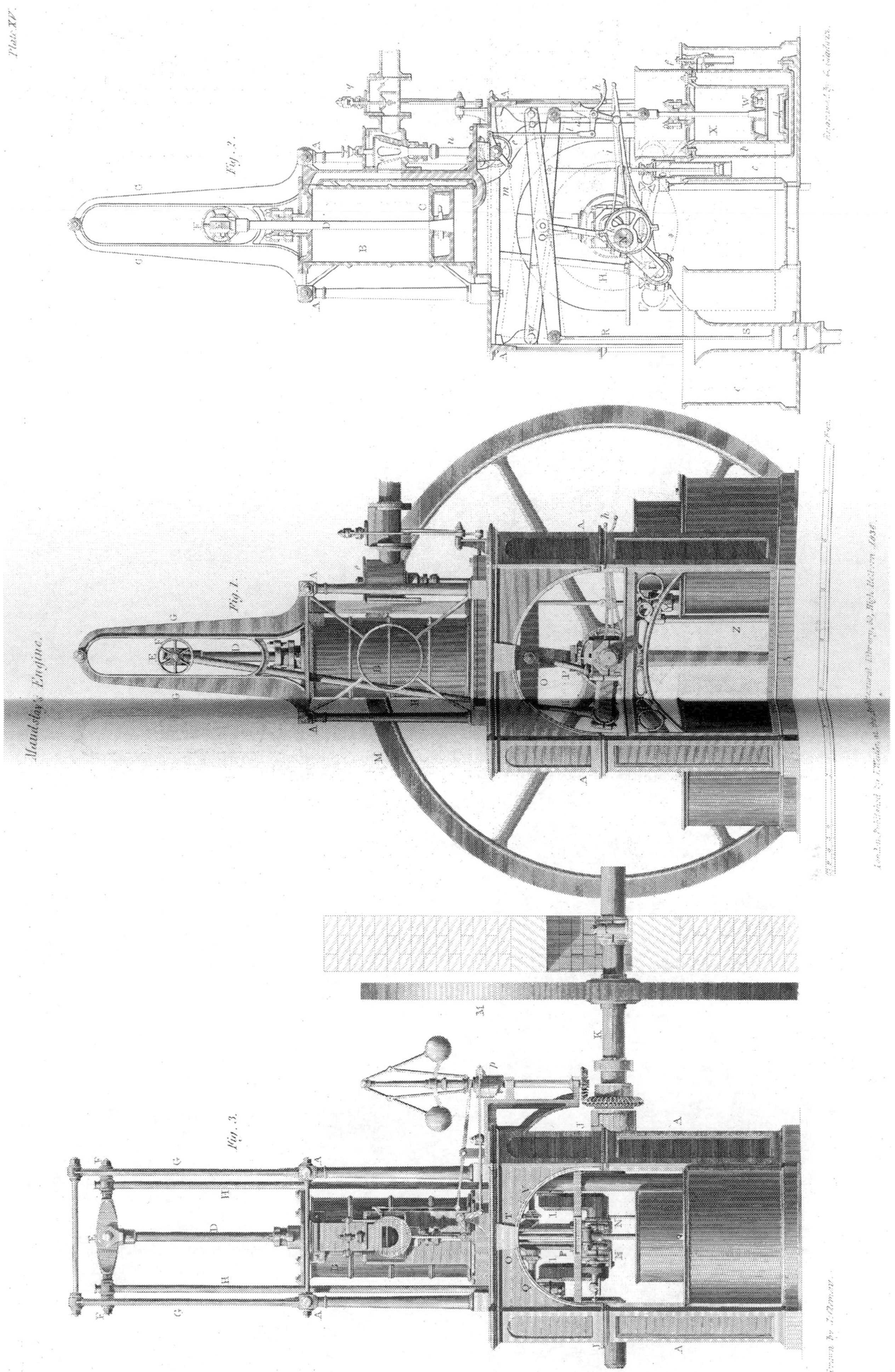

Plate XV.

Maudslay's Engine.

Fig. 2.

Fig. 1.

Fig. 3.

London, Published by J. Weale at the Architectural Library, 59, High Holborn. 1836.

Engraved by J. Ormont.

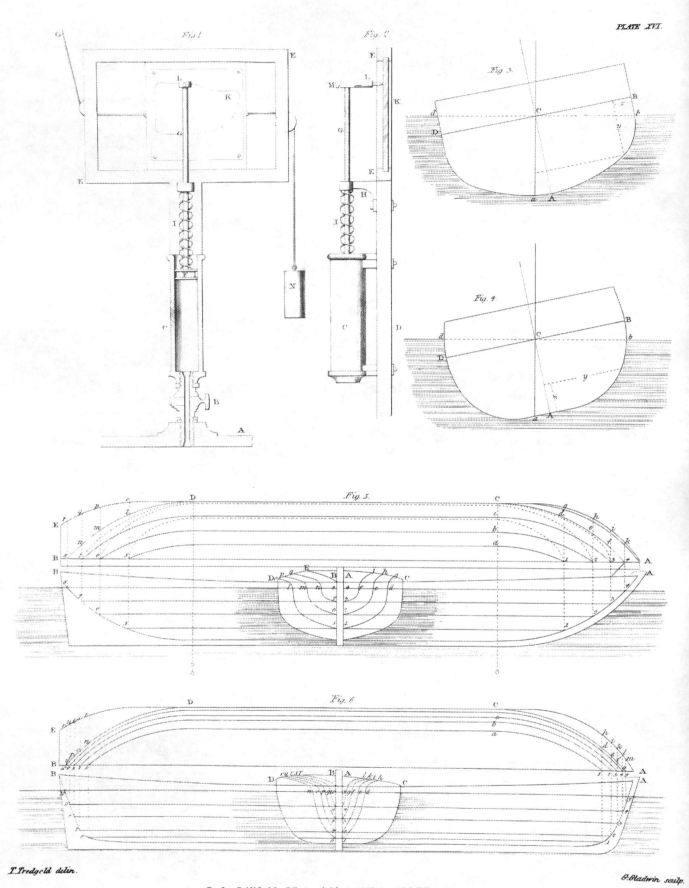

PLATE XVI.

Fig. 1.

Fig. 2.

Fig. 3.

Fig. 4.

Fig. 5.

Fig. 6.

T. Tredgold delin.

G. Gladwin sculp.

London. Published by J. Taylor, Architectural Library, High Holborn, 1827.

PLATE XVI.

Fig. 1. and 2. represent the instrument called the indicator for measuring the force of the steam in the cylinder of an engine. (See art. 560.)

Fig. 3. and 4. are diagrams to illustrate the comparative stability of opposite classes of forms for vessels. (See art. 599, 600.)

Fig. 5. If the motion of a vessel were always direct, its sides should be parallel, and one of the section Fig. 3, may be terminated by making both the extremities of the same figure, and formed by circular arcs; then if the section be similar, so that the stability may be equal throughout the length, (art. 599,) the water lines will increase in curvature towards the keel (they are shewn by dotted lines;) but the actual obliquity of the resisting surface by which these resistances are measured, decreases in descending. The objection I should make to this mode of forming a vessel is, that it would not have a sufficient tendency to keep in its course; and I think a better form would be obtained by conceiving the midship section to advance parallel to itself, and also towards the keel, in the same manner as is shewn in the next figure.

Fig. 6. If the section Fig. 4, be the midship section, and the plan of the load-water line be formed by arcs of circles, and the sections be all drawn by the same mould as the midship section, as far as the breadth at the part allows, then the form will be as Fig. 6; the water lines would be all of the same curvature, the capacity would be easily measured, and the construction would be simple.

But it is necessary to remark that parallel sides are best only for direct motion. In an oblique motion, such as that almost universally produced by wind, the vessel should diminish towards the stern; the oblique force of the wind then presses its side against the fluid so as to produce an effect similar to that of an inclined plane, if the sails be properly set, and I think the diminution should commence where the curvature of the fore part ends.

It is chiefly for direct motion that a steam vessel is intended, and where it is so, parallel sides have the advantage; but where sails are to be used with effect in addition to steam power, the direct resistance must be a little increased, or the capacity diminished, to get a clean run when the oblique force of the wind is available. Hence, it appears that a vessel adapted for one mode of action is not the best for another; and instead of theory being imperfect, it is evident that it only wants to be followed up by analyzing the different cases which occur in practice. It is difficult to conceive how much this subject has been neglected, or how much remains to be done.

PLATE XVII.

Fig. 1. is a section of a steam vessel with its boiler in two parts; the right part is shewn in a section across it through one fire place F, and its flues N, P; and through the cross flue L of the other fire place, and through the safety valve U, (art. 263;) shewing the dampers O, R, and the passages of the flues to the chimney. Of the left part the fire door end is shewn, with the fire doors D, D, the handles for clearing the clinkers B, B, the doors for cleaning the flues at E, E, E, and the gauge cocks G; also part of the chimney C, the steam pipe S, and a slide valve V, to shut off the steam from the engine. There should be as much space between the boiler and the sides of the vessel, as to admit a person to go round to examine it. The floor under the boiler should be rendered as strong as possible, and the boiler should rest on a plate of iron bedded on a layer of bricks or tiles laid over the floor in cement; in this manner a thin plate of wrought iron extending under the whole, being flexible, and brick a slow conductor of heat, is more secure than a much thicker cast iron plate.

Fig. 2. is a plan, shewing the arrangement of the two fire places F, F, and their flues. The fire door is at D, the fire on the grate F, the clinkers fall at H, and the smoke turns at L and returns along the flue N, rises at O, and goes back along a flue P over N. The boilers should be strengthened by internal frames disposed in triangles, and so as to afford supports for the flues.

Fig. 3. is a longitudinal section through the boiler, and one of the fire places: the same letters refer to the same parts in all the figures, (see art. 239—244;) for the fire and flue surface (art. 204;) for the capacity (art. 215—220;) for the area of the chimney (art 278;) for safety valves (art. 259—272;) for the strength of boilers (art. 525;) and for the management of sea water (art. 565.)

PLATE XVII.

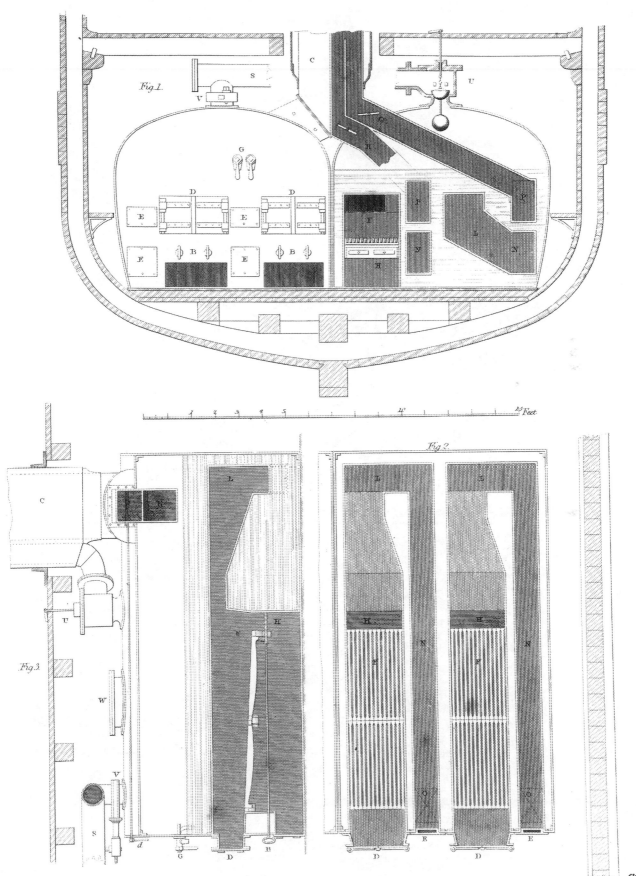

Fig.1.

Fig.2.

Fig.3.

Tredgold delt.

London, Published 1827, by J.Taylor, Architectural Library, High Holborn.

Gladwin sc.

PLATE XVIII

Steam Boat Engine.

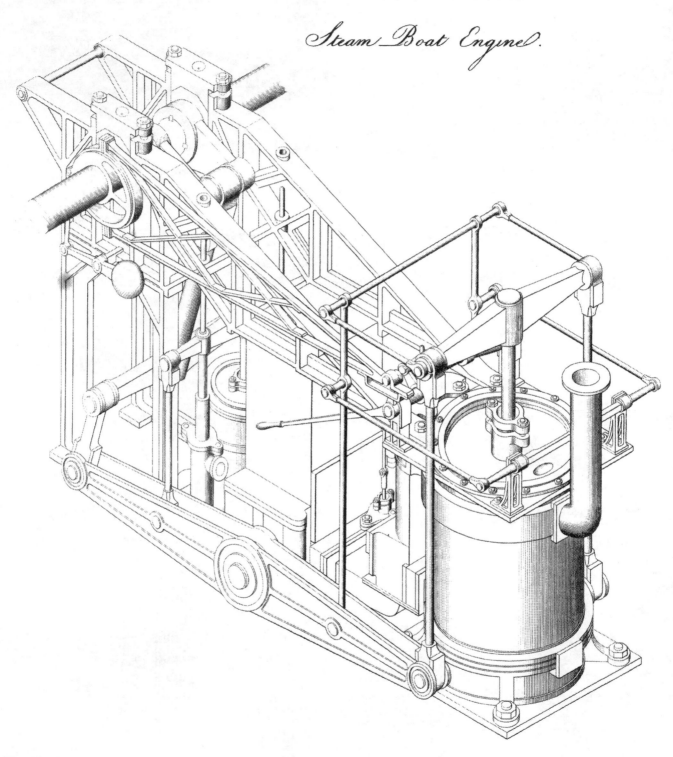

T. Tredgold delin.

G. Gladwin sculp

London, Published by J. Taylor, Architectural Library, High Holborn, 1827.

PLATE XVIII.

This is an isometrical projection* of a steam boat engine, in the manner they are arranged by Messrs. Boulton and Watt; and the same general principle of construction is followed by all the best manufacturers. The arrangement did not however originate from Boulton and Watt, but was adopted by them, in common with others, from the engines of the Clyde steam boats.

The steam comes from the boiler by the pipe in the front of the figure, and passes into the steam case and round the cylinder to the slide box, (see art. 146;) from whence it is let into the cylinder in a manner which will be more clearly understood by referring to the next plate: from the lower part of the cylinder a trunk proceeds to the condenser, which is below a square cistern; beyond which a part of the air pump is seen, and to the left of it the hot water pump to supply the boiler.

The motion of the parts commences at the cylinder: the piston rod is supposed to be descending, and by means of a cross bar (called a cross head) and two side rods, it depresses the ends of the side beams, these side beams moving on axes in the centre; the other ends rise and force a cross bar upwards, to the middle of which the connecting rod is fixed, by which the crank of the paddle wheel shaft at the upper part of the figure is turned; and also by the rising of the further end of the side beams the cross head of the air pump and hot water pump is raised by two side rods. The motion of the piston rod is guided by a combination of rods called the parallel motion, (see art. 495,) and the slide is moved by an eccentric wheel on the crank shaft; and to reduce the friction its weight would cause, it is balanced by a heavy ball acting on a lever below it.

Though the figures on the next plate are not a plan and section of this engine, yet the same parts, with the exception of the steam pipe, are nearly in the same places, and hence by comparison of the two the uses and action of the parts may be understood.

The weight of an engine of this kind is not exactly proportional to its power, but is nearly so; and that of a forty horse power engine, with proper duplicate parts, water, and other appendages, is about one hundred tons.

See Section X.

* Some account of this simple and useful mode of drawing, which was invented by Professor Farish, is given in Dr. Gregory's Mathematics for Practical Men.

PLATE XIX.

In this plate, Fig. 1 is a section, and Fig. 2 a plan, of a steam boat engine. A strong frame of cast iron supports the crank shaft I, and connects the parts of the engine; and the whole is sustained by two strong beams on the floor. The steam pipe S brings the steam from the boiler to the passages of the slide, and from thence to the top or bottom of the cylinder A, and it passes from the slide to the condenser B, where it is condensed by a rose jet playing constantly; the air pump C expels the air and water into the cistern D, from whence it runs out by a pipe. The motion of the piston is transmitted to the crank by means of the double side beams or levers E F, moving on the axis G; these are connected to the cross head of the piston, L, L, by two rods called the side rods, and the ends E are connected to the crank by the connecting rod H. The air pump is also worked by side rods from the side beams, and the hot water pump from the same cross head. The parallel motion is guided by the rod M N. (See art. 489.) The slide is on Murdoch's construction, (art. 447,) and is moved by a wheel on the crank shaft I, with a sliding frame P, (art. 481,) and by hand by the lever T; the slide rod being moved by slings from the arm R. The valve O is used to let steam into the condenser in setting the engine to work, and the air and water are blown out at the discharge valve into the cistern D.

For the strength of the parts, see (art. 496,) the proportions of the engine, (art. 419—422.) The parts of the slide should be counterbalanced, in order that its resistance may be equal in either direction.

Each vessel generally has two of these engines, placed parallel to one another, with a passage between them; and there is room left for working the fires between the cylinder and the boilers. The coals should be kept in iron tanks in the engine room, each to hold a given weight of coal, with the object of ascertaining with accuracy the consumption for any time; about four tons is a sufficient quantity to be together, and they should be kept out of the danger of taking fire.

The proportion of the power to the effect is given, Sect. X. (art. 655.)

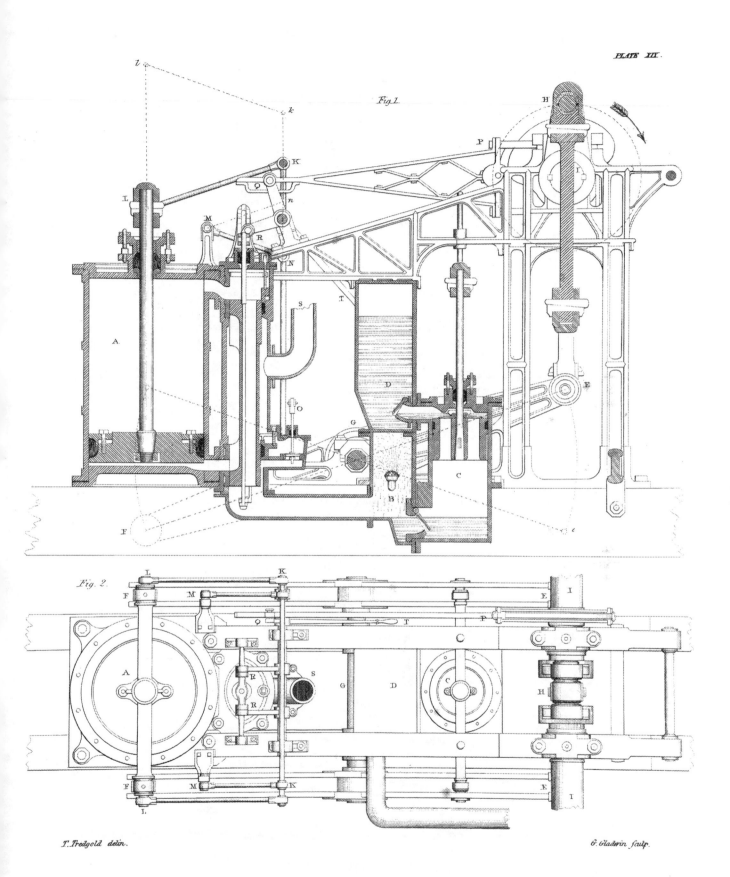

PLATE XIX.

Fig.1

Fig. 2.

T. Tredgold delin.

G. Gladwin sculp.

London, Published by J.Taylor, Architectural Library, High Holborn.1827.

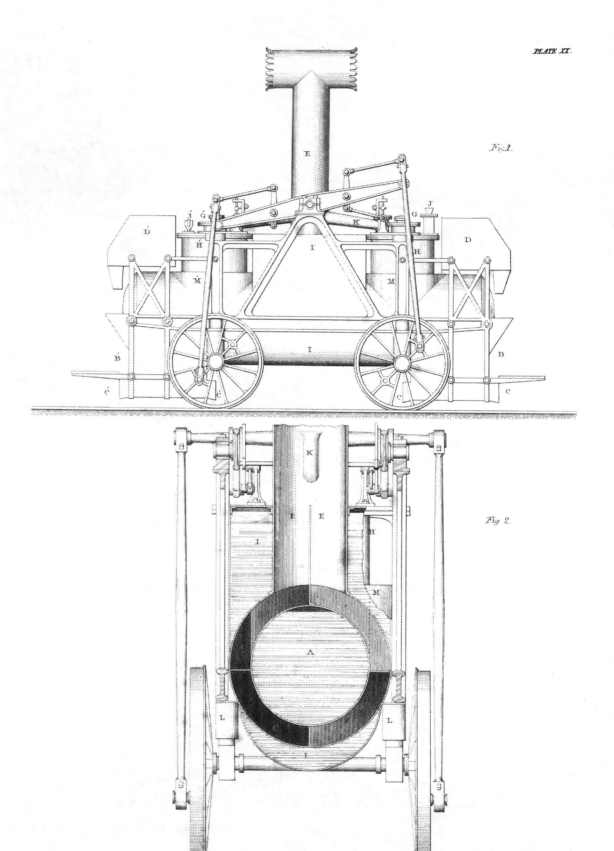

PLATE XX.

Fig.1.

Fig 2.

T. Tredgold delin.

G. Gladwin sculp.

London, Published by J. Taylor, Architectural Library, High Holborn, 1827.

PLATE XX.

Fig. 1. represents a side elevation of a steam carriage; and Fig. 2, part of a cross section to double the scale: the same letters refer to the same parts in both. The steam is generated in a cylindrical boiler A, and the fire and flues surround it; it is joined to two cylinders H, H, of the same diameter, intended as reservoirs for steam, and in these are inserted the engine cylinders, G, G'; the parts I, I, form a reservoir of water not exposed to the pressure of the steam, but surrounding the flues and chimney so as to be heated ready for injection into the boiler A by a small force pump.

In order to distribute the heat of the fuel so as to render it effective on a larger surface, there are two fire places, with fire doors at B, B', but fed with coals by hoppers from the boxes D, D'; the doors are used only to clear the bars, and should be kept open as little as possible. The fire flues meet at the middle; the one from the fire B' rises at F', Fig. 2, passes along on the upper surface of the cylinder A, round H at M, also round the end of the boiler, and returns on the opposite side, to ascend the chimney in the division E'; the other proceeds in the same manner in the opposite direction, and ascends at E. There are two apertures for air C, C, to each ash pit, both of which should be provided with registers, so that those may be open which either face a strong wind, or (in ordinary cases) those which face the direction of the motion of the carriage. For a like reason the top of the chimney E should have two apertures, that the motion of the air, or the motion in the air, may assist the draught.

The engine and boiler are supported by a frame, and this is supported by the axis; but to prevent the carriage resting on three wheels, there may be four spiral springs in the boxes L, L, and the cross heads must be connected to the piston rods by moveable joints, and all the bearings must be formed so as to admit of the motion which would take place by the sinking of one of the wheels in a certain degree. The waste steam passes from the slides to the chimney by the pipes K, and there should be two safety valves, one locked in a box at J, the other open for the use of the engine man at J'. (See art. 266—273.)

There will, I think, be some advantage in making the pistons act together, because the effect will be as great as by dividing it, supposing both methods to be perfect; and in acting together there would be less interference of the motion of the one with that of the other. The slide would be best moved from curved teeth on the beams. (See art. 481.)

For the proportions and construction of the boiler, (see art. 244, 227, 278 and 522—526:) for the engines, (see art. 271—380;) and for the power required, (see art. 590.)

CORRECTIONS AND ADDITIONS.

Page 10 line 38 *for* pump *read* steam.
—— 11 — 4 *for* cylinders *read* cylinder.
—— 34 *read* Dr. Edmund Cartwright, born 1742, died 1823.
—— 85 line 1 note, *for* to be liquid *read* to be from liquid.
—— 107 — 21 *for* (Sect. XI.) *read* Treatise on Warming, &c. (art. 220.)
—— 134 — 14 *for* 111 *read* 11.
—— 134 — 19 *for* 1·27 + 264 *read* 1·27 × 264.
—— 178 — 7 *for* twenty-one *read* 2100 and continue the correction
through the investigation, the result will be 3665·1 instead of 3670 lbs.
—— 201 — 2 *for* valve *o* *read* valve *c*.
—— 324 — last *for* $\sqrt{}$ *read* $\sqrt{l.}$
—— 329 — 17 *for* Wylem *read* Wylam, also 1240 lbs. which is under
Lightning should have been under Harlequin.
—— 333 — 23 *for* following *read* fallacy.

The phenomena observed by M. M. Clement and Désormes, respecting a flat plate being sustained by escaping air instead of being forced away by it, will not have a sensible effect on the escape of steam by a safety valve, if it produces any; for it appears, from some trials I have made, that the plate must be considerably larger than the aperture to produce the effect, and this is not the case in those valves. M. Hachette's illustration and explanation of the phenomena, has been given by Mr. Brande, in his New Series of the Quarterly Journal of Science, Vol. II. p. 193.

Page 333, (art. 660.) I am indebted to Mr. Edward Deas Thomson, who has just returned from America, for the following account of one of the latest and the best boats they have constructed in the United States. "The *North America*,' built by Kemble and Co. of New York, under the direction of Mr. Stevens, the owner,
Dimensions

Length of deck	178 feet
Depth of hold (from under side of beam to the keel) .	9 feet
Breadth (moulded)	28 feet
Breadth (extreme above water) . .	58 feet
Draught of water	4½ feet
Diameter of Paddle Wheels . .	21 feet
Breadth of Paddle Wheels . .	13¼ feet
Depth of paddles . . .	2 feet
Two engines with the diameter of cylinders . .	45 inches
Length of stroke . . .	8 feet
Number of strokes per minute . .	22 to 26

The usual force of the steam is nine inches of mercury above the atmosphere, the extreme force fourteen inches, and the two engines consume two cords of wood per hour. The passage from New York to Albany is about 160 miles, and is performed at an average in twelve hours, with a consumption of twenty-five cords of wood. The boilers are placed before, and the machinery abaft the paddle wheels."

I find by Bull's Essay on Fuel (Philadelphia 1827,) that a cord of wood weighs about 3826 lbs; and as the engines are of eighty-five horses' power, this gives forty-five pounds of wood per hour for each horse power. The ratio of coal to wood is (art. 190,) 8·22 to 22·6, according to which the equivalent quantity of coal would be 16·4 pounds.

INDEX.

————◆————

3 A

THE END.

Printed in the United States
By Bookmasters